FIRST EDITION

ORIGINS

THE STORY OF THE BEGINNING OF EVERYTHING

BY **Bahram Mobasher**

University of California—Riverside

Bassim Hamadeh, CEO and Publisher
Gem Rabanara, Project Editor
Alia Bales, Production Editor
Emely Villavicencio, Senior Graphic Designer
Trey Soto, Licensing Coordinator
Natalie Piccotti, Director of Marketing
Kassie Graves, Vice President of Editorial
Jamie Giganti, Director of Academic Publishing

Copyright © 2019 by Cognella, Inc. All rights reserved. No part of this publication may be reprinted, reproduced, transmitted, or utilized in any form or by any electronic, mechanical, or other means, now known or hereafter invented, including photocopying, microfilming, and recording, or in any information retrieval system without the written permission of Cognella, Inc. For inquiries regarding permissions, translations, foreign rights, audio rights, and any other forms of reproduction, please contact the Cognella Licensing Department at rights@cognella.com.

Trademark Notice: Product or corporate names may be trademarks or registered trademarks, and are used only for identification and explanation without intent to infringe.

Cover image copyright© by Babak Tafreshi. Reprinted with permission.

Printed in the United States of America.

ISBN: 978-1-62661-481-9(pbk) / 978-1-62661-482-6 (br) / 978-1-5165-7011-9 (al)

To my parents

"We delight with the beauty of the butterfly, but rarely admit
the changes it has gone through to achieve that beauty".

MAYA ANGELOU

AUTHOR

"The greatest obstacle to discovery is not ignorance, it is
illusion of knowledge".

DANIEL J. BOORSTIN

HISTORIAN

"It is the mark of an educated mind to be able to entertain
a thought without accepting it".

ARISTOTLE

PHILOSOPHER

CONTENTS IN BRIEF

CONTENTS *EXTENDED*

ACKNOWLEDGMENTS

Throughout my working life, I have had the honor of collaborating with a large number of truly outstanding scientists. Many of them became life-long friends. I am indebted to them all for teaching me to think deep, have focus and be ambitious in selecting my research topics. Among them, I am particularly grateful to Prof. Richard S. Ellis, Prof. Sandra M. Faber, Dr. Henry C. Ferguson, Prof. Adam Riess, Prof. Michael Rowan-Robinson and Prof. Nick. Z. Scoville. I have been inspired by them and greatly benefited from discussions with them over many years. I am truly grateful to my family, to my wife Dr. Azin Mobasher and children Armeen and Tara for their support, love and understanding.

I would like to express my deep gratitude to many people who helped me writing this book. In particular, my thanks go to Dr. Mario De Leo-Winkler for his valuable comments on the text and for designing and creating many of the original figures in this book. I am indebted to Ms. Olivia Barrlett and Cassandra Threadgill, who provided invaluable support in the editing, proofreading and design of the text. Finally, I am grateful to Ms. Gem Rabenera my project editor and Ms Alia Bales my production editor for being patient with me and tolerating me missing the deadlines.

PREFACE

I was ten years old when NASA astronauts landed on the Moon. My fascination by that event and its significance only increased as I became older. At every stage of my life I was inspired by it for reasons different from those a few years earlier. At the time my grandfather told me about his own experience that, when he was my age, it took him months to travel a distance that would only take a few hours today. During his lifetime, he had seen different means of transport- by foot, horse carriages, first cars ever made, first passenger planes and now the spaceship covering the Earth-Moon distance in three days. This shows how far us, human beings, have come in one generation. This event taught me a few things: the importance of thinking big, willingness to take up challenges no matter how difficult or impossible they may seem, a desire to push things to their limit for the sake of finding something new, and having a focused vision and working towards achieving it. This also showed me, as my grandfather also passionately stressed, that this achievement and science in general, belongs to humanity and not a particular nation, community or culture. I came to understand and appreciate all these much later in my life, well after I finished graduate school. In the words of Benjamin Franklin—*"The tragedy of life is that one gets too old too soon and too wise too late"*. We are living in a unique time where science and technology have progressed to a level that we can study from the details of the universe to the inside of a living cell. Our mission is to encourage people to move beyond everyday activities, to wonder, to ponder questions and be curious about the world they live in. The quest for knowledge and the joy of understanding the world around us, has no age, race, gender or cultural barriers. This is achieved by pushing our knowledge of the physical world to its limit by addressing the most fundamental questions. This book is an attempt to do just this- to use the combined knowledge we have gained in different disciplines and push them to their limit to address questions concerning the origin of everything we observe and experience in nature.

My aim in this book is to take the reader out of the usual activities of the daily life and introduce the joyful world of discovering the truth (or what we think is the truth) about the universe, atoms, cells, and life itself. The goal is not just to answer questions but to learn to ask new and deep questions. After all, science is not just trying to uncover the unknown but dig deeper in what we know and search for the right question to ask. This is addressed by adopting a reductionist approach: by reducing complex phenomena to their simple components and trying to understand them. The attitude throughout the book is that for every existing thing, there is an explanation and a reason as to why *it is like this and not like that*. The book also singles out the simple events that although insignificant, have played a major role in the development of the world around us. It shows how so many seemingly different things came together to make our very existence possible. I also hope this book would teach the reader the value of reason and objective thinking. Today, we have witnessed twisting of scientific facts by individuals to serve personal and preconceived views. We have the duty to provide the means to educate the public as the true wealth of a nation is measured by the awareness, knowledge and education of its citizens. In the words of American broadcaster Walter Cronkite: "Whatever the cost of our libraries, the price is cheap compared to that of an ignorant nation"

The book connects pieces of knowledge from different disciplines to build a coherent picture of the world we live in, starting with a historical review of the origin of scientific thinking and evolution of knowledge. At the beginning, everything was a part of philosophy. Once knowledge enhanced, independent disciplines were developed and advanced. Today, science has progressed so much that one could study the common areas between different fields. Later, the book explores the nature of space and time before studying the very early universe and the origin of particles and their mass. I delve into how the largest cosmological scales and the evolution of the universe are affected by physical properties of elementary particles. This is a fascinating story by itself, connecting the very small (particles) to the very big (structures in the universe). After discussing the present state of the universe, its content and its evolution, the origin of galaxies and stars is studied followed by the study of the origin of chemical elements. The origin of the planetary systems and our home planet, Earth, is discussed next. This is then followed by a discussion of the origin of continents, oceans and mountain ranges on the earth as well as the origin of the planet's atmosphere. I discuss how Earth became a habitable planet and how the ingredients needed for life became available. This then leads to the study of the origin of life, on earth and its evolution through geological times. I present the evidence of how our early ancestors formed cities, developed agriculture and invented languages by means of communication. "Origin" and "evolution" influence one another, the evolution of one thing leads to the origin of another thing. Therefore, alongside the study of the origin, I also discuss evolution. Due to its nature, the subject of this book is multidisciplinary, bridging between different independent fields.

I feel very fortunate to have had the opportunity of writing this book. Although not complete by any means, it has been an amazing learning experience for me. I was fascinated by how much we, as humankind, know about nature and how many questions are still unanswered. The following two images summarize my amazement. First, is the image of the Earth as seen from the Voyager I satellite after twelve years' journey when it turned its camera back towards the Earth and took a picture, before leaving the solar system. The Voyager I satellite was the first

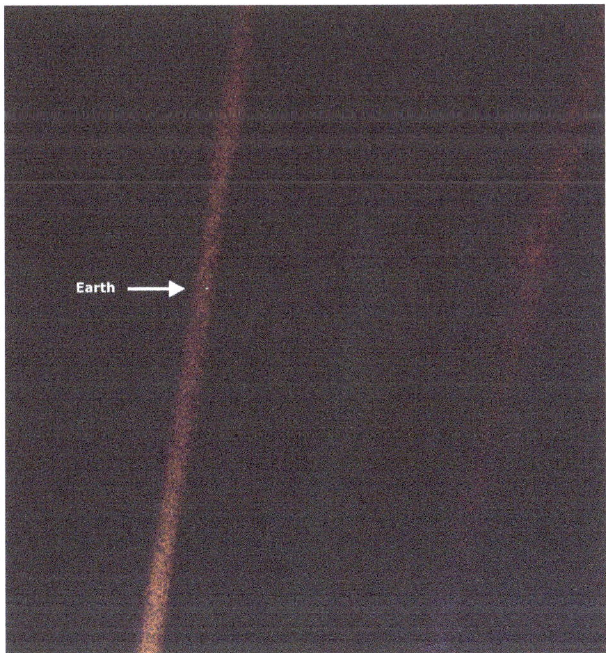

Figure A: Image of the Earth as seen from the edge of the solar system. This picture was taken by the Voyager I satellite on February 14th, 1990 from a distance of 6 billion kilometers (about 40 times the distance from Earth to Sun) after an almost 12 years journey.

Figure B: Image of the Hubble Ultra Deep Field taken at optical/infrared wavelengths by the Hubble Space Telescope, a result of 1 million seconds of exposure. This is the deepest image ever seen of the universe. It contains of the order of 10,000 galaxies, many at the edge of the observable universe (around 13 billion light years away).

manmade artifact to leave the solar system and venture into the space between the stars (figure A). This shows the Earth as an observer at the edge of the solar system would see it—*"a small speck of dust hosting seven billion people floating in infinite space"*, as Carl Sagan elegantly quoted. It is impossible to admire this image and not be amazed at how small, insignificant and lonely we are in this vast cosmos. The second image is that of the Hubble Ultra Deep Field (figure B), the deepest image (at optical/infrared wavelength) ever captured of the universe up to the present time. In this picture we see the boundary of the observable universe, the most distant galaxies formed just a few hundred million years after the beginning of the universe at a distance of over 12 billion light years. These two figures combined, demonstrate how we, the human race, despite occupying an insignificant volume of space, have reached the boundary of space and time to decipher secrets of the world we live in. It also reveals the immense number of things still waiting to be discovered.

This book covers material from a number of disciplines. I have attempted to make the subjects understandable to everyone independent of their background in the concerned field. For students, the book provides the means to encourage them to think deep and outside the box, and ask questions. For the general reader in the quest for knowledge, it provides the opportunity to discover a new world and share the fascination of discovery. For the educators, it provides the big picture of the world out there, to think and ponder about and to inspire the future generation of thinkers and innovators. As quoted by Albert Einstein: "Education is what remains after one has forgotten what one has learned at school." I hope the material in this book would become a part of your education.

FIGURE CREDITS

- Fig. A: Source: https://commons.wikimedia.org/wiki/File:PaleBlueDot.jpg.
- Fig. B: Source: https://commons.wikimedia.org/wiki/File:NASA-HS201427a-HubbleUltraDeepField2014-20140603.jpg.

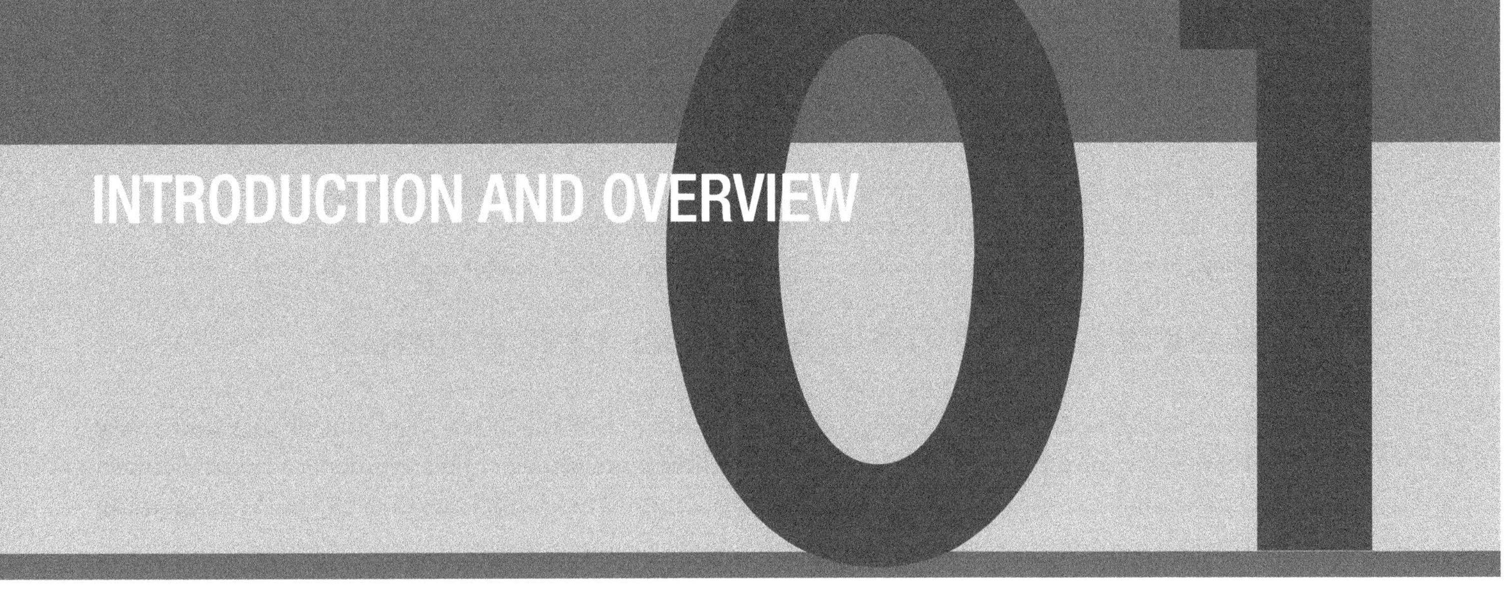

CHAPTER LEARNING OBJECTIVES

This chapter will cover:

- Summary of the history of the universe and life
- Timeline of the universe
- The fundamental constants in nature
- Definitions and measurement units
- Fundamental laws of nature
- Search for the origin of the laws of physics
- Symmetry in Nature

The title of this book is a guide to the material it contains—a multidisciplinary attempt to move between boundaries of science. While we move among widely different disciplines, the focus is the question of the "origin"—how everything we observe and experience in the world came to be the way it is? We observe natural phenomena and take them for granted. However, everything we observe in the physical world is likely to have had a beginning and somehow turned out to be the way it is today. Our aim here is to dig deeper to see where these all started and how these seemingly independent phenomena came together to provide the current state of the world we see and experience.

Science is the means and ways to explain nature through models contrasted with experiments and observations. This requires critical thinking to conceptualize an observed phenomenon and then try to explain it through established laws and test it through experiments. Through science, one could address human curiosity—from deep in space to inside an atom to the structure of a living cell, using the largest ground-based telescopes and space probes, powerful particle accelerators, and the most powerful electron microscopes. By exploiting the basic scientific principles, one develops and improves technology, which in turn is essential for more accurate measurements and observations and, possibly, new scientific discoveries. The study of the origins will help to know the world better and through that, ourselves and our own position in this world. The aim of this chapter is to put the book in context

"To myself I am only a child playing on the beach, while vast oceans of truth lie undiscovered before me"

- ISAAC NEWTON

"When it is not in our power to follow what is true, we ought to follow what is most probable"

- RENE DESCARTES

by presenting a very short synopsis of what will follow in the rest of the book. The chapter presents a summary of the history of everything, followed by the study of the fundamental laws of nature and the physical constants governing the world around us. It provides the general background needed for the rest of the book.

AN OVERVIEW OF THE HISTORY OF EVERYTHING

There is compelling evidence that our universe started around 13.8 billion years ago from a big explosion—the big bang. Space and time were formed at that instant. That was a moment of extreme density and temperature. The universe has been expanding ever since, due to the stretching of space (Hester et al. 2010). The first particles appeared a fraction of a second after the big bang, with the nucleus of hydrogen atom, the lightest of the nuclei, forming about one minute after the big bang. Once the temperature of the universe cooled down due to its expansion, electrons joined the existing nuclei, and atoms were formed. This led to the formation of matter, as we know it today. Under the force of gravity, and initial perturbations in an otherwise uniform matter distribution, the matter came together, forming structures consisting of galaxies and clusters of galaxies (figure 1.1). Galaxies contain, on average, a hundred thousand million stars with the observable universe containing roughly a hundred thousand million galaxies.

Throughout the age of the universe, properties of galaxies have changed because of the interaction with other systems or passive evolution of the stars they consist of. Today, we have precise estimates about the constituents of the universe, although we are less certain about the nature of different components governing it—dark matter (22% percent), dark energy (74% percent), and ordinary matter (4% percent). Dark matter attracts galaxies through the force of gravity, slowing down the rate of expansion of the universe, while dark energy repels galaxies, accelerating the rate of the expansion (figure 1.1) (Hester et al. 2010). What we see on the sky at night and all we see in the universe constitutes only 4% of the content of the universe.

After stars were formed from the collapse of cold gas in galaxies, planets appeared on the scene, orbiting the stars. For the first time in the history of civilization, we are now able to find and study planets outside our solar system that would help us understand the initial phases of formation of our own planet—Earth—around 4.6 billion years ago. The stars are the main factories for the production of heavy elements. They eventually run out of fuel after converting all their light elements to heavier elements. If a star is massive enough, it explodes as a supernova, spreading the heavy material produced in them into the space between stars, enriching this medium with heavy chemical elements. This is the source of the heavy elements found on Earth, the elements responsible for life.

Evidence for the first living things on Earth goes back to around 3.5 billion years ago, in the form of primitive cells (a cell without a nucleus or other membranes). Subsequently, more complex cells resulted that were able to perform multiple tasks (figure 1.2). These were all started deep in the oceans and moved to land when Earth's atmosphere formed. The first living organisms did not need oxygen to stay alive, and as a result, they released oxygen as waste. This led to the buildup of oxygen in the atmosphere and subsequent formation of the ozone layer, which produced a protective layer around the Earth, shielding it against intense ultra-violet radiation from the sun, making the land habitable (Bennett and Shostak 2005).

The evolution of life on Earth is a hugely complicated process. Only the systems that could adapt to their environment could survive and prosper. Mutation led to creation of the diverse spices of plants and animals we observe today. The history of first primates goes back to over a million years ago, while the closest of our ancestors, *Homo sapiens*, lived on the planet around 150,000 to 100,000 years ago. The primates evolved and adapted to their environment. Their brain size increased, enhancing their intelligence (Larsen 2014). Early appearance and evolution of mammalian life took place in Africa. A group of Homo sapiens then started to move out of Africa and into where is now Europe and Asia around 30,000 years ago. At some point, they started to form communities,

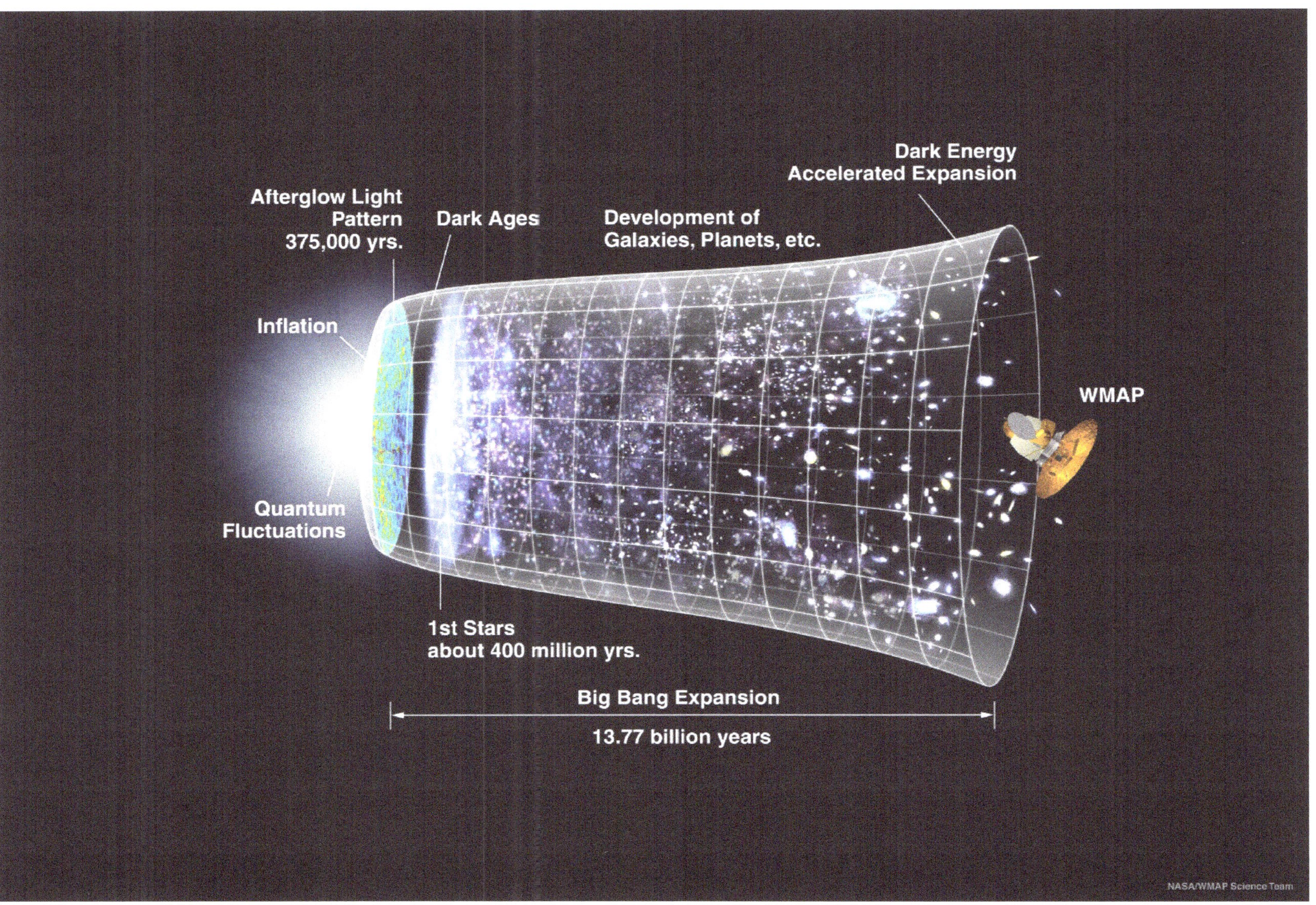

Figure 1.1. Different stages in the evolution of the universe, from the big bang to the formation of galaxies, stars, and planets.

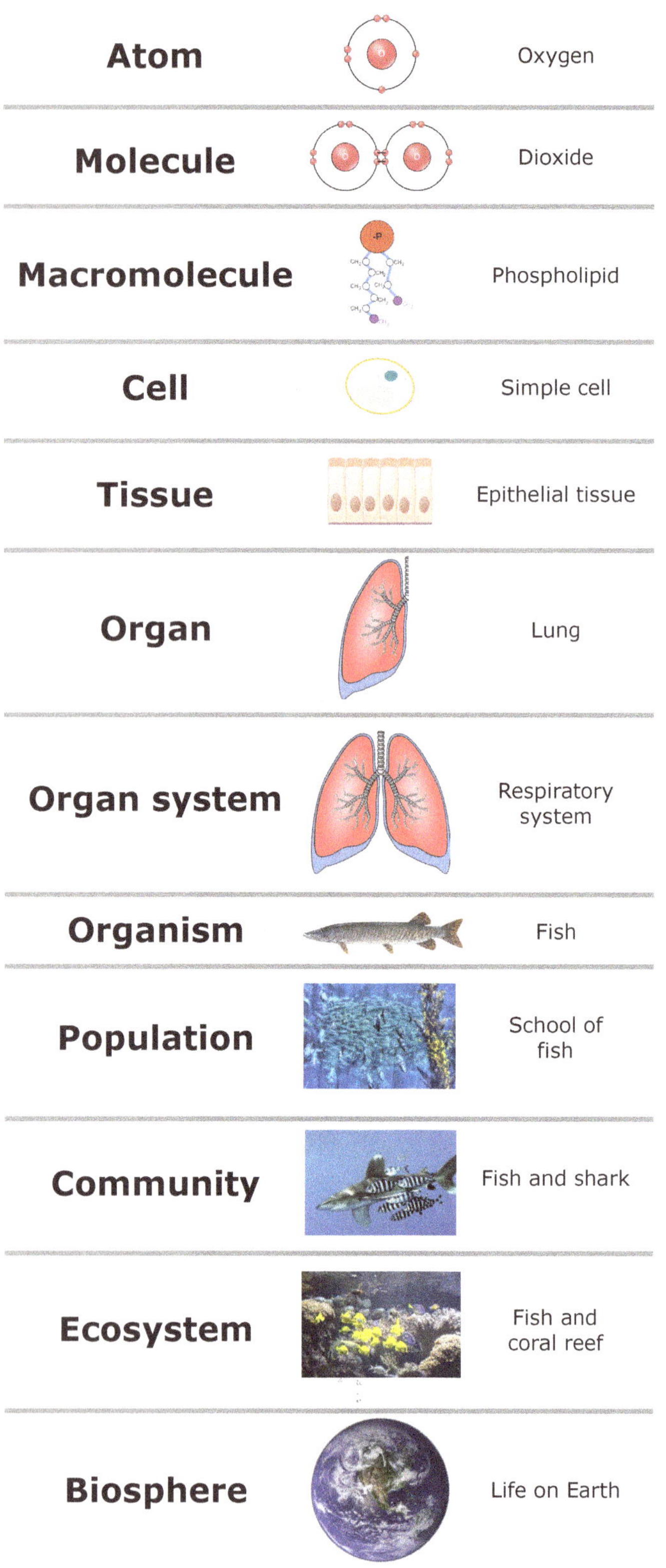

Figure 1.2. The development of complex living things, from a cell to more complicated systems.

communicating with one another and developing new ways to feed themselves. Around 10,000 years ago, agriculture was invented, and our ancestors learned to domesticate animals. Once they found the means to produce their food more efficiently and feed more, they found more time to spend on other things, like creating civil societies. Figure 1.2 shows the step-by-step development of life on Earth.

To give an idea about the relative time scales involved, it would be instructive to condense the history of the universe and life on Earth into one year (figure 1.3; Bennett and Shostak 2005). In this case, given an age of 13.8 billion years for the universe, each month corresponds to more than 1 billion years, while each day represents 40 million years, and each second around 400 years. Imagine the universe started at 12:00 a.m. on January 1. In this time-scale the Milky Way Galaxy was formed in May, with the solar system forming in early September. Primitive life on Earth started by late September, while more complex living systems were formed in November. An advanced form of life did not appear until mid-December. Fish were the first animals, coming to the scene on December 17, while land plants and animals appeared on December 20 to 23. Dinosaurs came to the scene around 25th–26th December. At 9:00 a.m. on December 31 hominids (our early ancestors) started to dominate Earth. Agriculture was developed 25 seconds and the Pyramids built 11 seconds before midnight. In this scale, the entirety of human history took place in the last 4 minutes in the history of the universe (figure 1.3).

PHYSICAL CONSTANTS

Why the universe is the way it is? How the conditions for life evolved? How our own planet was formed and evolved and became habitable? These, plus many more fundamental questions can be addressed by the value of the physical constants in nature. While it is not clear as to why

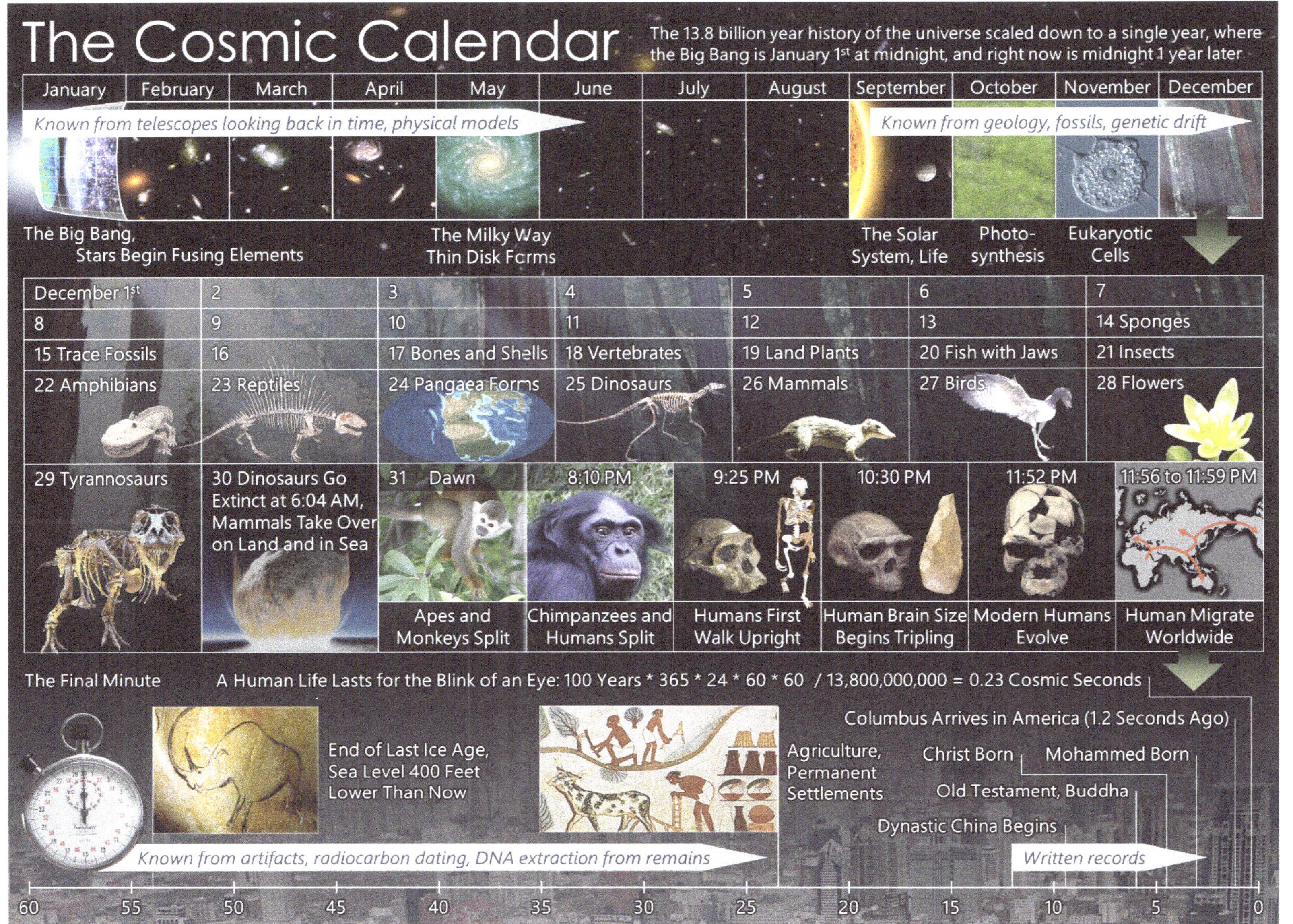

Figure 1.3. The history of the universe depicted in one year.

these constants have the values they have or what their origin has been, they are the most fundamental parameters in shaping the world around us.

Some of the most important physical constants include: the velocity of light (c), Planck's constant (h), electric charge (e), electron's mass, Newton's gravitational constant (G), and the constants responsible for the strength of fundamental forces in nature. The most fundamental constants have dimensions that are based on the basic quantities of mass, length, and time. There are also some physical constants that do not have any unit—that is, they are dimensionless. The example here is the fine-structure constant, e^2/hc. The value of this constant is 1/137.036, and it represents the strength of the electromagnetic interaction. The interesting point about dimensionless units is that they express facts about the universe completely independently from our choice of units. The universe would have been a very different place if these constants had different values.

The universe in its present form and the conditions to support life on Earth could only exist if the physical constants have the values they have. If these constants were slightly different, we would not have been here. In the following, I give a few examples to make this clear (Rees et al 2001), considering the four forces in nature: gravity (that governs large scale structure of the universe), electromagnetic (that keeps atmos together), week (responsible for particle decays) and strong (that is reponsible for keeping together particles inside the nucleus of atoms) (these will be discussed in more detail in Chapter 4). (1). If the strength of strong force that binds the nuclei of atoms (protons and neutrons) were 2% stronger, two protons would fuse to one another, making the next heavy nucleus after hydrogen, with two protons. In the present universe this is a very unstable element that immediately decays. However, a stronger nuclear force will make it stable. It will then stay around longer, allowing hydrogen (that existed in abundance at that time) to fuse into it. This would consume all the hydrogen in the early universe, seriously changing the nuclear processes in stars and therefore the production of chemical elements, making the world very different from what it is today; (2). The strength of electromagnetic force, as measured by fine structure constant described above is 10^{36} times that of the gravity. If this were slightly smaller, the universe would have been smaller and short-lived, leaving no possibility for life to evolve; (3). Two protons and two neutrons combine to form nucleus of Helium atom. However, the total mass of these only constitutes 99.3 percent of the mass of the helium nucleus. The remaining 0.7 percent (0.007 fraction of the mass) is released as energy, fueling our Sun or other stars. This is fixed by the strength of the strong force that keeps the nucleus of atoms together. If this number were slightly smaller (0.006 instead of 0.007), protons and neutrons would not bind together and the universe would only consist of hydrogen—no heavy elements would be produced and no life in the universe. Now, if this fraction were slightly higher (0.008 instead of 0.007), all the protons would be combined with neutrons, leaving no hydrogen in the universe—leading to a very different universe. There are numerous examples of how a slight change in the values of physical parameters could affect the evolution of the universe and life (Rees 2003). I will discuss some of this amazing "fine tuning" later in the book.

DEFINITIONS, MEASUREMENT SYSTEMS, AND UNITS

Galileo Galilei stated around four hundred years ago, "Measure what is Measurable and Make Measurable what is not so." Measurement is the process that quantifies scientific observations. Some measurements are independent, while some could be expressed in terms of others. For example, length is an independent quantity, while area is expressed by two length measurements and volume by three length measurements (length, width, and depth). Mass and time are independent measurements that can be used to define force/energy or velocity/acceleration. Measurements are expressed in terms of units. Apart from the natural units explained above, units are arbitrary entities, and the physical quantities have different values depending on their chosen units. Early on, when the need for units to express measurements was realized, units were defined that were all subjective. For example, the unit of length depended on body parts (inch and foot), mass units depended on the mass of a grain of wheat (for small masses) or stone (for larger masses), and time units based on division of day into hours, minutes, and seconds were created by the Babylonians. As a result, some of the units currently adapted have only historical significance with no real scientific meaning. Modern definitions of units are now used, and given the size of the systems measured—whether the size of the universe or a cell—different scales are adapted (table 1.1).

Table 1.1. Measurement units

Multiple	Prefix	Abbreviation	Meaning	Number of units
10^{12}	tera-	T	trillion	1,000,000,000,000 units
10^{9}	giga-	G	billion	1,000,000,000 units
10^{6}	mega-	M	million	1,000,000 units
10^{3}	kilo-	K	thousand	1,000 units
10^{-3}	milli-	m	thousandth	0.001 units
10^{-6}	micro-	m	millionth	0.000001 units
10^{-9}	nano-	n	billionth	0.000000001 units
10^{-12}	pico-	p	trillionth	0.000000000001 units

There are different units for astronomical distances, depending on the size of the system in question. Within our planetary system (the solar system), the *astronomical unit* (AU) is used; it is defined as the distance from Earth to the sun, which is 149,597,871 kilometers. Larger distances outside the solar system are measured in terms of *parsec (pc),* which is defined as the distance at which one astronomical unit subtends an angle of one arcsec. This corresponds to 3.26 light-years or 31 trillion kilometers. Another unit for astronomical distances is the light-year, which is the distance light travels in one year, corresponding to 9.46×10^{12} Km. Larger distances are measured by kiloparsec (Kpc, 10^{3} parsec) or megaparsec (Mpc, 10^{6} parsec).

The conventional unit of mass in everyday life is the kilogram (kg), which is the mass of a cylinder kept at the International Bureau of Weights and Measures in France. This is not a satisfactory definition, since it depends on an object and has no legitimate basis. A more physical definition for *mass* in terms of motion of bodies, is as follows: Objects are inclined to keep their state of rest or a uniform motion on a straight line. This property of matter is called *inertia*. The mass of an object is a measure of its inertia. Mass should not be confused with weight, which is the accumulated force of gravity on the object. The mass of the object is the same regardless of its location, while the weight of an object differs depending on its position on Earth or any other planet. The mass of celestial objects is often measured in a unit based on the mass of our sun—that is, 1.989×10^{30} kg.

The basic unit of time is a *second*. This was initially defined as 1/86,400 of a solar day. However, since the solar day is not fixed throughout a year, this was not a fixed number. As a result, a *second* is now defined as 9 billion oscillations of a cesium atom. This provides a definition for time units with an accuracy of several millionths of a second. Astronomical or geological time is often expressed in millions or billions of years.

The velocity of an object is defined as the change in its distance, *d*, per unit of time, *t*, expressed as $v = d/t$, with the unit meters per second. *Acceleration, a*, is defined as the change in velocity per unit of time. If a system changes its velocity from v_1 to v_2 in a time interval Δt, its acceleration is $a = v_1 - v_2/\Delta t$ (when negative, it is called decceleration). Acceleration is defined in meters per square second, m/s^2.

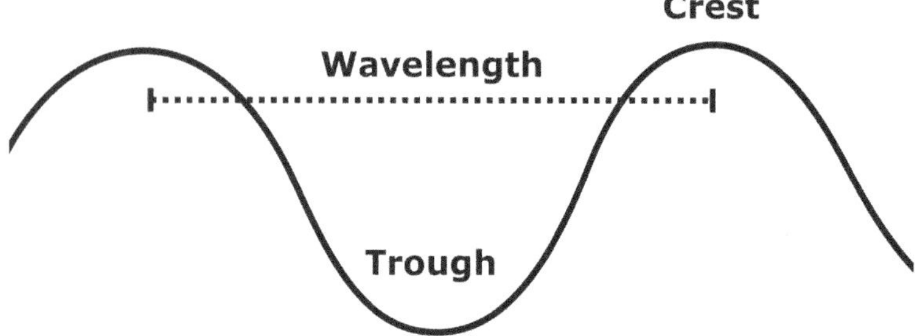

Figure 1.4. Wavelength, defined as the distance between two successive crests in a wave.

Temperature is defined as the average energy (speed) of molecules in an object or a system. In other words, temperature is a measure of the internal energy of an object resulting from the average velocity with which the molecules of the object move. Temperature is different from heat, which is defined as how energy is transferred from one system to the other. The unit of temperature used in scientific literature is the *Kelvin*. Zero Kelvin degrees is called *absolute zero* which corresponds to −273 degrees Celsius. At absolute zero, the atoms and molecules of an object do not move at all. There is no negative in the Kelvin temperature scale. To convert a Kelvin temperature to a Celsius unit, one needs to subtract 273 from the Kelvin unit. For example, 3 degrees Kelvin corresponds to −270 degrees Celsius.

For the elementary particles, mass and energy are equivalent through Einstein's famous relation $E = mc^2$ where E is the energy, m is the mass and c is the speed of light. Based on this, mass and energy are connected. The unit of energy is electron volt, defined as the energy gained by one electron (unit electric charge) when moving between two points with a potential difference of one volt. This is expressed in terms of million electron volts (MeV) or billion electron volts (giga-electron volts—GeV). Similarly, masses of elementary particles are expressed in terms of these energy units.

A concept that will be widely used in this book is the *wavelength*. This is defined as the distance over which the shape of a wave repeats. In other words, it is the distance between the two successive crests of a wave (figure 1.4). It is expressed in units of length and is often denoted by the Greek letter λ. *Frequency* is defined as the number of complete cycles of an oscillating system per second. It corresponds to the rate (number of times) at which the same sequence repeats itself within a unit of time. It is denoted by the letter f. The unit of frequency is the hertz (Hz). Larger units of frequency include kilo-Hz (KHz; 10^3 Hz), mega-Hz (MHz; 10^6 Hz), giga-Hz (GHz; 10^9 Hz), and tera-Hz (THz; 10^{12} Hz). For a wave moving with velocity V, the frequency (f) and wavelength (λ) are related by the equation: $f = V/\lambda$.

Light covers a range of wavelengths, from long (radio waves) to short wavelengths (X-ray and gamma ray). The visible wavelengths (the light our eyes are most sensitive to) only cover a short range. These are all electromagnetic waves and depending on their wavelengths, they are expressed in different length units (table 1.2)

Table 1.2. Electromagnetic waves and their corresponding wavelengths

Electromagnetic wave	Wavelength
Radio	> 1 mm
Microwave	1 mm to 25 μm
Infrared	25 μm to 2.5 μm
Near-infrared	2.5 μm to 750 μm
Visible	750 nm to 400 nm
Ultraviolet	400 nm to 1 nm
X-ray	1 nm to 1 pm
Gamma ray	< 0.001 nm

The commonly used units are nanometer (10^{-9} meters; *nm*), angstrom (10^{-10} meters; Å), and micron (10^{-6} meters or 10^4 angstrom; μm).

FUNDAMENTAL LAWS OF NATURE

Nature is governed by a set of physical laws that make it operate the way it does. The following subsections discuss the laws responsible for the motion of the planets around the sun and for the behavior of atoms and particles, as well as for daily operations of the world around us (Schneider and Arny 2015).

UNIVERSAL LAW OF GRAVITATION

All the objects on Earth are kept "down" because of the force of gravity. This is an attractive force between all the objects in the universe. The same way that Earth attracts us, we also exert to the Earth an equal force but in the opposite direction. The same forces operate between the celestial bodies, keeping the moon in orbit around Earth and Earth in orbit around the sun. This is the universal law of gravitation, first proposed by Isaac Newton in 1687 in one of the most famous books ever written, the *Philosophiæ Naturalis Principia Mathematica (Mathematical Principles of Natural Philosophy) known as the Principia*. This law states that an object with mass m_1 is attracted to another object with mass m_2, with a force proportional to the inverse square of their distance, r, expressed as

$$F = G\, m_1 \cdot m_2 / r^2$$

where G is the universal gravitation constant experimentally measured to have the value $G = 6.67 \times 10^{-11}\ \mathrm{N.m^2/kg^2}$.

LAWS OF MOTION

Newton's three laws of motion form the basis of the science of mechanics. These are summarized below.

The first law of motion: Every object remains at rest or moves with uniform speed in a straight line unless it is acted on by an external force. This is the law that introduces the concept of inertia, a tendency to resist a change in motion, as first proposed by Galileo.

The second law of motion: The acceleration of a moving object, a, is directly proportional to the net force acted on the object, F, and inversely proportional to the mass, m, of the object.

$$a = F / m$$

The third law of motion: When two objects interact, the force exerted on each of the objects is equal but opposite in direction to the force exerted on the other object. In other words, for any action there is a reaction equal in size but opposite in direction.

LAW OF CONSERVATION OF MOMENTUM

Momentum, p, is defined as the product of the mass of the object, m, by its velocity, v, expressed as

$$p = m.\,v.$$

It has the unit of *kg.m/s*. When an object is spinning or moving around itself, it has angular momentum, defined as the product of its mass, m, velocity, v, and size, r,

angular momentum = m.v.r.

The law of conservation of momentum states that the total momentum of interacting objects remains the same in the absence of external forces.

LAW OF CONSERVATION OF ENERGY

Objects in motion have a force, and once they collide with another object, they reduce their speed and transfer the force to the second object. The energy of motion is defined as kinetic energy (KE), and for an object with mass m moving with velocity v is expressed as

$$KE = \tfrac{1}{2}\, m.v^2.$$

Kinetic energy is measured in *joules*.

The energy an object has because of its position is called potential energy (PE). For example, when some mass is moved from the first to the third floor of a building, it stores potential energy. This type of potential energy is called gravitational potential energy since it is caused by gravitational attraction. For an object with mass m at distance h from the ground and under the gravitational attraction force g, this is defined as

gravitational PE = m.g.h

where mg corresponds to the *weight* of the object. Potential energy is measured in the unit of *newton meter* (*Nm*).

The law of the conservation of energy states that energy is never created or destroyed but converted from one form to another, with the total energy always remaining constant.

PLANCK'S LAW AND BLACKBODY RADIATION

Planck's law is the spectral energy distribution (the fraction of radiation energy emitted from an object at a given wavelength) of radiation emitted by a blackbody. The source of radiation is oscillating atoms, with their vibrational energy only having discrete values (being quantized). When an oscillator changes from an initial state of energy, E_1, to a lower state of energy, E_2, the amount of the energy released corresponds to the product of the frequency of radiation, f, and a constant value, h, called Planck's constant:

$E_1 - E_2 = h.f.$

The value of Planck's constant is $h = 6.626 \times 10^{-34}$ joules.

A blackbody is a hypothetical object that absorbs all the radiation energy falling on it until it reaches an equilibrium temperature. It then emits the absorbed energy at all wavelengths (Schneider and Arny 2015).

SEARCH FOR AN ORIGIN FOR THE LAWS OF PHYSICS

The order observed in the universe, the rising and setting of the sun for the last 4.6 billion years, the movement of the planets, the chemistry that governs biological processes in living subjects and everything else around us, all are manifestation of the laws of physics. These laws are expressed in rational and intelligible ways. It is not clear if the emergence of these laws was by accident. Similarly, the existing harmony between these laws, that they nicely complement one another without contradicting each other is among the most fundamental observations in science. Similar to what I discussed above about the physical constants, if the laws of physics were only slightly different from what they are, we would not have been here to study them. A straightforward explanation of these therefore comes from the *anthropic principle*. This states that the reason the universe is the way it is or the physical constants have the values they have or the laws of physics take the forms they do, is to accommodate and be compatible with the conscious life to observe them.

In order to explore the origin of the laws of physics, we first need to agree on a definition for these laws. The conventional definition is that a law describes patterns in nature and separates events that happen by chance from those that are there regardless of the conditions. A law is also able to provide reliable predictions. The rational order of the universe is manifested in the laws of physics. The task for scientists then is to take these laws as they are and apply them to explain natural phenomena, assuming they are independent of time and location in the universe.

Are the laws of physics a result of accidents or an amazing sequence of events that chose the best possible rules fine-tuned to govern the universe? We may never know the answer to this question. The laws of physics are absolute, unchangeable, and independent of the local conditions where they are applied—in the early universe, in the vicinity of our solar system, or in an atom. Physical processes do not have any effect on the laws governing them and the laws are completely independent of these processes.

Do the laws of physics serve a purpose? In other words, did they end up the way they are (out of many different possibilities) because this was the only way they could govern and sustain the universe without contradicting one another? In order to explain the universe by the laws of physics, these laws must have existed before the universe came into being, even before space and time existed. If true, the universe started in a deterministic way with all the events being predictable according to these laws. Would we ever be able to prove this? Attempts to unify forces in nature have opened the possibility that one day we may find where these laws came from and why they are the way they are. This explains how all the forces in nature today are manifestations of a single unified force from early in the history of the universe when density and temperature were at an extreme. It is therefore possible that the most fundamental laws were the ones at the very beginning of the universe that then led (probably by accident) to the laws we experience today. If this is the case, then how did those laws come about? One way to address this question is within the framework of the *multiverse* universe. In this scenario, it is hypothesized that a large number of universes exist and ours is only one of them. The laws of physics governing our universe were then originated in a different universe and were extended here. However, this only shifts the problem to a "different universe" that cannot be subjected to experimental tests. What is clear is that the laws of physics are only approximations to the truth. How accurately we can explain nature or predict future events is limited by the accuracy of our measurements (Primack and Abrams 2006).

SYMMETRY IN NATURE

The concept of symmetry plays an important role in the context of the laws of nature. The ability to repeat the same experiment at different places and times and come up with the same results relies on the invariance of the laws of nature under space-time transition. This characteristic gives the laws of nature an inherent regularity without which, it was impossible to discover them. An example of this is evident in characteristics of the particles responsible for the forces in nature, that are specified by field equations, predicting that the properties of them (i.e. the particles) are the same at any point in the field. We will review this in Chapter 4.

The expected regularity and symmetry in the laws of nature are sometimes hidden or broken by the arbitrary and unpredictable initial conditions. As we study the laws of physics at higher energies and smaller scales, we find more and more symmetries that are often hidden or broken at lower energies. Therefore, the symmetries are not revealed in the low energies of the present day universe but were present soon after the birth of the universe.

Imagine the very early universe when the temperature was enormous (of the order of 10^{32} degrees Kelvin or 10^{19} GeV—and here you could see the correspondence between temperature expressed in Kelvin or in GeV). The four fundamental forces in nature—electromagnetic, weak, strong and gravity—were indistinguishable at that time (I will explain these in detail in Chapter 4) with all the particles being massless. As a result, if you change one thing at that time by another, no one would notice, as everything was the same. Therefore, a symmetry existed at that time between the forces and particles. As the universe cooled down and the temperature dropped to $\sim 10^{22}$ degrees Kelvin (10^{14} GeV), due to some physical processes I will discuss later in the book, the symmetries broke. This resulted in particles gaining the mass they have today and the four forces becoming distinctive in nature. This continued and as a result of broken symmetries (because of the reduced temperature) all the different characteristics of particles and forces we observe in nature today, were revealed.

ORIGIN OF CONSERVATION LAWS

The conservation laws in nature are among the most fundamental laws. These must be obeyed in all the physical processes. Why the physical quantities (energy, mass, momentum) must be conserved? The first attempt to explain the origin of conservation laws was made by Emmy Noether, a mathematician, through what is now known as

Noether's theorem. The theorem states that the existence of conserved quantities in nature is a direct consequence of the symmetry of the laws of nature (the fact that the laws of nature do not depend on time). They are equally applicable to events in the past, present and future. Since the laws of nature are the same everywhere, momentum (measured by the mass and speed of the object) is also conserved. This means that if an object is at rest, its momentum is zero and will remain so unless an external force is acted upon it.

SUMMARY AND OUTSTANDING QUESTIONS

Everything around us in nature is likely to have had an origin. Study of their cause and their origin is essential for understanding how the world, as we see it, developed to its present form. This also provides the first step toward exploring changes with time in the observed phenomena—evolution. Such an adventure requires a multidisciplinary approach, reaching beyond the boundaries of specific disciplines. The aim of this book is to perform a scientific study of the origin of the observables in nature and to search for the reasons and causes that led the world to be the way it is.

In very early times, when the universe was much less than a fraction of a second old, processes at the microscopic level were dominant. At that time the interplay between particle physics and cosmology led the early evolution of the universe. Many of the phenomena we observe and measure in the universe today are the result of the interactions at that time. If things had gone slightly differently, everything would have been different today. The first generation of stars and galaxies were formed and then evolved through cosmic time. Atoms were then combined through the laws of physics, forming molecules and therefore the birth of chemistry. Through complex reactions between different chemical compounds, given the appropriate environments, biological molecules were formed; hence, the science of biology was born. This shows how seemingly different branches of science are connected. In the study of the origins, one needs to embrace and bridge between all these disciplines.

The laws of physics are symmetric—they are equally applicable in the past, present and future as well as anywhere in space. This symmetry has a consequence. This could explain the origin of the conservation laws. For example, why mass and energy in a system are conserved? This is explained by Noether's theorem as due to the laws of physics being independent of time. This symmetry in the very early universe, making the particles and forces indistinguishable, is now broken and this is the reason we experience different forces (with different characteristics) and particles in nature.

In order to explore the origin of the physical world, including the universe, stars, and galaxies, one needs to extrapolate the laws of physics back in time. Scientists do this by assuming that these laws apply at anytime and anyplace. The outstanding questions then are: What is the origin of the laws of physics? What is the origin of the fundamental constants? Why conservation laws apply so accurately? Once we have a framework to predict all details about the beginning of the universe, how could we verify it? Every corner of nature can be a subject to wonder. This is what inspires us as intelligent beings to find answers to these fundamental questions.

REVIEW QUESTIONS

1. Write the timeline of the universe from its beginning to the emergence of life.
2. Name some of the main events in the universe without which we would not be here today.
3. What was needed to start life on Earth?
4. How could life be sustained for some living creatures and not for others?

5. What is the biological process responsible for the diversity of living things?

6. Given that the age of the universe is 13.8 giga-years (billion years) and the age of Earth is 4.6 giga-years, if you were to accommodate the life of the universe in one year, when did Earth come into being?

7. What is the advantage of dimensionless units over those that have dimension?

8. What is the definition of the *fundamental physical constants*?

9. What is the difference between mass and weight?

10. What is the definition of *temperature*, and how does one convert Kelvin to Celsius degrees?

11. For the elementary particles, scientists sometimes use energy units instead of mass units. Explain how this is done and define the energy unit used.

12. What is the definition of *wavelength*? List units of length (distance) from atomic to astronomical scales.

CHAPTER 1 REFERENCES

Bennett, J., and S. Shostak. 2005. *Life in the Universe*. 2nd ed. Boston: Pearson/Addison-Wesley.

Hester, J., B. Smith, G. Blumenthal, L. Kay, and H. Voss. 2010. *21st Century Astronomy*. 3rd ed. New York: Norton.

Larsen, C.S. 2014. *Our Origins*. New York: Norton.

Primackm J.R. and Arams, N.E. 2006, The view from the center of the Universe, Riverhad books--Penguin Group, PLC.

Rees, M. J. 2001. *Just Six Numbers: The Deep Forces that Shape the Universe*. New York: Basic Books

Schneider, S.E., and T.T. Arny. 2015. *Pathways to Astronomy*. 4th ed. New York: McGraw-Hill.

FIGURE CREDITS

- Fig 1.1: Source: https://commons.wikimedia.org/wiki/File:CMB_Timeline75.jpg.

- Fig. 1.2c: Copyright © Veggiesaur (CC BY-SA 3.0) at https://commons.wikimedia.org/wiki/File:Phospholipid_Chemicalmakeup.png.

- Fig. 1.2d: Copyright © domdomegg (CC by 4.0) at https://commons.wikimedia.org/wiki/File:Simple_diagram_of_yeast_cell_(blank).svg.

- Fig. 1.2e: Copyright © OpenStax College (CC by 3.0) at https://commons.wikimedia.org/wiki/File:423_Table_04_02_Summary_of_Epithelial_Tissue_CellsN.jpg.

- Fig. 1.2f: Source: http://www.freestockphotos.biz/stockphoto/15174.

- Fig. 1.2g: Source: http://www.freestockphotos.biz/stockphoto/15174.

- Fig. 1.2h: Source: https://commons.wikimedia.org/wiki/File:Muskellunge_USFWS.jpg.

- Fig. 1.2i: Copyright © Elliott Lee (CC by 3.0) at https://commons.wikimedia.org/wiki/File:School_Of_Fish_(33018045).jpeg.

- Fig. 1.2j: Copyright © Peterkoelbl (CC BY-SA 2.5) at https://commons.wikimedia.org/wiki/File:Carcharhinus_longimanus_1.jpg.

- Fig. 1.2k: Copyright © Cliff (CC by 2.0) at https://commons.wikimedia.org/wiki/File:Coral_Fish_(2863778947).jpg.

- Fig. 1.2l: Source: https://commons.wikimedia.org/wiki/File:Earth_Western_Hemisphere_transparent_background.png.

- Fig 1.3: Copyright © Efbrazil (CC BY-SA 3.0) at https://commons.wikimedia.org/wiki/File:Cosmic_Calendar png.

DEVELOPMENT OF SCIENTIFIC THOUGHT: A HISTORICAL OVERVIEW

CHAPTER LEARNING OBJECTIVES

This chapter will cover:

- The search for reality from early to modern times
- The history of development of scientific thought
- First attempts to explain the world by objective reasoning
- The first world models
- The emergence of different scientific disciplines
- The scientific revolution and discovery of the laws of nature

The desire to know about our origin is an integral part of the intellectual curiosity that defines humanity. The main challenge here is to find rational explanations of the working of the world in search of reality and to address the question of why things are the way they are and what our role is, if any, in this vast cosmic landscape. In this journey, new ideas are subject to vigorous experimental verifications, and new hypothesis must conform to natural phenomena. This is the so-called scientific method that strives to explain observed phenomena through scientific, mathematical, and testable laws. An explanation is only acceptable if confirmed by experiments. Therefore, if an origin is assumed for anything in this world, it should first be identified and then verified empirically.

Often, in addressing deep questions like the origin of natural phenomena, we encounter abstract or speculative explanations. The quest to address these and the way philosophers and scientists have approached this question have evolved through the centuries. In the past the approach was abstract, but later, when humankind was able to perform observations of natural phenomena, it became possible to connect the models to the real world and to fully realize the hidden code for how things came to be the way they are. The most fascinating job is to decipher this code and uncover the truth behind it.

We humans have come a long way to understand the world around us. We should certainly be proud of this, but at the same time, we should acknowledge

"All Truths are easy to understand once they are discovered. The point is to discover them"

- GALILEO GALILEI

"The philosopher is in love with truth, that is, not with the changing world of sensation, which is the object of opinion, but with the unchanging reality, which is the object of knowledge"

- PLATO

that we have no unique place in this world. This has become clearer as we discover more about the universe. The ability that distinguishes us from other creatures is the power of our thought and imagination and the ability to learn from experience. This chapter briefly reviews the evolution of thought and different approaches taken to understand the hidden secrets of nature and, by extension, our origin. The pace of progress was very slow in the beginning but rapidly increased (Gleiser 2014).

Starting with Greek philosophers, this chapter summarizes the discoveries that changed humans' perception of the world, as knowledge was built up and different scientific disciplines emerged. This is followed by the realization of the power of observation, leading to the scientific revolution. The chapter discusses progress in modern science, the factors responsible for that, and how this progress changed our view of the universe and the world around us. When different scientific disciplines emerged after the scientific revolution, it was no longer possible to study one discipline in isolation. For example, the science of chemistry depends on the laws of physics, while biology is heavily dependent on understanding chemical processes. It is the relationships between such seemingly independent fields that one needs to discover in order to have a deep understanding of the question of the origins.

EARLY SEARCH FOR REALITY[1]

The first Greek philosophers attempted to search for reality through logic by moving from a faith-based to a knowledge-based approach. They were among the family of pre-Socratic philosophers who searched for a single and absolute principle for reality (Kirk et al. 1995). The first *pre-Socratic* philosopher was Thales (624–546 BCE)[2], who was the first to subscribe to the idea that all objects come from a single ultimate material: Water. Because of this, he believed in a unity of everything. Thales was the one who started the innovative approach to unify mathematics, astronomy, and philosophy and was the first philosopher to explain natural things by theories and hypothesis. A student of Thales, Anaximander (610–546 BCE), was the first science philosopher who believed that nature was governed by laws and tried to explain natural phenomena as a series of causes and effects. He conducted the earliest recorded scientific experiment. Although Babylonians are recorded as the founders of astronomy, Anaximander is often credited as the first to think about cosmology in a nonmythological way. He tried to observe and experiment with different aspects of the universe and its origin and to explore the mechanics of celestial bodies (Kahn 1994). Another notable student of Thale was Pythagoras (570–490 BCE), who was the first to call himself a philosopher (or lover of wisdom). He was the first pure mathematician and is known as the father of numbers. Pythagoras was the first to explain nature through mathematical methods and saw a beauty in the world that was expressed by mathematics. The one pre-Socratic philosopher who went beyond physical theory in search of metaphysical foundations and moral applications was Heraclitus (535–475 BCE). His ideas of a constantly changing universe with an underlying order and reason formed the foundation of later European worldview. Heraclitus lived around the same time as Parmenides of Elea (515–450 BCE), who was a very influential pre-Socratic philosopher and is known as the father of metaphysics. He followed the method of reasoned proof for assertions. In denying the reality of change, he started a turning point in the history of Western philosophy that significantly influenced the philosophers after him, including Plato.

Two early Greek philosophers who were seriously influenced by Parmenides's ideas were Anaxagoras (500–428 BCE) and Empedocles (490–430 BCE). Anaxagoras had insightful ideas in physical sciences that were quite

---

[1] The discussion of the early development of science by Greek philosophers makes significant use of following sources:
Internet Encyclopedia of Philosophy: http://www.iep.utm.edu
Stanford Encyclopedia of Philosophy: http://plato.stanford.edu/index.html
[2] "BCE" stands for "Before Common Era" which means the same thing as "BC" (Before Christ). "CE" stands for "Common Era" and is the same as Anno Domini (AD) meaning "the year of our lord"

revolutionary at their time. For example, he was the first to explain the cause of eclipses, and his ideas helped later development of atomism. Empedocles is credited as the originator of the cosmogenic theory of the four classical elements of the ancient world—earth, water, air, and fire—which became the standard belief for many centuries. One of the last pre-Socratic philosophers was Democritus (460–370 BCE), who developed a materialistic account of the natural world and significantly contributed to the establishment of the philosophical school of atomism. Although he was contemporary of Socrates, his views were closer to those of pre-Socratic philosophers (Kirk et al. 1995).

The above discussion summarizes the early search for reality that set the foundation for a scientific understanding of nature for the two millennia that followed. The fundamental achievement during the three hundred years from Thales to Democritus was that humankind could think about simple observations and decipher nature's message. This also helped the development of thought from initial hypothesis to final models in order to explain natural phenomena. This is how humankind started to think about its place in the universe and the question of the origin. The important feature during this time that contributed to the development of thought was the power of dialogue and conversation that was established by Socrates (469–399 BCE) and is known as the Socratic method of question and answer. This method played a major role in developing and contrasting ideas by his students and well after that when addressing the nature of reality.

FIRST WORLD MODELS

The first model for the universe was proposed by Socrates's student Plato (428–348 BCE). In his famous work, *Timaeus*, Plato gave his view of the creation of the universe and everything in it. In his view, the universe was not created out of nothing, as suggested by medieval theologians, but from the already existing elements fire, air, water, and earth, which constituted various compounds that formed the world. Plato's universe consisted of stars, planets, sun, and moon, all rotating around Earth in spheres. He concluded that the sphere of the moon was closest to Earth, followed by the sphere of the sun, and then the other planets farther away, followed by the stars as the most distant. He proposed that celestial bodies must be symmetrical and have a perfect shape and that this was the only possible way they could exist. Plato's student Aristotle (384–322 BCE) used observations to assert his models of natural phenomena and rational arguments to explain them. Aristotle's universe was not changing with time and had always existed in the same way. Aristotle's Earth-centric model for the universe was universally accepted as the only viable model for the universe for nearly nineteen hundred years, until the Copernican revolution. During this time, modifications were made to the Earth-centric model, but the basic principle of the model remained unchanged.

Although Aristotle established the philosophical foundations of the geocentric model of the universe, he didn't elaborate the details of his model based on astronomical observations. In the second century (Common Era), Ptolemy (90–168 CE) developed a standard geocentric model in which the observed planetary motions were explained by a combination of circular motions. However, to make his models compatible with observations, Ptolemy had to deviate from some Aristotelian principles. Despite many attempts to modify the Ptolemaic model, this uncertainty about the configuration of the planetary spheres remained unsolved until the mid-sixteenth century.

In his work *De revolutionibus orbium coelestium* (*On the Revolution of Heavenly Bodies*), Nicolas Copernicus (1473–1543) attempted to reconcile these discrepancies by proposing that the sun rather than Earth is at the center of the universe. He believed that the size and speed of each planet ultimately depended on its distance from the sun. This was a revolutionary concept, often referred to as the Copernican revolution, and although it faced opposition at its time, it paved the way for future discoveries (Brush and Holton 2001).

A WORLD MODEL BASED ON OBSERVATIONS

The Danish astronomer Tycho Brahe (1546–1601) was the last of the naked-eye observational astronomers. He performed accurate observations of the location of celestial bodies, their angular measurements, and their motions along the celestial sphere. By observing a supernova and a comet, he concluded that they were farther away from us than the moon. This measurement was very important because in Aristotelian and Ptolemaic models, comets and supernovas were considered meteorological or atmospheric phenomena. Brahe, therefore, provided one of the first observational evidence against the classical astronomy. He also suggested a new model in which the planets revolved around the sun while the sun (with all planets) orbited around the central Earth. Brahe made the most accurate measurements for the position of celestial bodies at that time (Almasi 2013). His assistant, Johannes Kepler (1571–1630), used these data and fitted them to mathematical models, deriving his three laws of planetary motion. Kepler was influenced by Neoplatonic thoughts and believed that "the geometrical things have provided the creator with the model for decorating the whole world." He was the first to discover universal laws governing the motion of the planets. His initial model assumed a circular orbit for the planets around the sun, but this revealed eight minutes of arc discrepancy with the accurate planet positions measured by Brahe. Instead of attributing this to observational errors, he modified his models and finally found the best fit when assuming an elliptical orbit for the planets around the sun, with the sun being in one focus (his first law of planetary motion), as well as the relation between the orbital velocity and orbital distance of a planet from the sun (his second law of planetary motion). By trying many combinations of models, he then arrived at his third law of planetary motion—the square of the orbital period of any planet around the sun is proportional to the cube of its semimajor axis. This was a perfect match to the observational data available at the time and mathematical models that were later confirmed by many independent experiments. The significance of Kepler's laws of planetary motion became clear in 1687 when they were used by Isaac Newton to derive his universal law of gravitation.

In 1610 the Italian astronomer, philosopher, and engineer Galileo Galilei (1564–1642), using his newly built telescope, observed that four celestial bodies orbited Jupiter—he had in fact discovered the moons of Jupiter. This contradicted Aristotelian cosmology that proposed that all celestial bodies orbited Earth. In the same year, Galileo noticed that Venus had the same phases as the moon. It was not possible to accommodate such observations within Ptolemy's geocentric model, as the heliocentric model predicted different phases for Venus. These observations rejected the Aristotelian view of a geocentric universe in favor of a heliocentric (moving around the sun) solar system. This affected fundamentals of science and philosophy and human perception of the world.

THE SCIENTIFIC REVOLUTION

Isaac Newton (1642–1727) was born the year Galileo died. In his book *Philosophiæ Naturalis Principia Mathematica* (*Mathematical Principles of Natural Philosophy*), published in 1687, he developed the foundation for the concept of universal gravitation and laws of motion. By combining Kepler's laws and the law responsible for centrifugal force discovered by Christiaan Huygens (1629–1695), Isaac Newton showed that centripetal force (the force that makes a body move on a circular path and whose direction is toward the center around which the body is moving) was indeed responsible for motion of the planets and that gravitational attraction between the sun and its planets decreases proportional to the square of distance between them. This was the birth of Newton's law of gravity that revolutionized science. Using this, he predicted the orbit of the comets, motion of the planets, and many other celestial phenomena. Newton's work confirmed beyond doubt the validity of the heliocentric model for the solar system and rejected the Aristotelian model for the world. Newton's law of gravity and his laws of motion have successfully explained the observations within the solar system for the past 350 years.

Apart from the law of gravity, another important development in the post-Newtonian era was the understanding that electricity, magnetism, and light are all manifestations of the same phenomenon. This was expressed in four equations developed by Scottish mathematician James Clerk Maxwell (1831–1879) between 1860 and 1871. This was the first expression of the unification of forces and was named electromagnetic force. He showed that the speed with which electromagnetic wave propagates is the same as the speed of light and that light is an electromagnetic disturbance propagated according to the laws set by Maxwell's equations. Together, Newton and Maxwell set out the foundation for modern science that continues to the present time.

Newton's views of the world were challenged at the beginning of the twentieth century. With the breakthroughs made in different branches of natural sciences, a new view of the world emerged. In 1905 Albert Einstein (1879–1955) presented his special theory of relativity, shattering humans' perception of the world up to that time. In developing his theories, Einstein reasoned based on Maxwell's theory of electromagnetism. He argued that the speed of light that appears in Maxwell's equations was a constant regardless of the speed of the source producing it. Further, he proposed that the laws of physics are the same regardless of their reference frame. These two concepts formed the basis of the special theory of relativity and modern physics.

EMERGENCE OF MODERN VIEWS FOR THE NATURE OF MATTER

In the early nineteenth century, British chemist John Dalton (1766–1844) noticed that elements always react in ratios of whole numbers. By measuring relative masses of different components (elements) of a compound, he realized that the elements enter into reactions in multiple numbers of discrete units, or atoms. French physicist Jean Perrin (1879–1942) experimentally confirmed Dalton's atomic theory, for the first time showing that matter consisted of discrete units called *atoms*. In an experiment with cathode rays, British physicist J. J. Thomson (1856–1940) noted the existence of particles eighteen hundred times less massive than atoms. It was soon realized that these "new" particles had negative electric charges and were responsible for electric currents in wires; therefore, they were called *electrons*. The discovery of electrons challenged the idea that atoms were the ultimate constituents of matter and indivisible entities. However, at the time it wasn't clear how the electrons were distributed within the atom. In an experiment in 1909, Thomson's student Ernest Rutherford (1871–1937) was studying scattering of positively charged alpha particles by thin metal foils when he noticed deflection of the particles by more than 90 degrees, contrary to what he had expected. Rutherford explained this observation by proposing a concentration of positively charged particles in a small nucleus in the center of the atom. Thomson and Rutherford proposed the first models for the atomic structure, the plum pudding model and planetary model, respectively. However, in 1913 another of Thomson's students, Danish physicist Niels Bohr (1885–1962), showed that Rutherford's model could not be stable and proposed the first model for atoms based on quantum concepts. In this model the nucleus was at the center of the atom, with electrons moving around the nucleus on separate orbits. The electrons could change their orbits by absorbing or emitting discrete packets of energy. Information of the structure of atoms led to the understanding of chemical bonds in atoms in 1916 and the long-standing problem of explaining emission and absorption spectra. This was explained as the interaction between the electrons in different atoms, with the chemical properties of the elements explained by American chemist Irving Langmuir (1881–1957) in 1919 as the special pattern in which the electrons were connected.

In the 1920s the world of physics was revolutionized by the development of quantum mechanics, changing humankind's view of the building blocks of matter. The quantum nature of matter was revealed by Max Planck's discovery of blackbody radiation in 1900 and Einstein's explanation of the photoelectric effect in 1905. The physical characteristics of elementary particles were first measured in an experiment in which beams of silver atoms, after passing through a magnetic field, were split, depending on the direction of the angular momentum (or spin) of the particles. In 1924 Louis de Broglie (1892–1987) proposed that all particles to some extent behave like waves. This model was used in 1926 by Erwin Schrödinger (1887–1961) to develop a mathematical formulation for atoms by considering electrons as waves rather than point particles. One consequence of the wavelike nature of electrons is that one could not measure both the velocity and the position of a particle at any given time; this is known as the *uncertainty principle,* developed by Werner Heisenberg (1901–1976) in 1926. Quantum mechanics gives deep insight into the working of matter at atomic scales (Collins 2007).

Today electrons are considered fundamental particles, while the constituents of the nucleus—*neutrons and protons*—consist of even more fundamental particles called *quarks* (each neutron and proton consists of three quarks). The existence of different types of quarks is experimentally verified. Also, given the wavelike nature of particles, their behavior is more conveniently explained in terms of fields (Chapter 4). The second half of the twentieth century witnessed enormous progress in the development of theories to explain the nature of elementary particles and behavior of quarks as well as the attempt to unify all forces in nature. As one searches deeper and finds more details, experimental verification of the models becomes more difficult, requiring more sophisticated experiments. Nevertheless, it is within our present technological abilities to test the existing models.

MODERN VIEWS OF SPACE AND TIME

In 1915 Einstein developed the general theory of relativity within a beautiful mathematical framework. In developing this, he changed the Newtonian concept of absolute space and time, relating mass-energy to the geometry of space and gravity. General relativity predicted many observable phenomena within the solar system and made predictions about the fabric of space and time. Many of these predictions were experimentally confirmed soon after the General Relativity was developed. These include the precession of the perihelion of Mercury, the bending of light due to gravity and more recently, the concept of gravitational waves produced by massive bodies. Extending the theory to beyond the solar system, solution of Einstein's equations predicted a dynamic universe, either contracting or expanding. In the lack of any observational evidence, Einstein introduced a mathematical term to his equation to counterbalance the attractive force of gravity, resulting in a static universe. This term, known as *cosmological constant*, produced pressure to counteract the force of gravity and therefore push galaxies apart. In one of the truly amazing encounters of theoretical prediction and observational verification, American astronomer Edwin Hubble (1889–1953) discovered the expansion of the universe in 1925 using the largest telescopes at the time, located at Mount Wilson, California.[3] After the discovery of the expansion of the universe, Einstein abandoned the concept of the cosmological constant. However, after the discovery in 1998 that expansion of the universe is accelerating due to a mysterious force known as *dark energy* (Chapter 10), interest in the cosmological constant revitalized. The force of dark energy has the same effect as the cosmological constant. The true nature of this repulsive force is unclear at present. This is one of the rare instances in which an observational fact is awaiting theoretical explanation.

The discovery of remnant radiation from the beginning of the universe, called *cosmic background radiation,* is another example of a theoretical prediction followed by observational verification. The existence of this radiation had been predicted by George Gamow (1904–1968) in 1953 and observationally discovered by Arno Penzias (1933–) and Robert Wilson (1936–) in 1964. This radiation fills the entire universe and provides an absolute frame of reference in space and time. Since its discovery, there have been detailed studies of this background radiation by dedicated space and ground-based missions.

---

[3] The 100 inch (2.5 meter) Hooker Telescope in Mount Wilson Observatory was the largest telescope in the world during 1917–1949 and was used by Edwin Hubble to discover the expansion of the universe.

With these observations, cosmology, defined as the study of the origin and evolution of the universe, was put on solid experimental grounds. This was the start of cosmology as an independent and scientific discipline (Brush 1992). Since then, the road to discovery continued, and with advances in technology, it emerged as a precision science. Over the past few decades, the progress has been phenomenal, moving us closer to understanding the reality of the world we live in.

FROM NATURAL PHILOSOPHY TO NATURAL HISTORY AND BIOLOGY

Pre-Socratic philosophers wondered about life, posing many questions in this regard. Hippocrates (460–370 BCE), known as the father of medicine, was the first person to believe that diseases were caused naturally and not through superstition and gods, separating the discipline of biology (the science of life) from religion. Hippocrates was a contemporary of Aristotle, whose work in biology was empirical, unlike his speculative writings in natural philosophy. Aristotle categorized many species of animals and believed that living things were all arranged in graded scales of perfection, from plants to animals. This view was shared by other scholars until the eighteenth century. After Hippocrates and Aristotle, Claudius Galen (129–216 CE) emerged as a notable biologist and physician; he established the basis of many biological disciplines today, including physiology and surgery. After Galen, despite documented evidence for Chinese, Mesopotamian, and Egyptian scholarly activities in biology, no major progress was made until the time of the European Renaissance, from the fourteenth to the seventeenth century (Lindberg 2008).

Andreas Vesalius (1514–1564) was one of the first physicians to replace abstract reasoning with factual empiricism in medicine and biology. He dissected and studied animals and wrote one of the most influential books on human anatomy, *On the Fabric of the Human Body*, in 1543, the same year that Copernicus published his revolutionary book about heliocentric astronomy. Vesalius's approach also helped in observing and studying plants. The invention of the microscope by Dutch microbiologist Antonie van Leeuwenhoek (1632–1723) in 1670s started a revolution in biology. This led to a huge increase in magnification and the discovery of microscopic life. English naturalist John Ray (1627–1705) was the first to classify plants in subgroups, giving scientific definition to the term *species* and starting the science of *taxonomy*. At about the same time, Danish anatomist and geologist Nicholas Steno (1638–1686) observed that remains of living organisms can be trapped in layers of rocks, producing fossils, which suggested an organic origin for fossils. This was not fully accepted by scientists at the time, due to philosophical differences as well as the influence of religion in questions like the age of Earth.

Early in 1700s, natural sciences started to diversify, with different disciplines growing independently from others. All the knowledge in natural sciences up to the seventeenth century was collected by French Naturalist Comte de Buffon (1707–1788) and published in thirty-six encyclopedic volumes called *Histoire Naturelle* (*Natural History*). Buffon's lifetime work was published the same year Immanuel Kant proposed his famous theory about Earth's formation. These works collectively set the foundation for the science of *geology*. This was complemented in the early nineteenth century by the work of Alexander von Humboldt (1769–1859), who analyzed the relationship between organisms and their environments following the quantitative approach of natural philosophy. Humboldt's work led to the study of the relationship between the spatial and temporal distribution of biological organisms that eventually resulted in the development of independent disciplines like geology, paleontology, and biogeography in the early nineteenth century. These attempts collectively laid the foundation for the study of evolution. Fundamental to this was the work by Georges Cuvier (1769–1832), who performed experiments comparing mammals with fossils and concluded that fossils were the remains of species that had become extinct. Later, with advances in science and technology, fossils became one of the main tools for studying natural history and eventually the history and evolution of our planet.

The science of evolution took a new turn in 1859 with the publication by Charles Darwin (1809–1882) of his book *On the Origin of Species by Means of Natural Selection*. Before Darwin, French naturalist Jean-Baptiste Lamarck (1744–1829), influenced by Comte de Buffon, proposed a different theory for evolution, with British explorer Alfred Russell Wallace (1823–1913) finding similar evidence for evolution as Darwin. What made Darwin's work distinctive was the volume and strength of the undisputed data he compiled and the scientific method he followed to deduce conclusions from those data. By the end of the nineteenth century, a growing number of scientists accepted the concept of evolution and a common descent for all living creatures. Although disputed for many years, huge compilation of data today, have put evolution on firm scientific grounds.

Meanwhile, another revolution was taking place. Advances in microscope technology led to the discovery of *cells* as the basic units of living organisms and the development of cell theory in 1839 by Theodore Schwann (1810–1882) and Matthias Schleiden (1804–1881). This discovery was based on the observations that new plant cells were formed from old plant cells and that the process was similar in animal cells. With the help of more powerful microscopes by the end of the nineteenth century, different components of cells were identified and studied. Having found the basic constituents of living organisms, biologists could now direct their attention to the origin of life. In a series of experiments, Louis Pasteur (1822–1895) showed that living organisms could not be produced from nonliving material, ending a debate that had started in Aristotle's time.

The discoveries made during this period heralded new disciplines in science, leading to the development of life sciences and their division to different subareas. This allowed scientists to dig deeper into understanding life itself and the origin of the materials that are responsible for life (Schrödinger 1992).

THE BIRTH OF ORGANIC CHEMISTRY AND MOLECULAR BIOLOGY

In the early twentieth century, scientists began to explain the behavior of living things through physical and chemical processes. The science of *organic chemistry* was born, attempting to separate organic and inorganic materials. The point of great importance was the understanding that presence of organic material is a requirement of any living organism, but not every organic material is a signature of life. This led to the study of the physical and chemical functions of living creatures. At the same time drug metabolism was discovered as well as *proteins* and *fatty acids*. By the 1920s scientists started to work on the metabolic pathways of life, leading to significant advances in *biochemistry*. During the 1930s many scientists applied techniques of physics and chemistry to biology. This led to the birth of *molecular biology*, paving the way for many outstanding discoveries that have continued to the present time.

Advances in biochemistry and genetics led to the discovery by Oswald Avery (1877–1955) that *nucleic acids* and not proteins are the genetic material of genes and chromosomes. Following this, in 1953 James Watson (1928–) and Francis Crick (1916–2004) proposed a model for the structure of genetic material—*dioxyribonucleic acid* (*DNA*). The fundamental point here was the special pairing of nucleic acids that suggested a copying ability for DNA, needed for any living molecule. This discovery was made possible due to advances in the then new technology of X-ray crystallography. Therefore, advances in one field started to have significant effects on progress in other disciplines (Watson 1980). In the following years, from the late 1950s to the early 1970s, molecular biology expanded rapidly, with the science of biology diversifying to a number of independent disciplines. This included evolutionary biology and later astrobiology, addressing the most fundamental questions as to the molecular basis of diseases, new branches in medicine and origin and evolution of life.

At no other time in the history of civilization has humankind made so much progress in understanding the basic functions required for life to begin and evolve as it has since the beginning of the twentieth century. Advances in technology allowed deeper questions to be asked and investigated, while other areas of natural sciences were brought into play to address the most fundamental questions. I will explore some of these in the following chapters in this book.

SUMMARY AND OUTSTANDING QUESTIONS

This chapter presented a brief (and incomplete) chronological summary of the development of science since the beginning of civilization. The aim was mainly to show the origin and evolution of scientific thinking and major discoveries throughout the centuries, as well as the development of the scientific method. Discoveries made at the beginning of civilization by ancient philosophers were as profound as those made in recent years using the most advanced methods and technologies. These collectively provided humankind with the opportunity to acquire a more accurate understanding of the truth.

In the early days, unlike the present time, scientific process was more abstract, developed by individual thinkers, and at times originated from faith rather than objective observations. However, at all times these thinkers had one thing in common—a quest for the truth. During this process many different questions came up. When seeking answers, many more questions arose, which were subsequently studied. All these questions and theories submitted themselves to experimental verification and were only accepted if they were supported by experiments. This chapter also covered the emergence of the disciplines that are now used to directly study the most fundamental questions—broadly speaking, how did the universe, galaxies, stars, planets and life come about and evolve to the state they are in now? One thing is clear from the discussions in this chapter: Every era had its own challenges and puzzles. Every discovery brought us closer to understanding the reality but it also generated many new questions and now, nearly 2500 years since the first philosophers pondered the question of the truth, we are as far away from it as we have ever been.

About the same time that Galileo was looking through the first telescope to study what was then considered to be the cosmos, van Leeuwenhoek looked through his newly built microscope to study the "micro-cosmos." Therefore, the microscopic world that led to biological revolution moved alongside the study of macroscopic phenomena, the cosmos (McClellan and Dorn 2015). These studies are dependent upon the laws of physics, that in turn, are responsible for chemical bonds forming the compounds that are needed in biology to generate the energy required by our cells and to transfer hereditary information across our ancestral lines.

To test the theories for the origin of the universe, scientists needed to simulate the conditions that prevailed at that time. This led to the construction of the largest particle accelerators, which accelerate particles to a speed close to that of light in two different directions; when the particles collide, they generate energies of the order that existed in the early universe. Through this, scientists have revealed how physics at microscopic scales could affect the state of the universe today. Galileo's telescope has now been replaced by large ground-based and space-borne observatories, which allow us to study the most distant parts of the universe or find other planets, while van Leeuwenhoek's microscope has turned to advanced electron microscopes and magnetic resonance imaging machines to study how cells develop, how they produce their energy requirements, and how the human mind works. This is the greatest achievement of humankind—to decipher the laws of nature.

The present development of modern science is very different from that during the time of Aristotle and Galileo. We have moved more toward an objective view of science based on experiment and observation. Similarly, the perception of reality is completely different now than it was centuries ago. Also, the pace with which science has developed is much faster. It took humankind thousands of years to remove Earth from the center of the universe, while over the past hundred years our entire understanding about the world around us has changed. Today the largest and best-equipped community of scientists that ever existed in history is working to solve problems once seemed unthinkable (Bowler and Morus 2005). If there is a time to scientifically study the origin of the universe and life, this is that time.

REVIEW QUESTIONS

1. Early Greek philosophers are divided into pre-Socratic and post-Socratic. On what basis is this division made?
2. Who was the first to explain natural phenomena (a) by theories and hypothesis, (b) as a series of causes and effects, and (c) by mathematical methods? Note the timing of these three events and explain if this initiated the process of scientific thought.
3. How did Parmenides of Elea influence the philosophers after him, and how long did this influence last?
4. What was the earliest model of the world?
5. Why did the Platonic model of the universe last for so long?
6. On what basis did Nicolaus Copernicus propose his model of a sun-centric universe?
7. Explain how the combined work of Tycho Brahe and Johannes Kepler led to the laws of planetary motion.
8. What observations by Galileo rejected the Aristotelian geocentric model of the universe and why?
9. What was the main difference between Newton's and Einstein's views of the world?
10. Explain the different steps that led to the final discovery of the first model for atomic structure.
11. Why was the cosmological constant introduced to Einstein's equations? Why was it abandoned and then introduced again?
12. Explain the early development of biology from Hippocrates to Aristotle and Galen.
13. When in the history of biology did technology start to play a significant role?
14. Describe the discovery of fossils, their interpretation as organic material, and how this discovery was received at the time.
15. Explain how the science of geology started.
16. Who was the first to conclude that fossils are remnants of extinct animals?
17. What makes Charles Darwin's work on evolution so unique and convincing?
18. During the 1920s and 1930s, organic chemistry, biochemistry, and molecular biology were developed as independent disciplines. Explain the fundamentals behind these developments.

CHAPTER 2 REFERENCES

Almasi, G. 2013. "Tycho Brahe and the Separation of Astronomy from Astrology: The Making of a New Scientific Discourse." *Science in Context* 26 (01): 3–30.

Bowler, J.P., and I.R. Morus. 2005. *Making Modern Science: A Historical Survey.* Chicago: University of Chicago Press.

Brush, S., and G. Holton. 2001. *Physics, the Human Adventure: From Copernicus to Einstein and Beyond.* New Brunswick, NJ: Rutgers University Press.

Brush, S.G. 1992. "How Cosmology Became a Science." *Scientific American* 267 (2): 62–68. doi:10.1038/scientificamerican0892-62.

Collins, G.P. 2007. "The Many Interpretations of Quantum Mechanics." *Scientific American*, November 19.

Gleiser, M. 2014. *The Island of Knowledge: The Limits of Science and the Search for Meaning.* New York: Basic Books.

Kahn, C.H. 1994. *Anaximander and the Origins of Greek Cosmology*. New York: Columbia University Press. Reprint, Indianapolis: Hackett.

Kirk, G.S., J.E. Raven, and M. Schofield. 1995. *The Presocratic Philosophers*. 2nd ed. Cambridge, UK: Cambridge University Press.

Lindberg, D.C. 2008. *The Beginning of Western Science*. Chicago: University of Chicago Press.

McClellan, J.E., III, and H. Dorn. 2015. *Science and Technology in World History: An Introduction*. Baltimore: John Hopkins University Press.

Schrödinger, E. 1992. *What Is Life?* Cambridge, UK: Cambridge University Press.

Watson, J. 1980. *The Double Helix: A Personal Account of the Discovery of DNA*. New York: Norton.

THE ORIGIN OF SPACE AND TIME

03

CHAPTER LEARNING OBJECTIVES

This chapter will cover:
- Space and Time: a historical view
- The nature of space and time
- The arrow of time
- Space, time, and gravity
- The nature of mass and energy

We live in space and feel the passage of time. These are the two most fundamental things in nature we deal with every single moment of our life. But, what is the nature of space and time? Did they always exist or come to being sometime in the distant past? If so, could we then imagine a world devoid of space and time? Are they the fundamental fabric of the universe or result of our perception? Can we study them by scientific means? Study of the nature of space and time has evolved throughout many centuries, from philosophical viewpoints to scientific and testable theories. Nevertheless, understanding the origin and nature of space and time remains elusive and among the most fundamental and intellectually challenging problems. While the change in our perception of space and time through history provides a fascinating story, we are still wondering what these entities actually are and how they relate to the world we live in. To perform such a study, one needs to contrast the abstract concept of space and time against the reality revealed from modern observations.

A related issue is the concept of mass and energy. Study of the relation between these quantities has challenged generations of scientists. In modern physics, the nature of mass and energy and their origin cannot be understood without understanding their relationship to space and time. At large scales, space and time are influenced by the force of gravity, which is itself dependent on the presence of mass and energy as was shown by Einstein's theory of general relativity.

"Nothing puzzles me more than the time and space; and yet nothing troubles me less"

- CHARLES LAMB

"Nothing exists except atoms and empty space; everything else is opinion"

- DEMOCRITUS

This chapter starts with a historical review of the concept of space and time, followed by a discussion of their nature. The reality of the arrow of time will then be discussed, and the question will be addressed as to whether space and time are separate or integrated entities. The influence of gravity on the fabric of space and time will be studied as well as the relationship between space-time and mass-energy.

A HISTORICAL VIEW OF SPACE AND TIME

The concept of space and time has evolved over centuries. Plato (428–348 BCE) in *Timaeus*[1] (dialogue) explained space as the entity in which things come to be and time as the period of motion of heavenly bodies. His student Aristotle (384–322 BCE), in his influential work in science and philosophy, *Physica Auscultationes*[2] (*Lectures on Nature*), defined space (the place of a thing) as "the first (i.e. innermost) motionless boundary of what contains" and time as "a constant attribute of movement that cannot exist by itself but is relative to motion" (Sachs 1995). Saint Augustine of Hippo (354–430 CE), in his autobiographical work *Confessiones*[3] (*The Confessions*), argued that the knowledge of time depends on the knowledge of movement, and therefore time cannot exist where there are no creatures to measure its passing. In this context, Saint Augustine connected time to the question of the creation. Later on, theologians rejected the notion of an infinite universe, as proposed by Greek philosophers. They argued that an actual infinite cannot exist and ascribed a beginning to the universe and hence a beginning for time. In his very influential work on the philosophy of space and time, *The Critique of Pure Reason*, Immanuel Kant (1724–1804) tried to explain time as a notion that, when combined with another notion called space, allows one to understand knowledge through sense, experiment, and data as opposed to pure reason (Kant [1781] 1999). The concept of absolute space and time—that is, the entities depending on themselves, as opposed to relying on other objects for their existence—was debated by Newton (1643–1727) in his book *Philosophiae Naturalis Principia Mathematica*[4] (*Mathematical Principles of Natural Philosophy*). In Newton's model, absolute space is needed to account for phenomena like rotation and acceleration that would not depend on other objects. In contrast, German philosopher Gottfried Wilhelm Leibnitz (1646–1716) argued that what we call between two objects is not space but the relationship between them. In other words, space has no meaning if objects do not exist, and motion is defined only as a relationship between these objects. In Newton's view however, space and frame of reference exist independently of the object in it, with the objects moving with respect to space itself. Newton's idea of absolute space prevailed for almost two centuries, until Ernst Mach (1838–1916), an Austrian physicist and philosopher, proposed his principle (known as Mach's principle) that the *inertia* (the tendency of an object at rest to remain at rest and an object in motion to continue in motion in the same direction) results from the relationship between an object with objects in the rest of the universe, however distant those objects are. In other words, inertia is caused by the interaction of different bodies in the universe.

---

[1] Plato's *Timaeus on Physics* is a monologue account of the formation of the universe. It is the collection of Plato's intellectual achievements. It depicts an orderly universe with the dialogues aiming to explain this. Persons in the dialogue include Socrates, Timaeus, Hermocrates, and Critias. Translated by B. Jowett in 1982.

[2] Aristotle's *Physica Auscultationes* is a collection of lectures in eight books dealing with philosophical principles of natural and moving bodies (living or nonliving). The book deals with the principle causes of motion and movement. This is a foundational work in physics, cosmology, and biology.

[3] Saint Augustine of Hippo's *Confessiones: The Confessions of Saint Augustine*. The original book was written by Saint Augustine of Hippo between 397 CE and 400 CE and contains an autobiography of Saint Augustine, explaining humanities' great concerns.

[4] Isaac Newton's *Philosophiae Naturalis Principia Mathematica* also called *Principia*—written and published in Latin in 1687 and translated to English by B. Cohen and A. Wittman, University of California Press, 1999. This is regarded as the most important book in the history of science and lays out the foundation of classical mechanics, the universal law of gravitation, and presents a derivation of Kepler's law of planetary motions.

Newton's views of absoluteness of space and time were challenged by Albert Einstein (1879–1955) in the beginning of the nineteenth century. Through some thought experiments, Einstein postulated that space and time are relative. Consider two astronauts in two different spaceships floating in space. If the spaceships are moving with uniform speed, there is no experiment any of the astronauts could do in their respective spaceships that could determine which one is moving and which is not. What one needs here is an absolute reference with respect to which they could measure their motion. But then, how do we know the "absolute" frame is not moving? This led to the conclusion that all motions are relative. This now leads to another important concept. Since none of the astronauts could perform experiments to detect their motion through space, the laws of physics must be the same in both spaceships. If this were not the case, experiments would have produced different results in the two spaceships, and the astronauts would be able to decide who is moving. This resulted to the *principle of relativity*, stating that the laws of physics are the same for all observers regardless of their reference frame, provided they are not accelerating (are in inertial frames). This led Einstein to arrive at the second principle that forms the foundation of the special theory of relativity, that the speed of light is the same for all observers independent of their motion. If the speed of light were not constant, then the astronauts could measure the speed of light in their spaceships and determine who was moving.

The above discussions all assumed a world without gravity. Now, how would this change by introducing gravity? Influenced by Mach's principle, Einstein argued that the gravitational force due to distant stars in the universe is responsible for acceleration and inertia. This led to the development of the *equivalence principle*, which states that the force felt by an observer in a gravitational field and that in an accelerating frame of reference are indistinguishable. In other words, the astronaut in a spaceship cannot distinguish between the force of gravity and the force produced due to acceleration of the spaceship. These led Einstein to propose that inertia, gravity, and acceleration are associated with the way space and time are related in what is referred to as space-time. Putting this a different way, the concept of space-time is associated with geometry and curvature. According to the general theory of relativity, the mass of an object affects the geometry of space-time around it, causing it to curve, and the curvature of space-time is responsible for acceleration/deceleration of mass and hence gravity.

Here I address three fundamental questions: What is the origin and nature of space and time? Could they exist independent of one another? How is space and time related to the fabric of our universe?

THE NATURE OF SPACE AND TIME

For over two hundred years, Newton's concept of absolute space and absolute time prevailed. Newton also postulated a medium required for light to travel—the so-called *aether* considered to be an absolute reference frame for motion and a frame with respect to which light moved. In his most intriguing equations explaining propagation of the electromagnetic fields, James Clerk Maxwell (1831–1879) suggested that light is a form of electromagnetic wave moving at a speed of 300,000 km/sec. However, in Maxwell's formalism, no medium was needed for the light. Based on this, Einstein argued that if there is no observable evidence for aether and if the Maxwell's equations don't accommodate it, perhaps, no such thing exists. There were also experiments designed to search for aether based on the fact that if one moves toward or away from an incoming light ray, in the presence of aether, the light speed must be respectively faster or slower than 300,000 km/sec. The speed of light was measured under different circumstances and was found to be constant, ruling out experimental proof for aether. But then, with respect to what is the light moving? Here Einstein proposed one of the most fundamental concepts of modern physics: Light is moving at a speed of 300,000 km/sec with respect to everything, regardless of the speed at which that "everything" is moving. Basically, he argued that no matter at what speed an observer is moving when measuring the speed of light, the same speed is always measured. This has deep implications. Consider a spaceship moving with a speed comparable to that of light. Within that spaceship an astronaut is trying to measure the distance between two events at points A and B in the spaceship. This requires sending light from point A to point B and measuring the time it takes, multiplying it by its velocity to get the distance. The astronaut aboard the spaceship always measures a shorter length between points A and B compared to a person on Earth who is not moving with the spaceship. Similarly, time moves slower for the astronaut on the spaceship compared to time for the person not moving with the spaceship. It turns out that the only way to keep the speed of light constant (and hence in line with the principle of relativity) is to accept the fact that length and time are different in the two reference frames moving with respect to one another. This leads to the conclusion that unlike what Newton proposed, that space and time were absolute, they are in fact relative and depend on the motion of the frame with respect to which they move. While both space and time are relative, the combined *space-time* is absolute and the same for all observers regardless of their frame of reference (Primack and Abrams 2006). This concept becomes clear in the next section.

THE REALITY OF PAST, PRESENT, AND FUTURE

Time flows from the past to the present and into future. This is measured by our heartbeats, the biology in our bodies, periodic appearance and reappearance of events, or atomic clocks. The laws of physics have no temporal dependence; therefore, they treat past, present, and future the same. We have a perception of these and deal with them every day in our lives, but the question is whether they have any reality or are concocted within our minds. Imagine a teacher standing in front of a class looking at the students. The voice of the teacher takes a fraction of a second to reach students (because speed of sound is finite). Therefore, what the teacher calls "now" is not "now" for the students but a fraction of a second before. The space between the teacher and students can be divided into a large number of time slices (depending on the resolution), and to each time slice a "now" could be associated that differs from the one before or after it. This means "now" is not an absolute concept but depends on the time slices between the teacher and students, with each slice having its own "now." Therefore, "now" cannot provide an expression of reality, since it depends on the position of the students with respect to the teacher and changes as students move in space. The only notion that provides reality is when all the events in space-time are assembled. As Einstein quoted: "The distinction between past, present and future is just an illusion. What is real is the notion of space-time" (Greene 2005, page 139).

THE ARROW OF TIME

For an object floating in space, up, down, back, and front does not have much meaning. The object could move freely in any of these directions. However, this is not true for time—there is only one direction the object could move in time—it is the future.

Every moment in our everyday life we experience events that unfold in one direction in time and not the reverse. When water is spilled from a jar on the floor, it cannot be collected and sent back to the jar, or when a glass is shattered or an egg is cracked, they cannot be returned to their original form. These sequences of events provide the concept of past and future, or before and after. We remember the past but know nothing about the future. This is what we mean by time having a direction—*the arrow of time*. The result is the inherent asymmetry in the time axis when considering space-time. However, the laws of physics show a complete symmetry between past and future. In none of these laws do we find a distinction that they apply only in one time direction and not the reverse. In other words, this arrow of time is absent from fundamental laws of physics. Basically, the laws of physics as we know them do not tell us why events unfold in one order and not the reverse. However, we see this happening in nature through breaking a glass or spilling water. How could this then be compromised? It seems that, in theory,

BOX 3.1: WHAT IS ENTROPY?

Entropy is the number of states accessible to a system. It is defined as heat added to a system divided by the temperature (and hence, the unit of calorie per degree). The entropy of a system is the sum of all the small increments. If the system is gradually cooled down, the entropy decreases.

Boltzmann showed that the entropy of a material is related to the number of different ways that molecules of that material can fill its volume. This is the number of ways the molecules in a fluid could spread out (how many molecules will be located in equal volumes). This is called the multiplicity, w, of any given state, which gives the probability of that state. The entropy can be formulated using the multiplicity: $Entropy = k\ log(w)$ where k is the Boltzmann constant.

if we reverse the velocity with which the water spills or the glass shatters, the water could return to its jar and the glass to its original form. This satisfies the predictions by the laws of physics.

Once you let water spill out of the jar, the degree of disorder among water molecules increases. In this case water will be confined to a larger space than the one in the jar, and hence molecules could arrange themselves in more configurations than they could when they were in a jar. The degree of disorder in a physical system is called *entropy* (Box 3.1). For cases of high entropy systems, a large amount of reordering of the constituents of the system (water molecules in the above example), will remain unnoticed, while for a low entropy system, a small amount of reordering will remain unnoticed. The number of ways the water molecules could arrange and rearrange themselves is significantly more when the water is not confined in the jar than if it were, indicating an increase in entropy (Box 3.1). Similarly, a physical system with many constituents has more ways to develop into disorder than order (it is subject to increasing entropy). This leads us to the second law of thermodynamics: *There is a tendency for a physical system to evolve to higher state of disorder, and therefore, higher entropy* (the first law of thermodynamics deals with the conservation of energy). This naturally explains the arrow of time for a physical system with a large number of constituents, since they are inclined to move toward higher entropy.

Given that the laws of physics are symmetric in time (and could equally well be applied to future events as well as the past), one could imagine scenarios in which entropy was higher in the past than at present or in the future. In other words, consider a chaotic and highly disorganized system in the past; given sufficient time, it could turn into an ordered system in the future. For example, we can see that our universe has acquired some order with all the structures—stars being born and dying, planets moving around their stars, the biological systems controlling our bodies, and neurons interacting with one another in our brains. This means the universe was less ordered in the past, with higher entropy. This implies that if we wait long enough (perhaps for an eternity), there is some likelihood that one would achieve order out of chaos. This is explicable by temporal independence of the laws of physics. Based on this discussion, it is possible to acquire structures and organisms from disorder and high entropy. In his book *The Fabric of the Cosmos*, Brian Greene (2005) argues that the source of all the order was the big bang from which the universe started. This happened with amazing order at the beginning, and we are now witnessing the gradual unfolding of that order in the universe. As we will see in the next chapters, at the beginning, the universe was hot with uniform gas consisting of only hydrogen and helium and a small fraction of lithium. The high density at the beginning provided the order. After a billion years, the structures were formed through the force of gravity, and this led to the formation of galaxies, stars, and finally planets like ours. Entropy is consistently increasing. Therefore, conditions at the beginning of the universe must have been critical in the presence of the time arrow today. As Greene (2005) states, "The fact that things started like this and ended like that but never start like that and end like this, began its flight in the highly ordered, low entropy state of the universe at its inception." The remaining question then is: How did the universe start in such an orderly manner?

Arthur Eddington was the first to explain the arrow of time in terms of the second law of thermodynamics. However, there is no experimental evidence supporting Eddington's hypothesis about the arrow of time. In the very early universe when it was all orderly because of its high density, the entropy was low and did not increase. If the arrow of time were indeed driven by the increase in entropy, there would have been no arrows. This means the time would have stopped. The consequence of this was that we would have never left that era. The expansion of the universe would have stopped. There would have been no structures in the universe, no galaxies, no stars, no planets and we would not be present here. This argues against the arrow of time being caused by the increase in entropy. The only justification of Eddington's explanation is that both time and entropy are increasing. However, this does not mean that one is causing the other-correlation is not a reason for causation (Muller 2016).

Are there alternative explanations for the arrow of time? For an event to explain the arrow of time, it must be unidirectional. Entropy was one of those, with other examples including a particle falling into a black hole with no possibility of returning or, a radioactive element decaying into lighter elements. Also, if we think of the universe in space-time, then why should the universe expand in terms of space and not space-time? Every second, we add a new second on time and the accumulation of those seconds generates the arrow of time. If this is the case, time could be continuously created, as is indeed the space.

THE REALITY OF SPACE-TIME IN THE PRESENCE OF GRAVITY

The discussion so far has been independent of gravity and was developed in reference frames that moved uniformly (with zero acceleration). However, in the real universe, there is matter in the form of structures. The general theory of relativity has produced an elegant way to explain the effect of matter on space-time. According to this theory, matter affects the geometry of space around it, generating a curvature in space. This "wrapped" space causes non-uniform (accelerating or decelerating) motion of any massive system trapped within it. This non-uniform motion is analogous to gravity and this is how a gravitational field is generated around massive bodies (Box 3.2).

According to the general theory of relativity, the force of gravity each of us experiences is the result of all the matter in the universe (and hence the geometry of the entire space-time), including the most distant stars and galaxies. For example, a free-falling object is subjected to the combined force of gravity from the rest of the objects in the universe. Therefore, if we remove all the matter from the universe, a free-falling body will not experience any force of gravity or acceleration. By removing all the rest of the matter, there would be no space curvature and hence no gravity. In this case the general theory of relativity will reduce to the special theory of relativity, as discussed in previous sections. Matter changes the geometry of space-time, and this would generate acceleration of moving bodies that in turn generates gravity. As a result, acceleration is defined with respect to all the matter occupying space-time, distributed throughout the universe (this is where general relativity and Mach's principle are reconciled) (Box 3.2).

German Philosopher Gottfried Leibnitz developed the relationalism concept in the 17th and 18th centuries. This holds that space arises from a certain pattern of correlations among objects. If two objects have similar properties, they will be located next to one another while if they differ, they will be located at a distance. This generates a pattern of connectivity. The relation between them are governed by laws of nature (i.e. quantum theory).

A new idea developed in recent years is *quantum entanglement*. According to entanglement, two particles created and moving in different directions remain correlated no matter how far they travel. The degree of correlation depends on the area of their interface (the geometry). In other word, measurements of different points in a field remain coordinated. Because of this, entanglement could result in a connection between the presence of matter and the geometry of space-time. This provides an alternative way for explaining the law of gravity (Box 3.2).

If two fields at their boundary are not entangled, they become a pair of uncorrelated entities, corresponding to two separate universes. There is no way to travel between these two universes. Now, when the two "universes" become entangled, it is as if a tunnel known as *wormhole* is opened up between them. As the degree of entanglement increases, the wormhole becomes smaller in length bringing the two universes together until they become one. According to this, the emergence of space-time is the result of the entangled fields. Observed correlations in the fields (electromagnetic or other fields) are the result from remains of the entanglement that binds space together. Every phenomena we observe, happens in the space-time domain. However, we never see space-time directly but infer its existence from every day experience.

THE CONCEPT OF MASS AND ENERGY

Any object in the universe possesses mass and/or energy. These manifest themselves as the "stored" energy hidden in an object at rest (not moving) and measured through Einstein's famous equation: $E = mc^2$, where E is the energy stored in a system, m is the total mass of the system and c is the speed of light, 300,000 km/sec, or as the kinetic energy for an object moving with a non-zero speed. The total energy of an object is the sum of these two components. The total energy is conserved. This means that if a system starts with certain energy, it will have the same energy at any given time even if some of its energy disappears or converts to other forms of energy (or mass). This is called the law of conservation of mass-energy.

According to Einstein's theory of relativity, all the energy that moves with an object contributes to the mass of the object. This increases the resistance against the acceleration (i.e. inertia). The kinetic energy of an object (measured by the mass and speed of the object) has different values in different reference frames, depending on the speed of the object with respect to that frame. Therefore, the total mass-energy of the object (the sum of its mass and its kinetic energy) changes while the mass is conserved.

What is the nature of mass? Basically, there are two concepts of mass. First, the inertial mass that is the property of an object that resists acceleration. Second, the gravitational mass that is the property of an object that determines how strongly it will be pulled by a gravitational field of a specific strength. The *equivalence principle* of general relativity postulates that the forces felt by accelerated motion and from a gravitational field are indistinguishable. In other words, there is equivalence between the concepts of inertial and gravitational mass. They are both the same thing.

SUMMARY AND OUTSTANDING QUESTIONS

Over the last century, we have come a long way to understand properties of the most fundamental things in nature—*space* and *time*. What started with Greek philosophers many centuries ago can now be explained in scientific terms and probed empirically. We have not yet deciphered the true nature of space and time but with each observation and discovery have made a leap toward that aim. We know that space and time are not absolute entities and are defined with respect to reference frames that are themselves not absolute.

Influenced by the Mach's principle—that the inertia of an object is the result of the relationship between that object and objects in the rest of the universe—Einstein developed the General Theory of Relativity in 1915. This heralded a revolution in science. According to this theory, gravity results from the geometry of space. As a massive body curves the space in its vicinity, objects moving in that curved space experience acceleration/deceleration that in turn, manifests itself as gravity. This is because there is no distinction between a system moving with non-uniform speed (accelerating or decelerating) and a system affected by the force of gravity. The conclusion is that objects with mass, generate "gravitational fields" and interact via those fields.

A long-lasting problem in physics is the explanation of the arrow of time—the reason that time only moves in one direction (i.e. towards the future). The laws of physics are symmetric with respect to time—they are equally applicable both to events in the past and the future. Arthur Eddington explained this in terms of increase in entropy (the second law of thermodynamics). To explain the arrow of time in terms of physical processes, one needs to identify those that are unidirectional (happen in one direction and not the reverse direction). Entropy is one such process (that only increases). Other possibilities are the falling of a material body inside a black hole (an object would be completely lost after it passes certain radius so-called the event horizon in a black hole). Another way to explain this is if time, like space, is being continuously created in the universe (Muller 2016). The time instantaneously created will be combined with that previously created and the one to be created, to produce the continuous flow of time, as it is observed. There is no experimental confirmation of any of these scenarios and the origin of the arrow of time is still an open question.

Despite impressive progress in understanding the fabric of space and time, there are still a number of questions outstanding. For example, how would space-time behave in an extremely entangled region surrounding very massive systems? Would we ever be able to understand the true nature of space and time? What causes the arrow of time? How accurate is the equivalence principle and how could be tested/measured?

REVIEW QUESTIONS

1. How was space and time defined in Plato's *Timaeus*?
2. How did Saint Augustine perceived the concept of space and time, and how did he connect them to the question of the creation?
3. Explain the concept of absolute space and time.
4. Compare Newton's concept of space and time with that proposed by Leibnitz.
5. Explain Mach's principle.
6. What is the principle of relativity?
7. What was the argument that led to the rejection of aether as the medium for the propagation of light?
8. Explain the arrow of time in terms of the second law of thermodynamics.
9. In a given reference frame, how could one simulate the effect of the gravity?
10. How does mass-energy relate to space-time and gravity?

CHAPTER 3 REFERENCES

Greene, B. 2005. *The Fabric of the Cosmos: Space, Time, and the Texture of Reality*. New York: Vintage.

Kant, I. (1781) 1999. *The Critique of Pure Reason*. Translated by P. Guyer and A. Wood. Cambridge, UK: Cambridge University Press.

Muller, R. A. 2016 Now—The Physics of Time, W. W. Norton & Company.

Primack, J.R and Arams, N.E 2006. *The view from the center of the universe*. Riverhead Boks. Penguin Group Inc.

Sachs, J. 1995. *Aristotle's Physics: A Guided Study*. New Brunswick, NJ: Rutgers University Press.

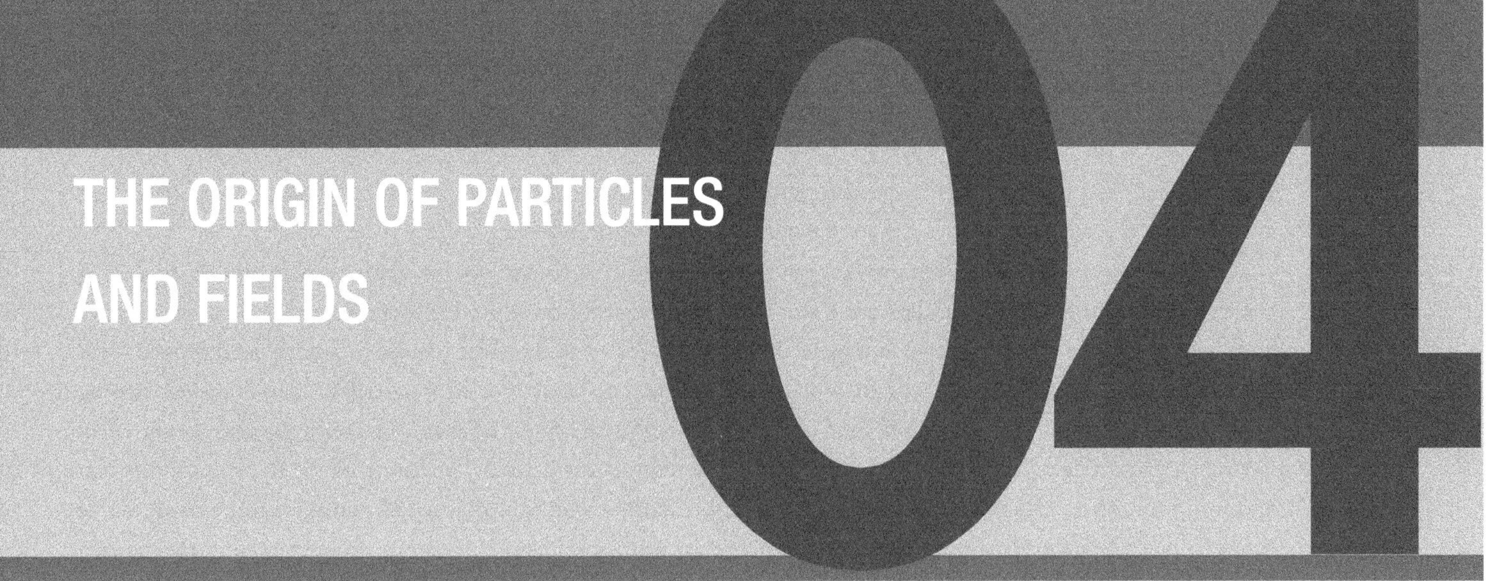

THE ORIGIN OF PARTICLES AND FIELDS

CHAPTER LEARNING OBJECTIVES

This chapter will cover:

- The constituents of matter
- Fundamental particles in the universe
- Interaction between particles
- The origin and nature of the forces in the universe
- Concept of fields
- The origin of mass
- Matter-antimatter asymmetry
- Unification of forces

"Science never solves a problem without creating ten more"

- GEORGE BERNARD SHAW

"An expert is a person who has made all the mistakes that can be made in a very narrow field"

- NIELS BOHR

The physical universe contains objects that span dimensions over 35 orders of magnitude, from sub-atomic scales (10^{-14} meters) to galaxies (10^{21} meters). In recent years, particle physicists and cosmologists working on the two ends of this scale started to converge on a picture that connects these widely different disciplines. Over the last fifty years, astonishing progress has been made towards understanding the nature of fundamental forces and particles and the laws of physics that govern them. Development of quantum mechanics at the beginning of the 20th century revolutionized our perception of matter and its constituents. This successfully explained the motion and interaction of sub-atomic particles and gave us a better understanding of how particles behave at lowest energies. Quantum mechanics showed us that energy and momentum are discrete entities (so-called quantized), objects can be found both as particles and waves (wave-particle duality) and there are strict limits as with what precision quantities can be measured (uncertainty principle). Rather than dealing with the exact position and momentum of a particle, quantum mechanics deals with the probability of them being at a given position and having a given momentum. Particles have "fields" associated with them and these "fields" have a value at each point in space and time. The examples are the "gravitational field" surrounding massive objects (like the Earth) or the "electromagnetic field" surrounding charged particles (like electrons and protons).

Interaction among particles takes place through their fields. For example, for two massive bodies to influence one another (e.g. the sun-earth system), their gravitational fields must interact.

One of the triumphs of modern physics is the development of the standard model for particle physics describing the interaction between forces and particles and the confirmation of these models by experiments. The models have explained the nature of the forces governing the universe at different scales and predicted new particles responsible for interaction between them. This naturally explains the origin and characteristics of particles today. Another exciting question in modern physics is how the fundamental forces in nature acquired their observed properties (their effective range and strength) and how they behaved in distant past. Knowledge of the nature of the elementary particles and forces allows us to decipher secrets of the universe and the conditions that led the universe to be the way it is today. This is where the physics of the very small (particles) and the very large (the cosmos) meet, resulting in the very first chapter of the book of the universe.

After a short review of quantum properties of matter and the new concepts, this chapter discusses the nature of the fundamental forces and particles and the interaction between them. It then studies the classification of particles in terms of their individual properties, the concept of fields, the origin of mass and unification of forces in nature. This chapter provides the necessary background in order to study the first fraction of seconds after the birth of our universe.

A QUANTUM VIEW OF MATTER

During the first half of the previous century, new physics was discovered, shaping our view of nature. In the early 1900s German physicist Max Planck (1858–1947) formulated the energy radiated by a blackbody (perfect absorber—see Chapter 1) and realized that electromagnetic waves (including light) could only be emitted in quantized forms. That is, in the form of discrete energy packets called *quanta* and not as a continuous set of values. Based on this discovery, Albert Einstein (1879–1954) explained the photoelectric effect by proposing that light has a dual nature, consisting of discrete bundles of energy (particles) called *photons* as well as behaving like electromagnetic waves. Danish physicist Niels Bohr (1885–1962) then postulated that electrons in atoms could only have specific energies (quantized) occupying different energy levels. French physicist Louis de Broglie (1892–1987) further showed that every particle is associated with a wave, with the wavelength depending on the momentum (mass and speed) of the particle (larger mass, shorter wavelength) (Box 4.1). These discoveries presented a world that is very different from that depicted by classical physics (Tillery et al. 2013).

The study of elementary particles took a new turn by the development of quantum field theory in 1950s. In particle physics, the interaction between two particles can be explained by the concept of the *field*, defined as a physical entity that has a value at each point in space and time. A field has energy and momentum and hence can be considered as a collection of particles of a given type. However, the idea of the field is not new. Isaac

BOX 4.1: QUANTIZED ENERGY AND WAVE-PARTICLE DUALITY

In a revolutionary discovery that changed the face of physics, Max Planck proposed that the energy emitted by a blackbody (an object that absorbs all the light radiated on it and emits at all wavelengths) could only emit electromagnetic waves (including light) in discrete (quantized) packets of energy. These energy packets are multiples of a constant, called Planck's constant (h), with the energy expressed by $E = h.\nu$, where E is the energy for a packet and ν is the frequency of the light.

Louis de Broglie proposed in 1924 that any particle has a wave associated with it. The de Broglie wavelength, λ, associated with a particle with momentum p, is given by $\lambda = h / p$.

Newton's law of gravity, with massive objects attracting one another, involves a field, the so-called gravitational field, as does Coulomb's law of electrostatic force, in which electrically charged particles influence one another. This can be explained by each particle generating a "field" around itself, affecting other particles and hence interacting with them. The energy in a field is quantized (exists in discrete packets), and this quantization can manifest itself as a particle. Therefore, particles can be understood as quanta of a field, making the concept of fields and particles interchangeable.

How do particles within a field or between two fields interact? A way to explain the interaction between two particles is by considering one to emit a virtual particle that is absorbed by the other. The virtual particle therefore transfers momentum between the two "real" particles. The fundamental forces in nature are explained through this exchange of virtual particles. For example, two electrons interacting through the electromagnetic force do so by the exchange of a virtual photon (figure 4.1). We will see later in this chapter that the photons are responsible for the electromagnetic interactions between two charged particles. Similarly, other forces in nature are associated with their respective virtual particles. This concept determines the origin of different forces in nature, with some particles mediating the interactions (called *force* carrying bosons, such as *photons*) and others forming the matter around us (called *leptons* and *quarks*). I revisit this later in this chapter.

THE NATURE OF FUNDAMENTAL PARTICLES

For many years scientists believed the most fundamental building blocks of matter were three subatomic particles—*protons*, *neutrons*, and *electrons*. An atom indeed consists of two parts, the nucleus (which contains protons and neutrons) and electrons that orbit around the

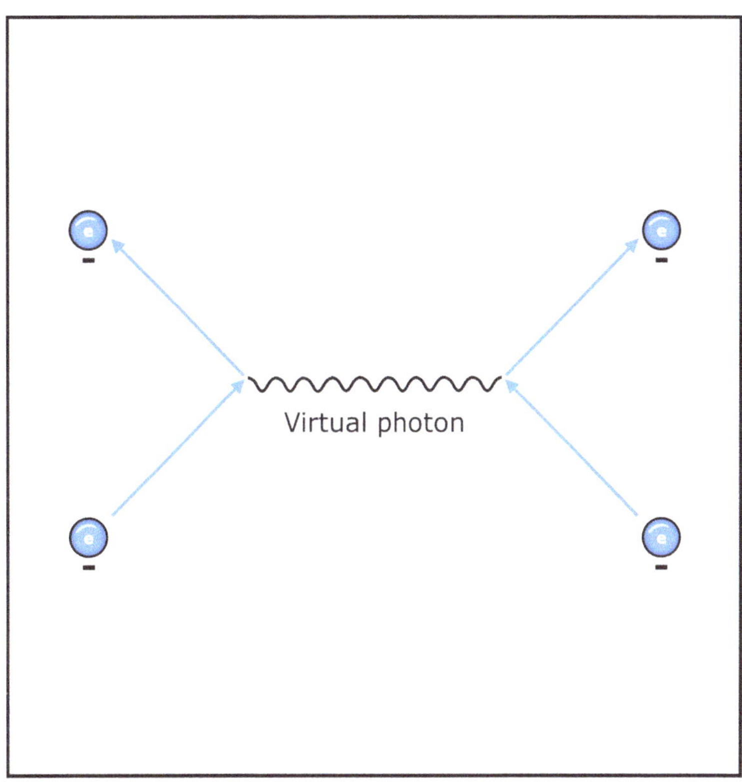

Figure 4.1. The exchange of a virtual photon between two electrons in an electromagnetic interaction.

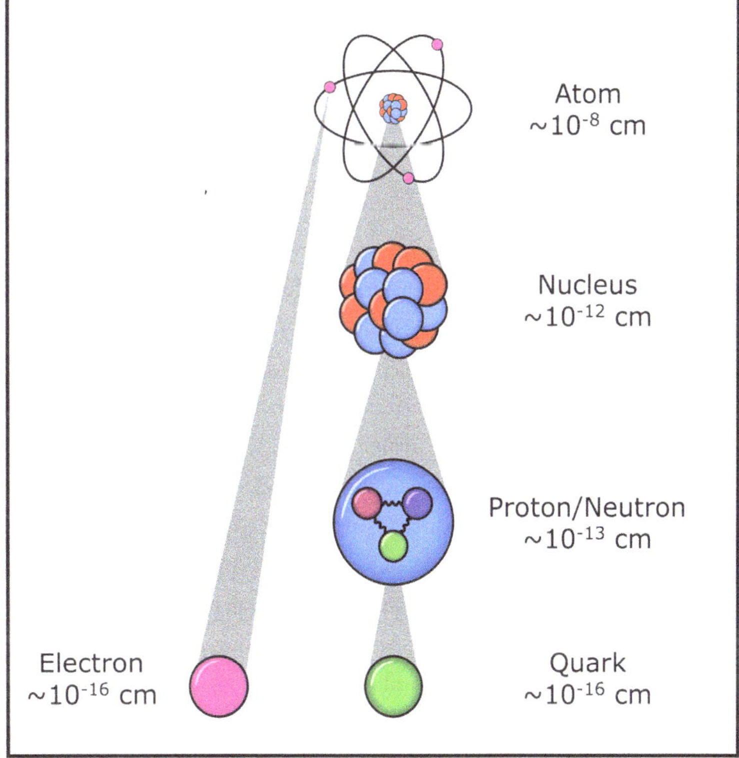

Figure 4.2. Different constituents of an atom. Protons and neutrons (collectively called nucleons) form the nucleus of an atom, while electrons orbit around the nucleus. Protons and neutrons each consist of three quarks.

Figure 4.3. Classification of the fundamental particles. Quarks and leptons are responsible for all the matter around us, while bosons mediate the forces (electromagnetic, weak, strong and gravity) in nature.

nucleus (figure 4.2). Scientists then realized that free neutrons decay to a proton and an electron through a process I describe later in this chapter. Considering the particles known at that time, they could not account for all the initial energy that entered this process, breaking the law of conservation of energy. Therefore, they introduced a new particle with no electric charge and extremely small mass to explain the missing energy. This particle was named *neutrino* (meaning "little neutral one"). Neutrinos only interact weakly, if at all, and as a result pass through matter without leaving any signals behind, which makes them very difficult to detect. Neutrinos were experimentally confirmed many decades after they were hypothesized.

The standard model of particle physics explains the nature and characteristics of the most fundamental building blocks of matter and the way their different components interact. According to this model, all the particles in the universe can be grouped into three families: *quarks, leptons,* and force carrier particles (figure 4.3). The most fundamental building blocks of all matter are quarks and leptons (figure 4.4). There are six different types for each of these particles. These interact with one another by exchanging force carrier particles (such as *photons*). Quarks are not found as free particles, due to a phenomenon called *color confinement*, but they combine to form composite particles called *hadrons*. Hadrons are further divided into *baryons* (consisting of three quarks) and *mesons* (consisting of a quark and an anti-quark) (figure 4.4). The most common type of baryons are protons and neutrons.

There are six types of quarks in the standard model, called *flavors*: up (u), down (d), strange (s), charm (c), top (t), and bottom (b) (figure 4.3). The existence of all flavors of quarks is experimentally verified. The up and down quarks have the lowest masses. They are therefore most stable and commonly exist in nature (the most stable particles are those with small masses). The other flavors have high masses and decay into the more stable up and down quarks. To summarize, quarks are found in triplets and doublets. Three quarks form a proton (uud) and a neutron (ddu) (figure 4.5), while a quark and antiquark ($\bar{u}$) form a *meson* (u$\bar{u}$). An antiquark has the same mass and spin angular momentum as the quark but with opposite charge. Quarks have intrinsic properties, including electric charge, mass, color charge, and spin. They have fractional electric charge and therefore, when combined in triplet or doublet, result in integer charges. For example, up quarks have an electric charge +2/3 (relative to a proton that has a charge of 1), while down quarks have an electric charge of −1/3 (relative to an electron that has a charge of −1). The combination of three up and down quarks (figure 4.5) therefore gives a proton a charge of unity and a neutron no charge.

Quarks are held together by the *strong* force that is mediated by particles called *gluons* and become stronger as they are pulled apart from one another (figure 4.5). This is like the endpoints of a string—as they are pulled apart, they tend to come together more forcefully. This confirms that quarks are not found as free particles except in the high temperatures and dense conditions of the very early universe. Quarks have color charges (unlike electrons and neutrinos) that allow them to take part in *strong interactions*. This is analogous to the baryons (protons and neutrons) that are also held together in the nuclei by the strong force, again

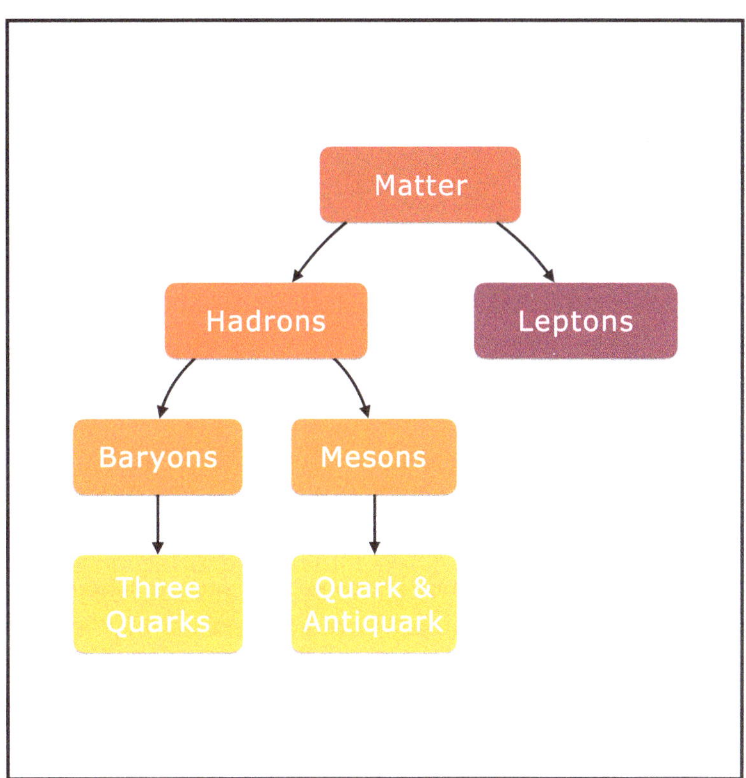

Figure 4.4. Ordinary matter classified into different types.

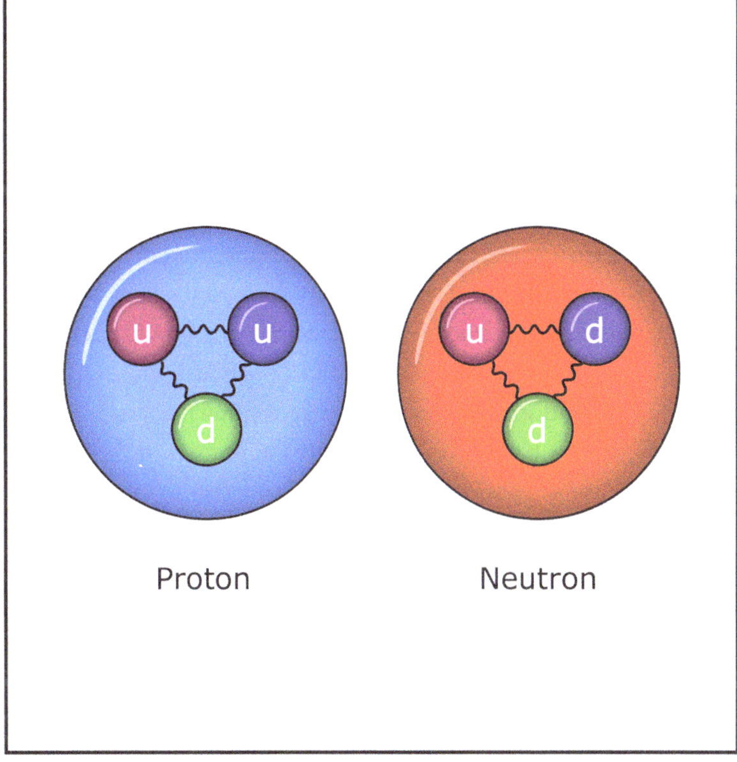

Figure 4.5. A combination of three quarks forming a proton (uud) and neutron (ddu). They are held together by the strong force. To satisfy the exclusion principle (as required by fermions), quarks are assigned three colors: red, green, and blue. These are not real colors but a way to assign them different characteristics to satisfy Pauli's exclusion principle.

BOX 4.2: BUILDING BLOCKS OF MATTER

There are two types of particles in nature: fermions (for example, electrons, neutrons, and protons) and bosons (photons, gluons, and W^+, W^- and Z^0 particles). The matter around us consists of fermions (with the most fundamental part being quarks). They follow a statistical law called the exclusion principle, according to which they cannot occupy the same state at the same time (in technical terms, two fermions cannot have the same quantum numbers). Bosons do not follow the exclusion principle. While fermions form the building block of matter, bosons are responsible for the interaction between the four forces in nature (gravity, strong, weak, and electromagnetic). The four bosons shown in figure 4.3 all have spin 1 and therefore are vector bosons. There are other kinds of bosons, including scalar bosons (Higgs bosons) and mesons (composite bosons made of quarks).

mediated by gluons. The concept of *color* was introduced to allow different quarks to reside in the same hadron (like a composite particle) without breaking the *Pauli exclusion principle* (Box 4.2). In other words, to allow three particles to reside in a baryon and satisfy the exclusion principle, a property with three values was needed, and hence three "colors"—red, green, and blue—were introduced (the combination of these three colors results in a "white" color and hence a colorless particle). These have nothing to do with real colors and only depict different quantum states, with only colorless or color-neutral particles allowed. Baryons consist of red, green, and blue quarks (the sum of these results in a colorless particle), while mesons consist of a quark (color) and antiquark (anticolor), making it color neutral (Bennett et al. 2007).

Another group of fundamental particles consists of leptons (figure 4.3), which are elementary particles with half-integer (1/2) spin that do not undergo strong interaction. Two types of leptons exist in nature: charged leptons (electrons) and neutral leptons (neutrinos). Like quarks, there are six flavors of leptons: electron (e^-) and electron neutrino (ν_e), muon (μ^-) and muon neutrino (ν_μ), tau (τ^-) and tau neutrino (ν_τ). Electrons have the smallest mass among the charged leptons and are the most commonly found. Leptons have intrinsic properties, including spin, electric charge, and mass. Unlike quarks, they do not have color charge, since they are not subject to the strong force.

Finally, force-carrying bosons are fundamental particles that mediate interactions (figure 4.3). The standard model accommodates the force-carrying bosons, called *gauge bosons,* mediating electromagnetic (photon), strong (gluon), and weak (W^+, W^- and Z^0) interactions (see next section). There is also the recently discovered *Higgs* bosons (to be discussed later in this chapter), and not-yet-discovered but hypothesized *gravitons* (mediating gravity).

The bulk of the matter in the universe is in the form of hadrons (composite particles consisting of quarks) and leptons. These are collectively called fermions (for example, electrons, protons, and neutrons are all fermions). These are distinctive in terms of their physical properties. The fermions (after Enrico Fermi, 1901–1954) have fractional spin and follow the exclusion principle, according to which they cannot occupy the same state at the same time and *bosons* (after Satyendra Bose, 1894–1974) have integer spin and do not follow the exclusion principle, meaning that they follow a statistical law that does not restrict them from occupying the same quantum state (see Box 4.2).

FUNDAMENTAL FORCES IN THE UNIVERSE

There are four fundamental forces governing the universe—strong, weak, electromagnetic and gravity. These are the forces responsible for all the interactions between particles and exchange of momentum among them, as well as for the building blocks of matter and formation of the structure in the universe (Box 4.3). As discussed earlier

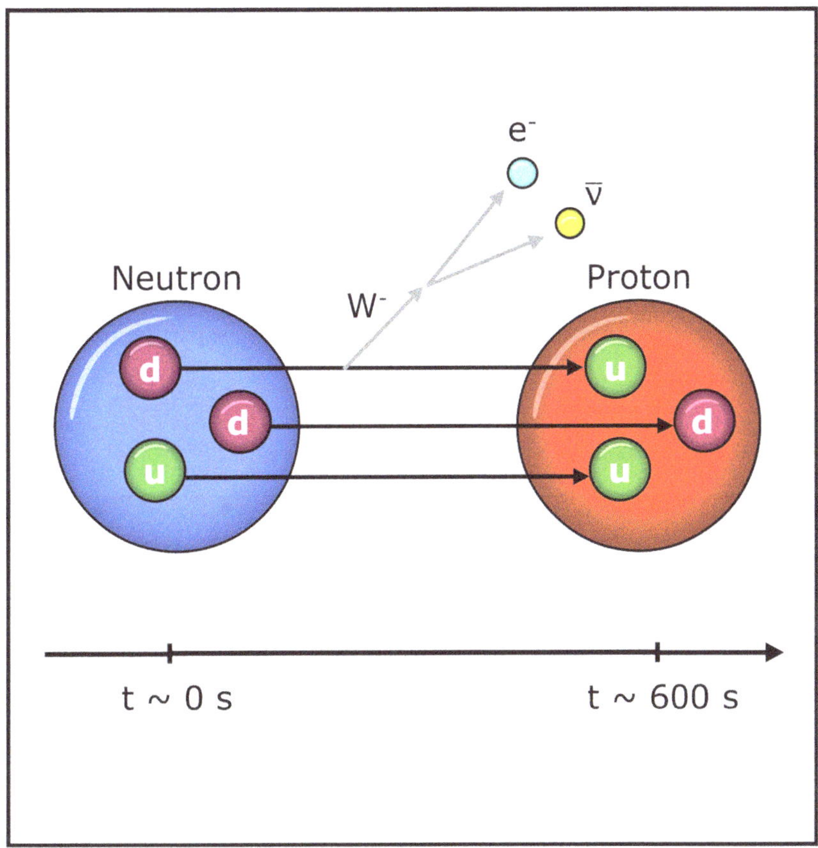

Figure 4.6. A free neutron decays to a proton by the exchange of a down quark to an up quark. This process is mediated by exchanging W⁻ particle. W⁻ boson decays to an electron and an anti-neutrino. The half-life of free neutrons is around six hundred seconds.

in this chapter, each force is mediated by the exchange of a virtual particle (force-carrying bosons), transferring momentum between any two real interacting particles.

The strong force is responsible for holding protons in the nucleus of atoms, as well as the quarks in the nucleons (figure 4.2). As protons have the same electric charges and since like-sign charges repel, strong force is needed to keep the nucleus together. In the nuclei with many protons (nuclei heavier than hydrogen), protons tend to repel one another because of the electrostatic force. The strong force counteracts the repulsion in the nuclei and is responsible for holding the nuclei together. The interaction of the strong force is mediated by the exchange of particles called *gluons* (which act like glue; hence the name). The quarks inside the nucleons (e.g. protons, neutrons, mesons) also interact via the strong force (Figure 4.5).

The weak force is also effective at nuclear scales. This is the force responsible for decay of particles. For example, neutrinos feel the weak force, and this is responsible for decay of free neutrons (figure 4.6). The weak force is mediated by massive particles: W^+, W^- and Z^0 bosons. These particles were experimentally discovered in the 1980s confirming that the same mechanism is responsible for both electromagnetic and weak interactions. These two forces are therefore manifestations of a single force—so-called *electroweak*. The strong and weak forces are only felt at the nuclear scale, and are very weak at the atomic scale (which is nearly one hundred thousand times larger than the nucleus)—(Table 4.1).

The electromagnetic force is responsible for keeping electrons around the nucleus to form atoms, as well as atoms to form molecules and complex chemical and biological structures that form the basis of life (table 4.1). The electromagnetic force is mediated by particles of light, the photons (figure 4.1). Therefore, electromagnetic is the only effective force for the interaction between atoms.

Gravity has an absolutely negligible effect at the dimensions which strong, weak and electromagnetic forces are effective and is the weakest of all forces (table 4.1). It is a long-range force and responsible for the Moon orbiting the Earth and the Earth moving around the Sun. An atom is electrically neutral since it contains similar numbers of positively charged (protons) and negatively charged (electrons) particles. Therefore, large and massive systems do not interact through the electromagnetic force, but because of their mass, they attract one another by the force of gravity. The strength of both electromagnetism and gravity declines proportional to the inverse square of the distance between the interacting bodies. The force of gravity increases with the mass of the objects experiencing it. Following the same formulation as for other forces, we expect gravity to be mediated by the exchange of virtual particles called *gravitons*. However, unlike the mediators for other forces, the particle responsible for gravity is not yet experimentally verified.

Table 4.1. Characteristics of the fundamental forces in the universe

Force	Strength	Range (m)	Particle
Nuclear strong — Holds the nucleus together	1	10^{-15}	Gluons, π nucleons
Electromagnetic	0.007	∞	Photon Mass = 0 Spin = 1
Nuclear weak — Neutrino interaction induces beta decay	10^{-6}	10^{-18}	Intermediate vector bosons (W^+, W^-, Z^0) Mass > 80 GeV Spin = 1
Gravity	6×10^{-39}	∞	Graviton? Mass = 0 Spin = 2

There are four distinct forces in nature. Very early in the history of the universe, these were unified in one single force. As the universe expanded, it cooled down, and these forces were separated and attained the distinct identities they have today. The fundamental forces are listed and briefly explained below:

Gravity is the weakest of all four forces and the first that acquired its distinct identity at 10^{-43} seconds after the birth of the universe (at a temperature of 10^{32} degrees Kelvin). It is a long-range force responsible for attraction between massive bodies. This is the reason we can feel gravity and not other forces, which are only effective over a short range. Gravity is explained by Einstein's general theory of relativity.

Strong force is responsible for holding quarks together to form protons and neutrons and also for keeping protons and neutrons together to form the nuclei of atoms. This was separated from the weak and electromagnetic forces at 10^{-35} seconds after the birth of the universe (at a temperature of 10^{27} degrees Kelvin). The particles responsible for mediating strong force are called *gluons* (table 4.1). This is a very short-range force and only affects distances of the order of 10^{-15} meters. The theory studying the strong force is called *quantum chromodynamics*.

Electromagnetic force is much weaker than the strong force but acts over much longer distances. It obtained its distinct identity by separating from the weak force when the universe was 10^{-12} seconds old (at a temperature of 10^{15} degrees Kelvin). The particles that carry the electromagnetic force are particles responsible for light (table 4.1). This force explains the interaction between charged particles. The theory of the electromagnetic force is *quantum electrodynamics*.

Weak force acts within the atomic nucleus and is responsible for radioactivity (beta decay), neutron decay, and interaction between neutrinos. It has a very short range and is very weak. The particles that carry the weak nuclear force are called W^+, W^- and Z^0 *bosons* (table 4.1). The standard model for the weak force is the electroweak model.

THE CONCEPT OF FIELDS

I introduced the concept of "fields" earlier in this chapter. This is of extreme importance in modern physics and here I take it a step further. All the four fundamental forces in nature—electromagnetic, gravity, weak and strong—have a field associated with them. For each field, there are particles that carry that field. We know from previous discussion that the electromagnetic field is carried by the photons, with the weak nuclear force being carried by the W^+, W^- and Z^0 particles and the strong force carried by the gluons. Just as photons transmit electromagnetic field, the particles responsible for gravitational field are hypothesized to be gravitons. These are not yet discovered but within the framework that explains other forces, it is natural to expect gravitons to carry the gravitational field.

Now, we could extend this framework to other particles and, in general, to matter. As I explained earlier in this chapter, every particle can also be considered as a wave, representing the probability of that particle being at any given location. At these small scales, the motion and position of particles are calculated by equations of quantum mechanics. It is not possible to fix the exact location of particles but only the probability that a particle may be at a certain position. To give an example, an electron is a particle but is also considered in terms of a wave (this is proven experimentally by observing the interference patterns produced by two electron beams after passing through two slits). In this case, an electron's probability wave is closely associated with an *electron field*.

Apart from the force fields and matter fields discussed above, there is also another field called the *Higgs field* (named after Scottish physicist Peter Higgs). It is believed that the entire space is filled with the Higgs field, a relic from the fraction of a second after the birth of our universe. It is the Higgs field that gives particles the properties they have and, as a result, properties of matter that constitutes the entire universe. The Higgs field is associated with a particle called *Higgs boson* acting like other force mediating particles (figure 4.3). The force mediated by Higgs boson is universal, as it interacts with all particles and especially with massive particles. According to the law of conservation of energy, mass is not generated by the Higgs field but is given to the particle through its interaction with the particle via the Higgs bosons.

THE ORIGIN OF MASS

Fields respond to temperature as the ordinary matter does. Soon after the birth of the universe, when it was around 10^{-43} seconds old with a temperature of 10^{32} degrees Kelvin, all the fields rapidly fluctuated. At that very early stage and under extreme temperature, all the fields had the same properties and were indistinguishable. As the universe cooled down with time (discussed in the next chapter), the initial matter and radiation density dropped and field fluctuations reduced, with the value of the fields becoming close to zero. At this point the Higgs field behaved differently from other fields because of the shape of its potential energy curve (figure 4.7). Once the universe cooled down and reached below a certain temperature, the Higgs field assumed a nonzero value (as steam is condensed to liquid water when temperature drops) throughout the entire space (i.e. the Universe). This corresponds to the lowest energy level (the potential energy is zero but the value of the Higgs field is nonzero, as shown from the potential energy curve in figure 4.7) called *vacuum*. This energy level (a level with zero potential energy but nonzero Higgs field) prevails through the entire universe. This results from the certain shape of the Higgs potential energy curve (figure 4.7). The process of a Higgs field assuming a nonzero value throughout space is called *spontaneous symmetry breaking*.

Now, suppose a particle moves in this uniform Higgs field. The field exerts certain amount of resistance or drag on the particle (figure 4.8). This causes the object to resist against acceleration. The entity that fights acceleration is the inertial mass of the particle. This is the origin of the inertial mass. In other words, inertial mass for a particle is generated because of the interactions between that particle and the uniform Higgs field. The degree to which the Higgs field resists a particle's

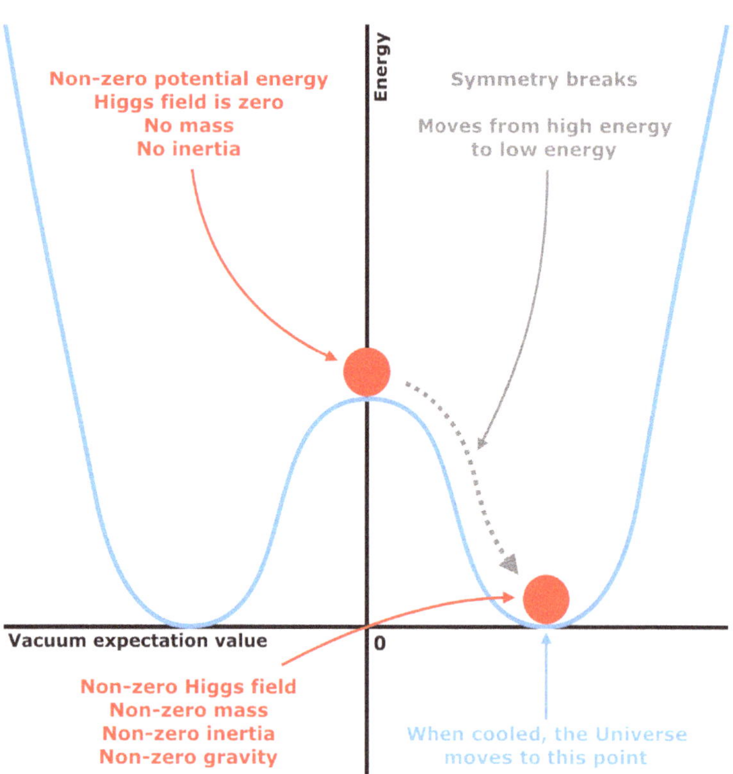

Figure 4.7. Shows the shape of the Higgs potential energy curve. Consider a ball located at the top of the "potential hill". At that point it has nonzero potential energy but zero Higgs field. It seeks to minimize its energy. It rolls down the hill and comes to rest somewhere at the bottom. That is a zero potential point but nonzero Higgs field. In the process of rolling down it interacts with the uniform and nonzero Higgs field and attains its mass.

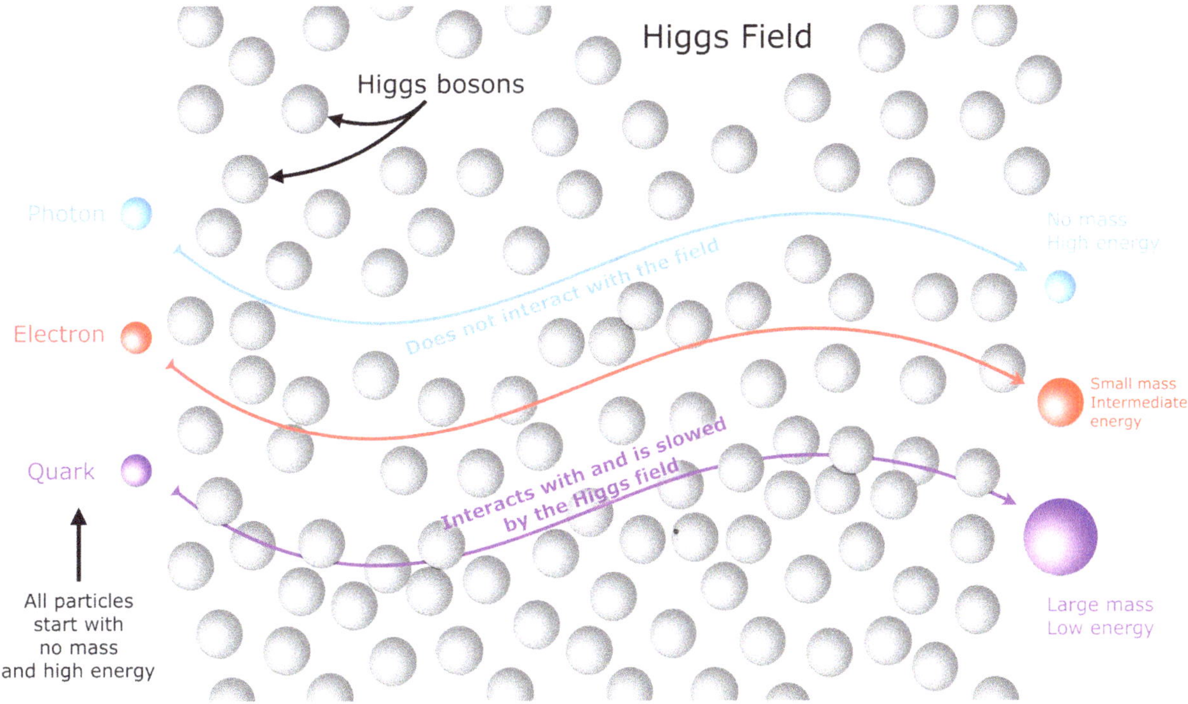

Figure 4.8. Schematic diagram shows the interaction of particles with the Higgs field.

acceleration varies depending on the type of the particle. The more strongly the particles interact with the Higgs field, the more massive they are (figure 4.8). This is the reason particles have different masses (e.g., the mass of electrons and quarks are different) (Box 4.4). If the Higgs field did not exist, all particles would be massless, like photons. In this case the difference between various particles would have disappeared.

BOX 4.4: WHY PARTICLES HAVE THE MASS THEY HAVE?

Every particle passes through a hypothetical energy field, called the Higgs field, that is present in the entire universe. By interacting with this field, particles acquire their mass. The strength with which different particles interact with the Higgs field determines their mass. If they interact more strongly, they are more massive. Particles that do not interact with the Higgs field are massless and move with the speed of light, like photons. Once mass is given to a particle its speed is slowed down. What gives the Higgs field its specific characteristics is the shape of its potential energy curve (Figure 4.7).

The Higgs field is associated with a particle called Higgs boson (figure 4.8). This is a force-mediating particle used by the Higgs field to interact with other particles. The Higgs particle is not stable (with a mean lifetime of 1.56×10^{-22} *sec*). The Higgs field was zero soon after the beginning of the universe but as the universe expanded and its temperature fell below a critical value, the Higgs field grew stronger so that any particle that interacted with it acquired mass. The Higgs field cannot be observed and could only be manifested through Higgs particles. These particles are massive (125 GeV) and short lived. Since it interacts with all the other particles (except massless particles), it can be created in high-energy collisions in accelerators (Box 4.7).

UNIFICATION OF FORCES

At the earliest stages of the universe when the temperature was extremely high ($\sim 10^{15}$ degrees Kelvin), all differences between various particles disappeared and all forces were unified into one. This was because at this extreme temperature, the Higgs field had zero value. Without Higgs field, there were no resistance to particles undergoing acceleration implying that all particles had the same (zero) mass. This was a highly symmetric state in that, if one particle mass had changed to another, nothing would appear different. As the temperature dropped below 10^{15} K, particles suddenly acquired mass depending on the degree of interaction between them and the Higgs field. Because of the nonzero and non-equal masses for particles, the symmetry between the masses was lost. This is what scientists mean by spontaneous symmetry breaking (Figure 4.7). Before the Higgs field acquired its nonzero value, not only all the matter particles were massless, all the force particles (particles that mediate forces in nature) were also identical (massless) (Box 4.5). This led to another symmetry. The implication of the symmetry between force particles was that in the absence of the Higgs field, all the forces in nature were the same.

In late 1800s James Clarke Maxwell realized that electricity and magnetism, although separate forces, were in fact different manifestations of the same force—electromagnetic. Sheldon Glashow, Steven Weinberg and Abdus Salam demonstrated in 1960s that at the very high temperatures of the early universe, the photons, W^+, W^- and Z^0 particles were all identical. In other words, there were symmetries between these force particles and therefore, symmetry between the forces associated with them. The implication was that, in the absence of the Higgs field, the electromagnetic and weak nuclear forces were part of a single unified force—*electroweak*. The unifying symmetry between these forces, existing at very high temperatures and absence of the Higgs field, were broken when the photons and W^+, W^- and Z^0 particles acquired their individual characteristics caused by their interaction (or non-interaction in the case of photons) with the Higgs field (Box 4.5). The result is that the two very different forces in nature— electromagnetic that is responsible for light and electricity, and weak nuclear force that is responsible for radioactive decays—were in fact a single force. Only in the absence of the Higgs field the symmetry in nature becomes apparent.

Is it possible that the other non-gravitational force, the strong nuclear force, could be unified with the electroweak force within the same framework that unified the electromagnetism and weak forces? If so, at temperatures around 10^{28} degrees Kelvin and 10^{-35} seconds after the birth of the universe, another phase transition must have happened, distinguishing this from other forces. Before that time, the force carrying particles- photons, W^+, W^- and Z bosons and gluons- must have had the same characteristics, freely interchanging with one another. This was the result of a complete symmetry among these three non-gravitational forces. This is called *Grand Unification*. When the temperature of the universe dropped below 10^{28} degrees Kelvin, a different spices of the Higgs field

---

BOX 4.5: PROPERTY OF THE HIGGS FIELD

As a region of space becomes more and more sparse from matter and radiation, the energy in that region becomes lower. Given this, the emptiest region of space would be a region with the lowest energy—so-called vacuum. For ordinary fields, this is when the energy is zero. This means that we consider the emptiest regions of space to be devoid of everything—they have zero energy with their associated fields having zero values. However, for Higgs fields things work differently. Because of a certain shape of its potential curve, it must have enough energy to jump up from the potential well to have zero potential. In other word, a Higgs field with no energy will slide to the bottom of the potential well, away from the zero energy state and therefore, has nonzero value (figure 4.7).

(grand unified Higgs) condensed to a nonzero value, breaking the symmetry. This process then gave gluons their mass. Because the grand unified Higgs has a different effect on gluons than other force particles, it only affected the strong force and not others. However, as yet, there is no experimental evidence to support this version of the theory. I will return to this in the next chapter when studying the first stages in the evolution of the universe.

ANTIPARTICLES

For every lepton and quark, there is an antilepton and antiquark that has the same mass but opposite charge. The antiparticle of an electron is called a *positron* and is identical to an electron but with a positive charge. Therefore, there are a total of twelve fermions (quarks and leptons; and the same number of anti-fermions (figure 4.3)). When a particle and antiparticle collide, they disappear and their entire mass turns to energy, according to Einstein's $E = mc^2$ equation (with E being the energy, m the mass, and c the speed of light).

The process of conversion of matter-antimatter to energy could also work in reverse (Perkins 2003). Under certain conditions, energy can convert into a pair of particle and antiparticle. Whenever an electron comes into existence, a positron will also appear through a process called *pair production*. Throughout this process, all the conservation laws (energy and charge) are satisfied. Therefore, in the very early universe, when the temperature and energy were extreme, particle-antiparticle pairs were continuously created through the pair-production process and disappeared into energy through the process of annihilation.

Why the present universe contains ordinary matter and not antimatter? A universe with an equal number of particles and antiparticles, continuously created and annihilated, would end up only with energy (unless through some yet-unknown physical process, the particles and antiparticles are separated). This is because as the universe cools down with time, the particle-antiparticle pairs cease to be created (since large amounts of energy is needed to initiate the pair production process) but the remaining pairs annihilate, turning to energy. However, everything around us today almost completely consists of matter. Somehow the balance between matter and antimatter must have broken in favor of matter. It is likely that through an as yet unknown process a fraction of a second after the birth of the universe, a small proportion of matter—one particle per billion—survived. Particles transform to their antiparticles with a very high rate. To generate this excess of matter over antimatter, it is hypothesized that because of an intervening process, particles were more inclined to decay to matter than to antimatter, leading to the present matter dominated universe.

Are the laws of nature the same for particles and antiparticles? Particle physics experiments have shown that particles called *D-mesons* change from being a particle to an antiparticle and reverse. This happens at *different* rates, depending on whether the meson is being converted to an antimeson or the reverse. Such process breaks the symmetry between particles and antiparticles and indicates that laws of physics are different for matter and antimatter. This is called matter-antimatter asymmetry. This asymmetry is seen where a particle and its antiparticle both decay with different rates (like B_0 mesons and its antiparticle $\bar{B}_0$ that decay with different rates). I will return to this after discussing the concept of the parity in the next section.

Charge-Parity (CP) violation can be considered in the following context:

1) Consider a particle decay process; 2) Now look at the reflection of the process in a mirror (parity conjugation) and, in that mirror image, replace all particles by their antiparticles (charge conjugation); 3) If the image so modified occurs in nature with the same probability as the original process, then CP is conserved otherwise, it is not.

PARITY

The parity transformation of a physical system replaces the system with its mirror image. This means inversion of the spatial coordinates of the system (changing signs) relative to the origin. The conservation of parity states that the laws of physics (such as decay rate of a particle) are the same for both the particle and its mirror image. Applied to particles, parity symmetry means that the equations of particle physics are the same under mirror inversion. This predicts that the mirror image of a reaction (a chemical reaction or radioactive decay) occurs at the same rate as the original reaction.

The theoretical symmetry between particle and antiparticle is often expressed in terms of charge (C) and parity (P). This is the product of two symmetries: charge, when transforming a particle to its antiparticle, and parity, which creates the mirror image of the particle. If nature treats particles and anti particles the same, CP is symmetric (the same for both). If not, CP is violated (not the same for a particle and its antiparticle) (Perkins 2003) (See Box 4.6). The conservation or nonconservation of CP has great implications toward explaining the origin of the existing matter-antimatter asymmetry in the universe. It is found that CP is conserved in electromagnetic and strong interactions but is violated in weak interaction (Perkins 2003).

Parity asymmetry was experimentally discovered in the weak interaction involving beta decay of cobalt 60 (cobalt with sixty protons and neutrons in its nucleus). It was detected that the inverse reaction did not occur as frequently as the original reaction. CP violation first confirmed through decay of K-meson (kaon). It was shown that weak interaction not only violates charge and parity symmetries individually, but also their combination. The discovery of CP violation created serious confusion in particle physics and cosmology, particularly the fact that CP is conserved in electromagnetic and strong interactions but not in weak interaction.

Apart from charge and parity, there is also a third operation—the time reversal (T) that refers to reversal of motion. Time reversal symmetry means that when a motion is allowed by the laws of physics, the reverse motion is also allowed. The combination of charge, parity, and time reversal (CPT) is always conserved in all reactions. CPT conservation implies equal mass values and lifetimes for a particle and its anti particle.

SUMMARY AND OUTSTANDING QUESTIONS

The standard model of particle physics has been very successful in explaining the nature of fundamental particles and forces. While many questions have been answered, many more are waiting for future experiments or new theories. The elegant classification of the elementary particles to different groups and their respective characteristics is an indication that nature can be expressed in organized and simple ways. We know that fermions (electrons, neutrinos, and quarks) are the constituents of all the matter around us, while bosons (photons, gluons, and W^+, W^- and Z^0 particles) are the virtual mediators responsible for the forces in nature. The origin of the difference between the two classes of particles is in the statistical laws they follow, their spin angular momentum, and whether they obey Pauli's exclusion principle.

The fundamental interactions between particles take place by the exchange of virtual particles grouped into force-carrying bosons (this is not yet confirmed about the gravity). The strong force keeping the quarks in hadrons and protons inside the nuclei is mediated by gluons and follows a theory that accommodates color forces to satisfy the Pauli exclusion principle (that only applies to fermions), the so-called *quantum chromodynamics*. The standard theory of particle physics successfully explains the weak force and the exchange particles responsible for this, W^+, W^- and Z^0 bosons. This was experimentally confirmed with these mediator particles discovered in

1983 at the European Organization for Nuclear Research (CERN) (Box 4.7). The mediating particles for the electromagnetic force are the photons. The behavior of the electromagnetic force, responsible for interactions at the atomic and molecular scales, is explained by the theory of *quantum electrodynamics*. This is one of the most successful theories in particle physics, with its predictions experimentally confirmed to high degrees of accuracy. The weak and electromagnetic forces are unified into the framework of the electroweak theory that is now universally confirmed. Finally, gravity is the only force that acts at large scales and is responsible for the movement of planets, stars, and galaxies in the universe. We know much less about this force than other forces, and it is significantly weaker than the electromagnetic or strong force (by a factor of 10^{-43}). The only theory for gravity is Einstein's *general theory of relativity*, which is a classical theory (it doesn't involve quantum properties of matter). While general relativity successfully explains the available observations and provides testable predictions, it cannot be used to explore the conditions very early in the history of the universe (at a time less than the Planck time, 10^{-44} seconds after the birth of the universe). It has proved to be extremely difficult to combine gravity with the other three forces to come up with a grand unified theory of all forces in nature. This is the subject of ongoing research.

The triumph of the current models is manifested by confirmation of many of their predictions. This shows the confidence with which we could use our theories to predict new particles. For example, positrons (electrons antiparticles) were predicted by Paul Dirac in 1928 before being experimentally discovered in 1932 by Carl David Anderson, who was studying cosmic rays. Many flavors of quarks were predicted before they were found in laboratory experiments during the 1960s and 1970s (the last flavor, top quark, was discovered at the Fermi National Accelerator Laboratory near Chicago in 1995). This process gives us confidence to extrapolate our theories and search for their predictions. However, we must note that a theory that is impossible to verify through experiments, however mathematically elegant, cannot be accepted as a viable theory.

For every particle in the universe there is an antiparticle with the same mass and spin but opposite charge. However, there is an asymmetry in the universe, with more matter than antimatter. It is not clear how this arose, and this is a mystery in physics today. It is likely that this resulted from processes very early in the universe, soon after it was born. If the particles and antiparticles were created in equal numbers, the universe would have been filled with energy and nothing else today. We would certainly not be here. To answer this question, one should look at extremely high energies of the order what existed in the very early universe.

We know that fermions, consisting of leptons and quarks, build everything around us. Today we have confirmed evidence for twelve types of quarks and antiquarks and twelve types of leptons and antileptons, a net total of twelve fundamental particles that constitute all the matter around us. At higher energies obtainable with a new generation of particle accelerators (Box 4.7) over the next few years, a new generation of massive particles may be created that might provide hints of new physics or yet again change our view about the universe.

The bosons and quarks obtained their mass through interaction with a hypothetical field, the Higgs field, mediated by Higgs bosons. This is considered as the process that gives particles their masses. Before that, all particles had the same mass (zero mass). Because of the special potential associated with the Higgs field, the symmetry between particles breaks at lower energies and therefore, particles attain the mass they have. The amount of mass any given particle has depends on how strongly the particle interacts with the Higgs field. The discovery of Higgs boson in 2013 at the Large Hadron Collider (Box 4.7) at the CERN Laboratories in Geneva confirmed Higgs scenario.

An outstanding question today is the reason for baryon asymmetry or, in other words, why the universe mostly consists of matter and not antimatter. The seed for this was planted very early in the history of the universe. A related question is why CP conservation is violated in weak interactions but not in electromagnetic or strong interactions.

BOX 4.7: PARTICLE ACCELERATORS

A particle accelerator uses an electric field to propel and accelerate electrically charged particles in a desired direction. An example is the Large Hadron Collider in Switzerland, which uses an oscillating electric field to accelerate two beams of proton in opposite directions over circular paths with a circumference of 27 km. The two beams are accelerated to a speed of 0.999999991 times the speed of light and collide head-on when they acquire an energy of 7 trillion electron volts (an electron volt is defined as the kinetic energy acquired by a single electron when accelerated across a potential of 1 volt). The collision produces an energy of 14 TeV, large enough to generate very massive (but short-lived) particles. The more energetic the accelerated particles are, the more deeply they could probe the structure of matter leading to the discovery of new particles.

Also, how could gravity be unified with the other three fundamental forces? Given recent discovery of gravitational radiation from colliding massive systems, could we one day detect gravitational signals from the early universe?

REVIEW QUESTIONS

1. Explain the nature of light.
2. What is the definition of a *field*, and what do scientists mean by quantization of fields?
3. What are hadrons? Name the two types of hadrons.
4. What are the main characteristics of quarks?
5. Name the particles that mediate the four forces in nature.
6. What is the basis for introducing color terms to quarks?
7. Explain the characteristics of the four forces in nature.
8. Explain the concept of parity and what we mean by the symmetry between particles and antiparticles.
9. Under what conditions were the four forces in nature unified?
10. Explain the concept of symmetry breaking in nature and what we mean by spontaneous symmetry breaking.
11. Define CP violation and its significance.
12. Briefly explain how particles attain their mass?

CHAPTER 4 REFERENCES

Bennett, J., M. Donahue, N. Schneider, and M. Voit. 2007. *The Cosmic Perspective: The Solar System*. 4th ed. Boston: Pearson/Addison-Wesley.

Gross, D.J. 1996. "The Role of Symmetry in Fundamental Physics." *Proceedings of the National Academy of Sciences* 93 (25): 14256–14259.

Perkins, D.H. 2003. *Particle Astrophysics*. Oxford, UK: Oxford University Press.

Tillery, B.W. E.D. Enger, and F.C. Ross. 2013. *Integrated Science*. 6th ed. New York: McGraw-Hill.

THE ORIGIN OF THE UNIVERSE

CHAPTER LEARNING OBJECTIVES

This chapter will cover:

- The evidence for an origin for the universe
- The big bang singularity
- The Planck units
- The inflationary universe
- History of the early universe
- Particles in the very early universe

The concept that the universe had a beginning and started at a time in the past, as opposed to a universe that always existed with an infinite age, was first proposed in 1927 by Belgian priest Georges Lemâitre (1894–1966). Finding a solution to the equations of general relativity, he predicted that the universe started from an extremely dense and hot condition sometime in the past, resulting from a violent explosion. This event was named the *big bang* by British astronomer Fred Hoyle (1915–2001) in 1953. In a classical paper in 1948 Russian cosmologist George Gamow (1904–1968), who later immigrated to the United States, and his student Ralph Alpher (1927–2007) predicted that hydrogen, helium and a tiny fraction of heavier elements were produced in right proportions in the big bang model. They also argued that the energy released at the time of the big bang led to an extremely high temperature for the universe soon after it was born. They predicted that remnants of this radiation must still be around in the present universe.

In 1929 American astronomer Edwin Hubble (1889–1953), using the telescope at the Mount Wilson Observatory in California, discovered that the extragalactic nebulae (galaxies) are all receding away at speeds proportional to their distances from us. This was the first observational evidence confirming the expansion of the universe. If the universe is expanding, it must have been smaller in the past, and therefore an extrapolation back in time would lead all the matter in the universe to concentrate in a small volume and an extremely dense state. Following the discovery of the expansion of the universe, other scientists—notably Arthur Eddington (1882–1944), Willem de Sitter (1882–1934), and Albert Einstein (1879–1955)—who had been working on a static (nonevolving) model for the universe abandoned those models. The big bang has now been accepted as the standard model for the origin of the universe. This was firmly established by the discovery by Arno Penzias and Robert Wilson in 1968, finding the remnant radiation left over from the big bang in the form of cosmic background radiation that had been predicted by George Gamow's team thirty years earlier.

This chapter first presents the observational evidence for the big bang as the origin of the universe before carrying out a step-by-step study of the evolution of the universe and dominant processes in the first few fractions of a second after its birth. A study of the formation of first atomic nuclei will then follow.

EVIDENCE OF A BEGINNING FOR THE UNIVERSE

There are three main observations strongly supporting that the universe started 13.8 billion years ago by a violent explosion called the *big bang*:

Observation 1: Discovery of the expansion of the universe by Edwin Hubble in 1929 showed that galaxies move away from one another with their spectrum shifted from shorter (blue) to longer (red) wavelengths (redshift), stretched due to the expansion of space over the years the light traveled (Box 5.1). Hubble also found that the distances of galaxies (D; in megaparsec) are directly proportional to their recession velocities (V; in km/sec). This relationship is known as Hubble's law:

$$V = H.D$$

where H is the Hubble constant in units of km/sec/Mpc. This means that the speed of a single point (galaxy) increases by 1 km/sec for every megaparsec increase in distance with respect to the frame of reference, in this case our own Galaxy (see Chapter 1 for the definition of megaparsec—Mpc). The fact that distances between galaxies increase proportional to their velocity implies that at some point in the past, they were all very close to one another, starting from a single point (figure 5.1).

Observation 2: The extreme temperature and density of the early universe provided grounds for the fusion of chemical elements heavier than hydrogen. For example, after protons and neutrons were formed very early in the history of the universe, they were combined to form the nucleus of deuterium (also called heavy hydrogen), while fusion of two deuterium nuclei produced helium. These processes are called *big bang nucleosynthesis* and were first proposed in 1946 by George Gamow, who predicted that in the high temperatures of the early universe, from a soup of elementary particles (protons and neutrons), primordial elements were formed (deuterium, helium and

BOX 5.1: THE DOPPLER EFFECT AND REDSHIFT

By monitoring the spectrum (the relative light intensity at different wavelengths) of the light emitted by a galaxy, astronomers indicate whether the galaxy is moving toward or away from us. Compared to the spectrum of the object at rest, if the *observed* spectrum is shifted toward shorter wavelengths (blueshift), the galaxy is moving toward us, the same way as the sound of an approaching car has higher pitch (shorter wavelength). Similarly, if the spectrum is shifted toward longer wavelengths (redshift), the galaxy is moving away from us, just as the sound of a car moving away from us has lower pitch (longer wavelength). This phenomenon is known as the *Doppler effect*.

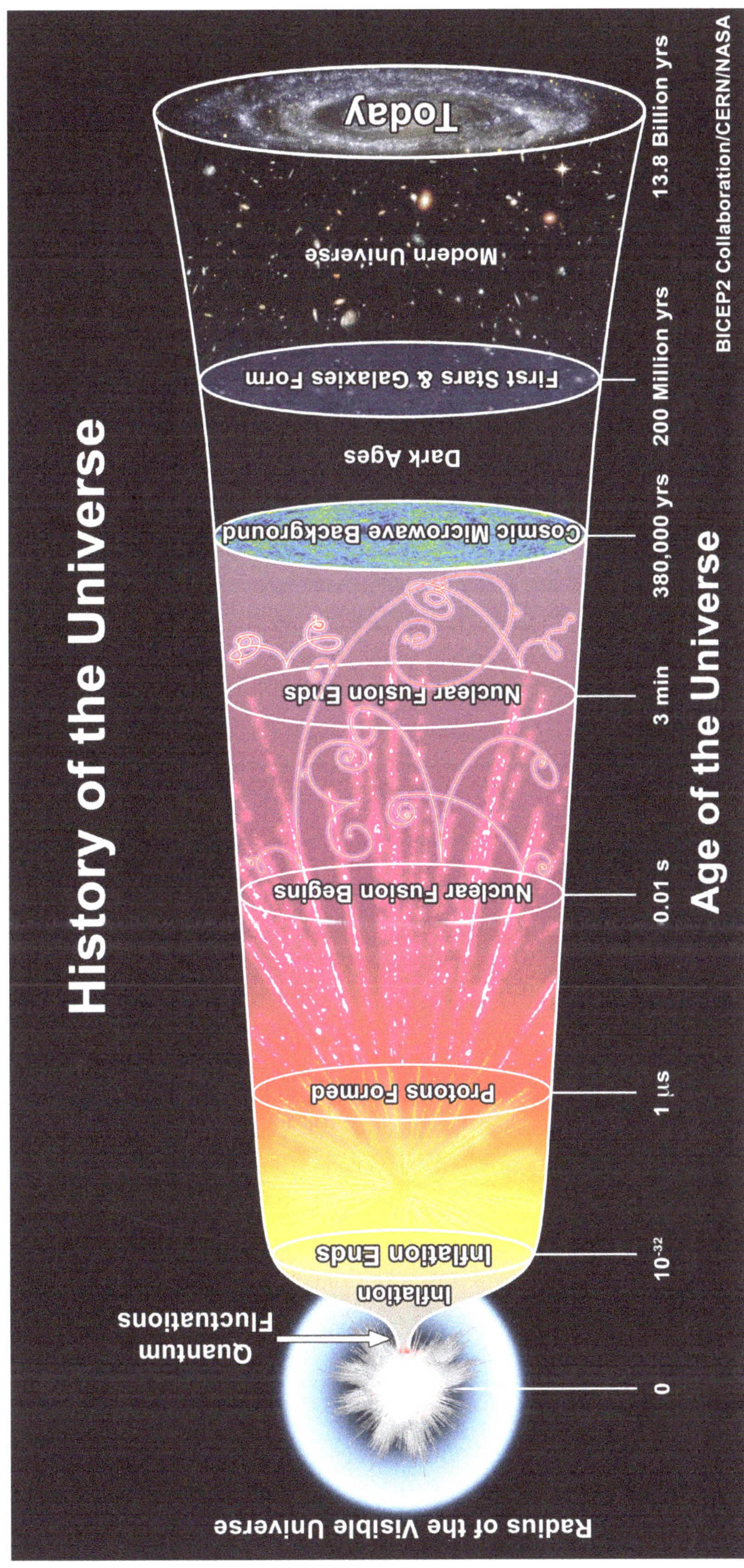

Figure 5.1. The history of the universe from the big bang to the present time.

traces of lithium and beryllium). The majority of these elements had existed before stars were formed. Therefore, the current abundance of light elements provides an indication of what formed in the early universe. In mid-1960s, based on big bang nucleosynthesis scenario, astronomers predicted that roughly 25 percent of the mass of the elements was locked into helium while the rest (75 percent) were in hydrogen. This closely agrees with observations revealing a helium abundance (by mass) of 20 percent to 30 percent. We also know that most of hydrogen nuclei (consisting of a single proton) were produced during the first minutes of the birth of the universe, independently and are not from its heavier isotopes like deuterium or tritium. Indeed, deuterium was not produced in stars but at the beginning of the universe, and it is only destroyed throughout the history of the universe, indicating that the abundance of deuterium today provides lower limits on its primordial abundance.

Observation 3: The strongest observational evidence in favor of the big bang model comes from the cosmic background radiation (figure 5.1). An afterglow of the big bang, this is a low energy radiation uniformly filling the entire universe. Due to the expansion of space over the past 13.8 billion years, the temperature of the radiation has lowered to 2.73 degrees Kelvin, with its wavelength shifting to longer *millimeter* bands. Earlier on in cosmic history, this radiation was in constant contact with matter in the universe and was freed (decoupled) from the interaction with matter roughly 380,000 years after the big bang. Before this time the temperature of the universe was so high that protons and electrons existed as free particles scattering cosmic background photons and making the universe opaque to radiation (that is, radiation could not penetrate through the proton and electron clouds that were distributed in space). When the temperature of the universe fell below 3,000 degrees Kelvin, helium nuclei (consisting of two protons and two neutrons) formed. Soon after this, hydrogen and helium nuclei (that have positive electric charges) captured electrons (that have negative electric charges), neutralizing the electric charge (recombination). After that, the universe became transparent to light. The intensity of this microwave background radiation follows a blackbody (an idealized object that absorbs all the electromagnetic radiation falling on it), consistent with a radiation that has been brought to equilibrium with its environment—just what is expected if this radiation were relic of the big bang. The cosmic background radiation contains three hundred photons per cubic centimeter, constituting 99 percent of the total number of photons in the universe.

THE VERY EARLY UNIVERSE

A number of events took place at the earliest time in the history of our universe that helped to shape the world we live in today. The forces responsible for holding matter together (at the microscopic and macroscopic scales), the particles that constitute the atoms around us, and the process responsible for the origin of mass all resulted from conditions in the very early universe. These events happened at different time intervals in the history of the universe and made major contributions to its future development (Figure 5.1). Furthermore, they did not happen in isolation and took place in sequential orders, one leading into the next. In the following sections I discuss a step-by-step history of the universe from its birth to when it was ten seconds old, when hadrons and leptons were formed and combined to generate matter and anti matter.

THE SINGULARITY

In Chapter 3 I discussed the concept of space and time and argued that they are related entities, although they are manifested differently in the present universe. Furthermore, the geometry of space is influenced by its matter content that in turn is responsible for the force of gravity resulting from it (due to matter affecting the geometry of space in its vicinity). This is the force that governs the universe at large scales. The relationship between gravity and space-time is explained by Einstein's theory of general relativity. This predicts that at the time of the big bang when all the mass in the universe was concentrated in an infinitesimally small volume and the density of the universe was infinite, there were no distinctions between space and time. This is called the *singularity*, predicted by the solution of Einstein's equations extrapolated back to the beginning of time, $t = 0$. To explain the very early universe, we therefore need to understand the physics of singularity.

While the behavior of matter at large scales is explained by gravity through the general theory of relativity, at infinitesimally small scales (the scale of atoms), matter has entirely different characteristics, explained by quantum theory. Therefore, the nature of gravity at scales the size of an atom (soon after the universe was formed) can only be explained by a new theory that incorporates quantum theory into the general theory of relativity—the theory of the very small and the very large. Scientists have not yet come up with an acceptable (or testable) quantum theory for gravity, making any prediction of the nature of singularity impossible. The time immediately after the birth of the universe when quantum properties of gravity were important is called the *Planck epoch*, corresponding to 10^{-43} seconds after the big bang (Box 5.2).

THE PLANCK EPOCH ($0 < t < 10^{-43}$ sec)

This is the time from the big bang to the Planck time ($t_\mathrm{p} = 5.39 \times 10^{-43}$ seconds), defined as the earliest epoch in the history of the universe after which one could use known laws of physics to study the behavior of the universe. At the Planck time, the size of the universe was about one Planck length (l_p)—the distance light travels in one Planck time (1.6×10^{-35} *meters*) (Box 5.2). Due to the extraordinarily small size of the universe during this time, quantum effects of gravity were prominent, and all the fundamental forces—electromagnetic, weak, strong, and gravity—had the same strength and characteristics. Our understanding of the state of the universe at this epoch is vague.

THE GRAND UNIFICATION EPOCH ($10^{-43} < t < 10^{-36}$ sec)

The grand unification epoch follows the Planck time and is the period when the temperature of the universe was around 10^{27} degrees Kelvin (corresponding to an energy of 10^{15} GeV). During this period, three of the fundamental forces—electromagnetic, weak, and strong—unified as one single force (figure 5.2). Gravity was also unified with the other three forces but was separated at the end of the Planck epoch (10^{-43}) seconds after the big bang. The grand unification ended when the strong force was separated from the electromagnetic and weak (electroweak) forces, at a temperature of 10^{27} degrees Kelvin or 10^{-36} seconds after the big bang (figure 5.2).

BOX 5.2: PLANCK UNITS

Using the universal constants of nature—gravitational constant (G), the speed of light (c) and the Planck constant (h)—a fundamental set of units are defined through dimensional analysis. These were first derived by German Physicist Max Planck, as listed below:

Planck time: $t_\mathrm{p} = (h\,G/c^5)^{1/2} = 5.391 \times 10^{-44}$ sec

Planck length: $l_\mathrm{p} = c\,t_\mathrm{p} = (h\,G/c^3)^{1/2} = 1.616 \times 10^{-35}$ meter

Planck mass: $m_\mathrm{p} = (h\,c/G)^{1/2} = 2.176 \times 10^{-8}$ kg

These units define the limits where quantum effects become dominant in explaining the behavior of matter.

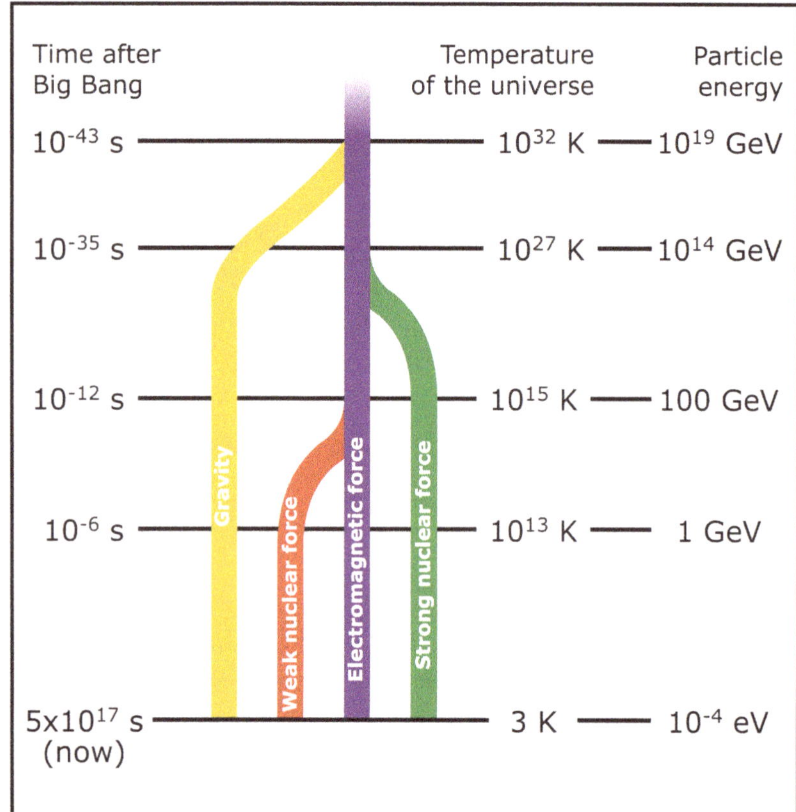

Figure 5.2. Unification of the four forces in the very early universe. As the universe expands and cools, the forces find their individual identities and separate from the single unified force.

The unification of electromagnetic, weak, and strong forces are predicted and experimentally verified by the discovery of the virtual particles mediating these forces (Chapter 4). To create the conditions under which these particles could show up, one needs extreme energies comparable to those of the very early universe (~100 GeV). The energies that can produce traces of the fundamental forces have now been generated in particle accelerators, resulting in the creation of particles that mediate these forces (Box 4.3). Since these are often massive particles, they decay into other components in a very short time. The exception is the gravity that is an elusive force, and efforts to unify it with other forces have not yet succeeded. Around 10^{-35} seconds after the big bang, when the strong, weak, and electromagnetic forces were still unified (figure 5.2), the universe experienced a rapid expansion in size in just a small fraction of a second. This is called the *inflationary* phase.

THE INFLATIONARY EPOCH ($10^{-36} < t < 10^{-32}$ sec)

There are two outstanding observations of the present-day universe that could not be accommodated within the standard big bang model—the *flatness* problem and the *horizon* problem (Schneider and Arny 2015). In what follows I give a brief explanation of these problems:

Flatness problem: Current observations show that there is a fine balance between the negatively curved (open) universe and positively curved (closed) universe, implying that the mass-energy density of the universe is very close to the critical value needed to close the universe (see Chapter 9 and Box 9.2). This requires extreme fine-tuning and is called the *flatness* problem. The problem was more serious at the very beginning of the universe, as any small deviation from this balance would greatly increase later, leading to significant consequences. For example, only a very small excess in the density of the universe over its critical value (the matter density needed to make the expansion of the universe stop and the universe collapse on itself) when the universe was 1 billion years old would have resulted in a universe that would have collapsed by now.

Horizon problem: The homogeneity observed in the temperature distribution of the *cosmic microwave background radiation* (see Box 7.2) suggests that the universe is the same in all directions. The only way this could happen is for different parts of the universe to be close enough that they could exchange information (to bring the

BOX 5.3: THE HORIZON DISTANCE

The horizon distance is defined as the distance light travels during the age of the universe. The fastest information could move is the speed of light (c). The longest known time is the age of the universe (t_u). Therefore, the longest distance one could receive information over the age of the universe is the horizon distance: $d_h = c.t_u$. The regions with distances larger than the horizon distance cannot exchange information and be connected. They are *causally disconnected*.

temperature into equilibrium). The fastest information could travel is the speed of light. Therefore, if two regions were so widely separated from one another that light hasn't had enough time to travel between the two, they are considered to be beyond each other's *horizon distance* (Box 5.3). The question therefore is: as the universe expands and more space moves within our horizon, how these well-separated regions are so similar to those already viewed by us (as revealed from the isotropy and homogeneity of the cosmic microwave background radiation). This is the *horizon* problem (Figure 5.3).

The inflationary scenario was first postulated in 1982 by Alan Guth as a solution to these two problems (Guth 1998). This proposes a phase of accelerated exponential expansion for the universe at about 10^{-35} seconds after the big bang, during which the size of the universe increased by a factor of 10^{26} and its temperature dropped by a factor of 10^5 during an extremely short time (from 10^{-35} to 10^{-32} seconds after the beginning of the universe; figure 5.3). After the inflationary period, the speed of expansion slowed down, and the universe continued to expand until the present time. Because of the inflation, the regions of the universe that were in contact (and equilibrium) with each other before the inflation were expanded away after that, solving the horizon problem. Also, any curvature or structure in the very early universe was smoothed out and largely removed because of the extreme expansion and increase in the size of the universe, solving the flatness problem.

The cause of the inflation is likely to be a negative form of vacuum energy density or positive vacuum pressure (vacuum is the lowest state of the energy of a system) resulted from the separation of strong force

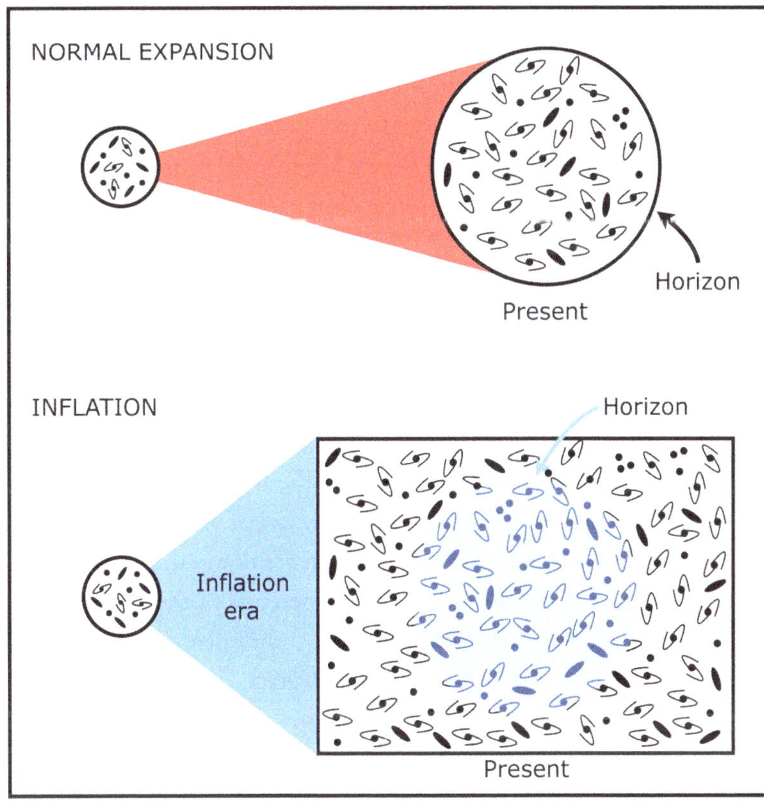

Figure 5.3. Comparison of the horizon between the normal expansion of the universe and expansion due to inflation. In the inflationary model, different parts of the universe were in contact before the inflationary expansion, solving the horizon problem.

from other elementary forces as briefly discussed in Chapter 4 and shown in Figure 4.7. This caused a symmetry breaking, leaving the universe with much more energy than it could otherwise have and hence a rapid outward force (antigravity), which would also create enormous quantities of particles in a very short time (Chaisson and McMillan 2011).

THE ELECTROWEAK EPOCH ($10^{-32} < t < 10^{-12}$ sec)

At around 10^{-32} seconds after the big bang, the potential energy that had driven the inflation was released, *reheating* the universe. The temperature of the universe that had dropped by 10^5 times during the inflation, increased to its value before the inflation. This energy is responsible for the formation of many particles, including hot *quark-gluon plasma* that filled the universe at that time[1] (Box 5.4). The temperature of the universe at this epoch was around 100 GeV (10^{15} Kelvin), enough to unify electromagnetic and weak interactions (figure 5.2). Particle interactions at this epoch were strong enough to create W^+, W^- and Z^0 particles and Higgs bosons, for which there is convincing experimental evidence now. As the universe further expanded and cooled, production of W^+, W^- and Z^0 particles was halted (around 10^{-12} seconds after the big bang). The existing W^+, W^- and Z^0 particles then decayed quickly, and the weak interaction became a short-range force. At this epoch, the particles did not yet have mass.

THE QUARK-GLUON EPOCH ($10^{-12} < t < 10^{-6}$ sec)

At $t = 10^{-12}$ seconds, the four forces of nature were separated, taking their present identity (Figure 5.2). However, the temperature of the universe was still high enough to prevent quarks binding together to form hadrons (protons and neutrons). During this epoch the universe was filled with hot and dense quark-gluon plasma. The quark-gluon epoch ended when the universe was 10^{-6} seconds old and when the average energy of particles fell below the binding energy of the hadrons, and hence quarks came together to form the hadrons. Since then, quarks ceased to exist as free particles (quarks follow a rule that their binding becomes stronger as they are pulled away from one another- like the two ends of a spring when it is stretched out. As a result, they cannot be found as free particles) (Box 5.4).

THE HADRON EPOCH ($10^{-6} < t < 1$ sec)

At this epoch, as the universe expands, the quark-gluon plasma becomes larger, with the distance between the quarks increasing. As the quarks moved apart, the elastic force due to gluons became stronger. When the distance between two quarks became around 10^{-15} meters, the gluon force between them broke, producing two new quarks, one at each end of the break. As the universe further expanded and its temperature dropped to 10^{10} degrees Kelvin, there was no longer enough energy to break the gluon elasticity, and each quark was tied (permanently) to its neighbor. The result was the production of hadrons (protons and neutrons) and antihadrons (Box 5.4). At this point, the temperature of the universe was still high enough to allow hadron and antihadron pairs to be continuously created. As the universe further expanded and the temperature dropped, the production of hadron-antihadron pairs was halted. These were subsequently annihilated and were removed, leaving a small fraction of hadrons still in the universe. The hadron epoch ended at *1 second* after the big bang, with protons and neutrons created as a result of it. The baryon asymmetry observed in the universe today has its origin in this period, when more hadrons than anti-hadrons were left over due to some unknown physical process.

---

[1] Soon after this, because of the symmetry breaking, particles acquired the characteristics they have today, getting their masses through interactions or non-interactions with the Higgs field.

THE LEPTON EPOCH (1 < t < 10 sec)

At the end of the hadron epoch, the energy density in the universe dropped so much that no more hadrons could be produced. The remaining energy density would then turn to lighter particles known as leptons (electrons, muons, and neutrinos) and antileptons (positrons, antimuons, and antineutrinos). As electrons and positrons annihilated, they converted their energy to photons, and in turn, photons collided and created electron-positron pairs (through the pair-production process). When the temperature dropped to 10^{10} degrees Kelvin, the photons could no longer produce electron-positron pairs. However, the lepton-antilepton pairs continued to annihilate with each other. The number of electrons and positrons remained the same. They were then attracted to protons and antiprotons through the electrostatic force, generating matter and antimatter.

BOX 5.4: PARTICLES IN THE VERY EARLY UNIVERSE

Quarks

In the present universe, quarks are only found inside other particles (protons, neutrons, and mesons). No free quarks are found, but due to the extreme temperature in the early universe, quarks likely existed as free particles. There are six types of quarks: up, down, top, bottom, strange, and charm. Quarks have fractional electric charges (in unit of the charge of an electron or a proton). They are bound in the nucleus by exchanging virtual particles called *gluons*. To satisfy the exclusion principle, a color is assigned to each quark (red, blue, or green) to allow them to be in different states (figure 4.5). Different types of quarks have different masses.

Gluons

These are carriers of the strong force. Gluons are particles but act as elastic bands binding quarks together (figure 4.5). When quarks are close to each other, the force produced by gluons is weak. This was the case in the early universe and the reason why quarks were found as free particles at that epoch. As quarks are separated, the gluon force between them becomes stronger, as an elastic band being stretched.

Hadrons

Quarks combine together to produce the particles we see in nature as independent entities. Three quarks (up and down) combine to produce protons and neutrons. These are called *baryons*. A quark and an antiquark make a *meson*. Baryons and mesons, grouped together, are called *hadrons*. Neutrons do not have a charge and are heavier than protons. Free neutrons are unstable, decaying to a proton and an electron (figure 4.6).

Bosons

These are particles that mediate forces. They all have integer spins. W^+, W^- and Z^0 bosons are the particles that mediate the weak force. For example, a neutron decays to a proton, an electron, and electron antineutrino by exchanging a W boson (figure 4.6). These are heavy particles with masses 80.38 GeV (W) and 91.10 GeV (Z^0) and hence decay soon after created. W bosons have +/– electric charge, while Z particles have no charge. Photons are particles responsible for the electromagnetic radiation. They have zero rest mass and no electric charge. Higgs bosons are the particles associated with the Higgs field, which is responsible for giving particles their mass. They were predicted in the standard theory of particle physics and were recently discovered. They have a mass of 126 GeV, no electric charge and zero spin.

SUMMARY AND OUTSTANDING QUESTIONS

The timeline of the very early universe and the most important events during each epoch are listed in table 5.1. The universe began 13.8 billion years ago by an explosion, the big bang, and started to expand from an infinitesimally small volume with high density and temperature. With this explosion, space and time were created, as neither existed before the big bang. Space expanded like the surface of an inflating balloon, with matter riding on this stretching space. We do not have information from the singularity that preceded the birth of the universe, when space and time did not exist as separate entities. Because of the extreme gravity at the time of singularity, concentrating in an infinitesimally small volume, the known laws of physics break down. Therefore, there is as yet no convincing answer to the question of how the big bang happened and what was responsible for it. Indeed, we have no information about the evolution of the universe up to about 10^{-44} seconds from its birth, known as the Planck time, when the radius of the universe was about 10^{-35} meters (the Planck length).

Because of the extreme energy at this very early time, the four forces of nature known today—gravity, electromagnetic, weak, and strong—were unified into one grand force. As the universe expanded, the forces were separated and attained their individual identities. First, gravity was separated (at 10^{-43} seconds and a temperature of 10^{32} degrees Kelvin), followed by the strong force (at 10^{-35} seconds and temperature of 10^{27} degrees Kelvin) and finally weak and electromagnetic forces (at 10^{-12} seconds and a temperature of 10^{15} degrees Kelvin). After the Planck time, the laws of particle physics explain the behavior of the universe, within the framework of a self-consistent theory to explain unification of the strong, weak, and electromagnetic forces. The only long-range force, gravity, is still elusive—yet to be unified with other forces. At present, there is no physical or mathematical framework that would accommodate this with experimentally verifiable predictions.

Before the strong force was separated at 10^{-36} seconds, the universe went through a phase of extremely rapid expansion—called the *inflation*. This was proposed to address some of the current puzzles in the present universe—why the universe is homogeneous and why its mass density is just on the border of that needed between an open and closed universe. Because of the inflation, the radius of the universe increased by a huge factor, with its temperature decreasing sharply. What went on at these early times fixed the future evolution of the universe up to the present time.

The outstanding question here is how the initial density enhancement that led to the structure formation in the universe happened at this very early stage. The energy released because the inflation reheated the universe (it is called reheating because it first started with a hot universe with the temperature dropping substantially because of its expansion). The released energy was responsible for creation of the first particles.

The bosons that are responsible for mediating fundamental forces were formed as these forces acquired their separate identities. These include gluon (for the strong force), W^+, W^- and Z^0 particles (for the weak force), and photons (for the electromagnetic force). Quarks were also formed around this time but could not combine to form hadrons as the temperature of the universe was very high, and as a result, the hadrons disintegrated as soon as they were created. This continued until 10^{-6} seconds after the big bang, when quarks finally combined and formed protons, neutrons, and mesons. That was the last time in the history of the universe that quarks were found as free particles.

The fascinating thing about the study of the early universe is the direct interplay between the two seemingly unrelated scales. For example, at this point the rate of particle interaction competes with the rate of expansion of the universe. As the universe ages and the temperature drops, photons lose energy and could no longer destroy the protons and neutrons as they are created from the quarks, and hence hadrons form (around 10^{-6} seconds after the big bang), followed by creation of leptons around 1 second after the big bang.

Epoch	Time (seconds)	Temperature	Description
Planck	$< 10^{-43}$	10^{32} K	Established laws of physics do not apply. Quantum properties of gravity were important.
Grand unification	$< 10^{-36}$	10^{29} K	Electromagnetic, weak, and strong forces were unified.
Inflation	$< 10^{-32}$	10^{28} K	Inflation expands the space by a factor of 10^{26}. The temperature reduces from 10^{27} to 10^{22} K.
Electroweak	$< 10^{-12}$	10^{22} K	The strong force separates from the electromagnetic and weak forces.
Quark-gluon	$< 10^{-6}$	10^{12} K	Energy in the universe is too high for the quarks to form hadrons. A quark-gluon plasma is formed.
Hadrons	< 1	10^{10} to 10^{9} K	Energy becomes low enough for quarks to form hadrons. A small matter-antimatter asymmetry results in a universe dominated by matter.
Leptons	< 10	10^{9} K	Leptons and antileptons were present and in thermal equilibrium. Neutrinos were decoupled.

Through annihilation of photons, their energy is converted to a pair of particle and antiparticle (that is, electron and positron) that collide and turn to light. An outstanding question here is the reason for the existing asymmetry between particles and antiparticles. In other words, why are there significantly more particles in the present universe than antiparticles (so-called baryon asymmetry)? The answer, although not clear yet, seems to lie in the physics of the very early universe, when the balance between the two was broken, favoring an excess of one over the other. If there were a complete balance between the two, we would not be here today. Therefore, our very existence is due to the breaking of the matter-antimatter symmetry. Finally, the physical nature of singularity is unknown and will be an extremely important topic to study.

REVIEW QUESTIONS

1. Explain the observational evidence for the big bang.
2. What was the consequence of pair production in the early universe?
3. Describe characteristics of space-time at the singularity.
4. Describe the Planck units and explain their significance.
5. What do physicists mean by unification of forces?
6. What are the observational facts that are explained by the inflationary scenario? Explain how the inflation solved these.
7. Describe the horizon distance.
8. At what epoch in the history of the universe did the particles attain their mass?
9. When were protons and neutrons formed, and why was their formation delayed?

10. Explain "reheating energy" and its significance.

CHAPTER 5 REFERENCES

Chaisson, E., and S. McMillan. 2011. *Astronomy Today*. New York: Pearson.
Guth, A.H. 1998. *The Inflationary Universe: The Quest for a New Theory of Cosmic Origins*. New York: Basic Books.
Schneider, S.E., and T.T. Arny. 2015. *Pathways to Astronomy*. 4th ed. New York: McGraw-Hill.

FIGURE CREDIT

- Fig. 5.1: Source: https://commons.wikimedia.org/wiki/File:The_History_of_the_Universe.jpg.

THE ORIGIN OF LIGHT ELEMENTS

CHAPTER LEARNING OBJECTIVES

This chapter will cover:
- The formation of light elements in the universe
- The main parameters responsible for synthesis of light elements
- The origin of the imbalance between matter and antimatter
- The initial conditions that led our universe to be the way it is

Knowledge about the origin of light elements in the universe is the first step toward understanding the formation of baryonic matter. This is the ordinary matter around us including galaxies, stars, planets, and everything else we observe. Light elements are collectively defined as hydrogen, deuterium, lithium, helium, and beryllium. These elements were formed because of the intense heat of the early universe (causing high speed for elementary particles) allowing fusion of the elementary particles (protons and neutrons) to form the nuclei of the lightest of the chemical elements. The light elements were then fused in stars to synthesize heavier elements. The process of formation of chemical elements from lighter elements in the early universe is called *big bang nucleosynthesis* (Alpher et al. 1948). However, as the universe expanded, its temperature decreased and eventually reached a level that was no longer sufficient to form elements heavier than beryllium through the fusion process. Elements heavier than this were then formed in stars through the process of stellar evolution (chapter 14). Due to the small size of the universe at the time the light elements were formed, and the fact that only a tiny fraction of these elements could form later in stars, the density of light elements is uniform throughout the universe. In other words, whatever light elements (mainly hydrogen and helium) are found in the universe today, are left over from the time of the big bang nucleosynthesis and hence are primordial. Therefore, a measure of the abundance of these elements today will give clues about conditions at the very beginning of the universe (when the universe was about ten seconds old). Modern

"We, all of us, are what happens when a primordial mixture of hydrogen and helium evolves for so long that it begins to ask where it came from."

- JILL TARTER

"You see, the chemists have a complicated way of counting. Instead of saying one, two, three, four, five protons, they say hydrogen, helium, lithium, beryllium, boron"

- RICHARD P. FEYNMAN

observations confirm that the observed abundance of hydrogen and helium in the present universe is similar to that predicted from the hot big bang model.

This chapter will discuss the state of the universe before light elements were formed. This follows by the study of the origin of light elements. The initial conditions that led to the current abundance of these elements will be addressed, as well as the question of how these influenced the present state of the universe.

THE UNIVERSE BEFORE FORMATION OF LIGHT ELEMENTS

In the very beginning (about one second after the big bang), the universe was filled with a plasma of elementary particles, consisting of protons and neutrons as well as electrons and their antiparticle positrons, neutrinos (massless and noninteracting particles with neutral charge) and photons. In the small volume constituting the universe at that time, these particles interacted with each other through the electromagnetic and weak forces (electroweak). As a result, particles continually decayed or interacted to create new species of particles. This led to a state of equilibrium in the universe, with the number of particles of each species and their respective energies staying the same. This condition was acquired by the same particles being destroyed and created again. For example, protons (p) and electrons (e^-) interacted to create a neutron (n) and an electron neutrino (v_c):

$p + e^- \rightarrow n + v_c.$

Similarly, neutrons and positrons (e^+) combined to create a proton and an antineutrino ($\bar{v}_c$):

$n + e^+ \rightarrow p + \bar{v}_c.$

Therefore, in the very early universe, the number of protons and neutrons were kept the same because of these reactions. If the reaction rate between particles is high, equilibrium can be attained more quickly, whereas if the reaction rate is low, it will take some time before the equilibrium is reached. *The fundamental factor affecting the equilibrium at this stage is the competition between the rate of expansion of the universe and the rate of nuclear reactions.* The result of the equilibrium at this time is uniformity of the temperature throughout the universe. However, the total temperature changes due to the expansion of the universe.

Until *0.1 seconds* after the big bang, the rate of weak interaction was high enough to keep the particles in equilibrium. As the temperature fell below 10^{11} degrees Kelvin, the rate of weak interaction between neutrinos and electromagnetic radiation field became so slow that they no longer interacted and hence decoupled and continued to behave independently. Another prominent process at this time was collision of electrons and positrons, generating photons that contributed to the electromagnetic field. This was balanced by the reverse process of the generation of electron-positron pairs by the photons in the electromagnetic field (called pair production; figure 6.1). As the universe continued to expand, the temperature of the electromagnetic field decreased so that it could no longer generate new electron-positron pairs. This led to an increase in the temperature of the electromagnetic field (due to collision of electrons and positrons and generation of photons and the absence of the reverse process) but not of the neutrinos that were already decoupled. This led to the emergence of a neutrino background in the universe that still exists (Weiss 2006).

The only way to attain equilibrium at this time was through the weak reactions shown above. However, there are very few electrons or positrons left to allow these interactions. The expansion rate of the universe was much faster than the rate needed for these to keep up the equilibrium. As a result, the weak interactions froze out. The only weak reaction that could take place at this time was decay of free neutrons to protons (figure 4.6). This is independent of temperature and only depends on the radioactive decay rate of neutrons (this has a half-life of about ten minutes, defined as the time needed for half of any bunch of neutrons to decay to protons). The race between the rate of expansion of the universe and the nuclear reaction rate fixes the number and type of the atomic nuclei formed at this time. For example, if the universe had expanded faster that it did, neutrons would remain free and hence decay. As a result, all neutrons would have been converted to protons, with no chance of building elements heavier than hydrogen. As it happened, the rate of the expansion of the universe was just slow enough (compared to the rate of nuclear reactions) to allow neutrons and protons to combine to form light atomic nuclei.

FORMATION OF LIGHT ELEMENTS

At around *one second* after the big bang, the temperature of the universe was about 10^{10} degrees Kelvin (corresponding to an energy of 1 MeV per particle). At this epoch nuclear reactions were fast enough (compared to the rate of expansion of the universe) to bring the universe to equilibrium, generating the conditions needed for the formation of light elements (forcing

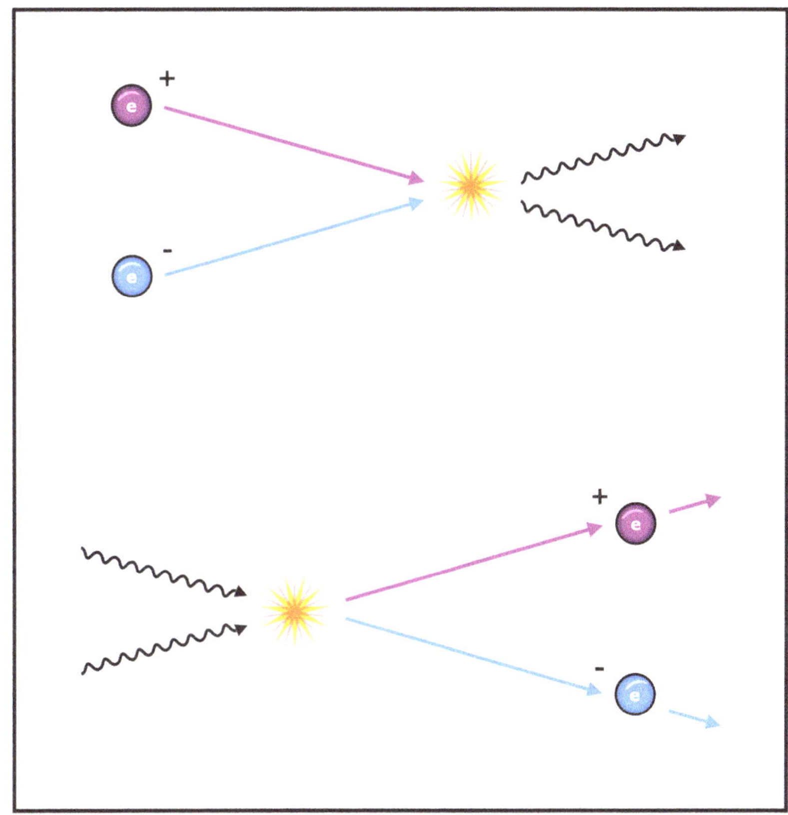

Figure 6.1. Collision of an electron-positron pair generates energy in the form of photons. The reverse reaction is when photons collide and generate an electron-positron pair, a process called pair production. These processes were dominant in the universe when it was less than a second old.

neutrons to combine with protons before they decay). The ratio of neutrons to protons remained constant at one to six (for every neutron there were six protons, due to the faster decay rate of neutrons—Box 6.1), as the age of the universe at this time was not long enough for a significant fraction of neutrons to decay to proton. This continued until one to two minutes after the big bang, when the temperature of the universe reached ~8 × 10^8 degrees Kelvin, needed to start deuterium (or heavy hydrogen, ^{2}H; number two here means one proton and one neutron[1]) synthesis by combining a proton and a neutron (figure 6.2). This time is not negligible compared to the mean lifetime of free neutrons, which is 890 seconds. Therefore, by the time conditions were suitable for deuterium formation, neutron decay decreased the neutron to proton ratio to one to seven (one neutron for every seven protons). If the expansion of the universe had halted at this time, it would have given enough time for neutrons to decay, and the universe would have ended up with a ratio of one neutron for every seven protons (Box 6.1), completely unable to form elements heavier than hydrogen (Perkins 2005).

Deuterium is fragile and, as it is produced, is destroyed by energetic photons. This delays formation of elements heavier than deuterium and is called *deuterium bottleneck*. As the universe expands and cools, photons lose their energy, and around three minutes after the big bang, deuterium nuclei were formed without being destroyed by photons (figure 6.2). After this time, the process of the synthesis of elements proceeded rapidly (figure 6.3). After

---

[1] This is the mass number of the elements defined as the number of protons and neutrons. The mass number is shown on the upper left of the chemical symbol of an element.

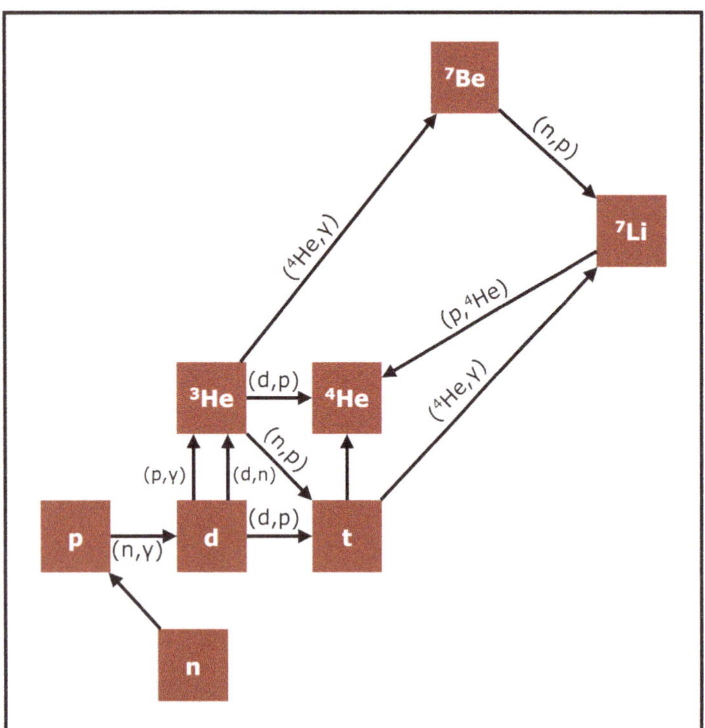

Figure 6.2. Shows different processes (neutron and proton capture) that lead to formation of light elements. There are no stable elements with mass numbers 5 and 8.

a deuterium nucleus is formed and survives, it captures a neutron to become tritium (^{3}H; denoted as *t*) (figure 6.2). The end product of this reaction is ^4He, with two protons and two neutrons (figure 6.2). There are different routes to form ^4He: (1) fusing two deuterium nuclei or (2) capturing a neutron by ^3He or capturing a proton by ^3H (figure 6.2; Weiss 2006).

The process of big bang nucleosynthesis continues after the production of ^{4}He, generating even heavier elements (however, see below). ^4He is one of the most stable nuclei. If for some reason the expansion of the universe had halted at this point, ^4He would have formed in abundance, with all the nucleons (protons and neutrons) ending up in their nuclei. Given the fast rate of cosmic expansion and slow rate of proton decay to neutron, this did not happen. Instead, the fast and efficient process of proton-neutron capture took place, leading to the formation of more ^4He nuclei (figure 6.2).

After ^4He formed, the progression of the elements to heavier nuclei was halted. ^{4}He is very tightly bound and is stable (with two protons and two neutrons—atomic number of 4). However, there is no stable nucleus with atomic number 5. Fusion of a proton or neutron to ^4He will not produce any elements, as ^5He and ^5Li are both unstable. Therefore, ^4He is resistant to fusion with protons and neutrons. Similarly, the synthesis of nuclei with atomic numbers larger than 7 is hindered by the absence of stable nuclei with atomic number of 8. By the time the universe was ten minutes old, with a temperature of $\sim 4 \times 10^8$ degrees Kelvin, the big bang nucleosynthesis was essentially over. Nearly all baryons were in the form of free protons (hydrogen nuclei) or ^4He nuclei. The small number of free neutrons decayed into protons with small amounts of deuterium, tritium, and ^3He left over (figure 6.3; Weiss 2006). Changes in the fraction of light elements formed during Big Bang nucleosynthesis is shown in figure 6.3.

BOX 6.1: HYDROGEN AND HELIUM ABUNDANCE IN THE EARLY UNIVERSE

At the equilibrium state, the ratio of protons to neutrons at the time of formation of light elements was seven to one. Now, consider sixteen nucleons (total number of protons and neutrons in a nucleus), of which two are neutrons and fourteen are protons (and hence preserving the seven to one ratio). From this, we could create one ^{4}He nucleus (two neutrons and two protons) with a mass number of 4. The remaining twelve protons will form hydrogen nuclei each with a mass number of 1 (a total mass number of 12). Therefore, the ratio of helium to hydrogen mass is four to twelve. This means, by mass, there are three times more hydrogen than helium, leading to a fraction (by mass) of 75 percent hydrogen and 25 percent helium (figure 6.3). This prediction very closely agrees with the observations and is one of the strongest proofs supporting the big bang scenario.

WHAT DRIVES BIG BANG NUCLEOSYNTHESIS?

An important parameter in the cosmology of the early universe, seriously affecting the nucleosynthesis, is the ratio of the number density of baryons to photons. This is measured from the total number density of protons and neutrons (baryons) and the photons in the cosmic background radiation (413 photons cm^{-3}). The baryon to photon ratio is $\sim 5.5 \times 10^{-10}$, meaning that for each baryon, there are about 1 billion photons in the universe. A high baryon-to-photon ratio increases the required temperature for deuterium synthesis, giving an earlier start to the nucleosynthesis. The consequence of this is that the process will be more efficient in producing ^{4}He, leaving less deuterium and ^{3}He, as the density and temperature of the universe drops.

The baryon-to-photon ratio cannot be as small as 10^{-12}. If it were, the process of big bang

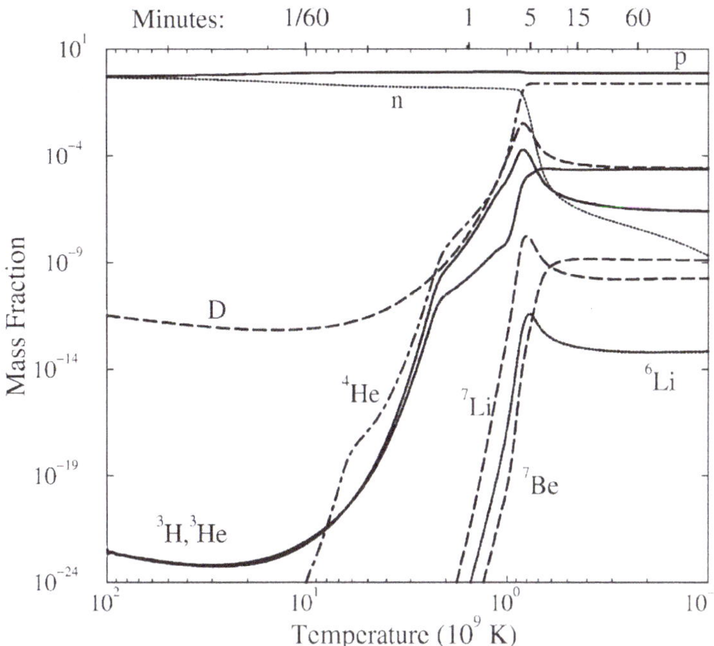

Figure 6.3. Changes in the abundances of elements with cosmic time. Helium started to synthesize around one hundred seconds after the big bang. Also, Beryllium (^{7}Be) only started to form at three hundred seconds after the big bang, delaying formation of elements heavier than atomic mass of 7 (beryllium bottleneck).

nucleosynthesis would be very inefficient, and we would expect only a very small amount of helium to be produced. Also, the baryon-to-photon ratio cannot exceed 10^{-7}. In this case the nucleosynthesis would have happened much earlier (before neutrons had a chance to decay), and the universe would have been free from deuterium, with the ^{4}He abundance being at its maximum. Therefore, observations of the abundance of light elements can constrain the baryon content of the universe and hence its matter density (Perkins 2005) (Box 6.2).

BOX 6.2: LIGHT ELEMENTS AND THE DENSITY OF THE UNIVERSE

The baryonic density expressed in terms of the number of neutrons and protons in a given volume in the universe directly affects the outcome of the nucleosynthesis. For example, if the density of the universe is high, nuclear reactions occur more frequently, leading to faster production of ^{4}He and less time for neutron radioactive decay and hence less deuterium and ^{3}He. Therefore, if we measure the relative abundance of ^{4}He and ^{3}He elements and find the proton-neutron density it corresponds to, one could constrain the baryonic density of the universe.

WHY IS THERE NO ANTIMATTER IN THE UNIVERSE?

The laws of physics predict the presence of baryons and antibaryons consisting of quarks and antiquarks respectively (because of their symmetry). However, we know that there is a significant preference for baryons over antibaryons in the present-day universe. At the high temperature of the early universe, the quarks were not confined in the baryons.

Therefore, in the very early universe, quarks and antiquarks were rapidly created through the pair-production process and annihilation generating photons. At this time the number of quarks, antiquarks, and photons were similar. As the universe expanded and cooled, quark and antiquark pairs were no longer created because there was not sufficient energy to produce them. Instead, they annihilated and created photons. Now, if there were a tiny excess of quarks over antiquarks in the very early universe, after annihilation of these particles, a small excess of quarks would remain. These were then combined, forming baryons as the universe expanded and cooled down (there were not many antiquarks to produce antibaryons). Due to the annihilation of quark and antiquark pairs, more photons were generated, while the quarks (building blocks of baryons) were removed. This explains the very small baryon-to-photon ratio in the universe. The cause of the baryon asymmetry in the very early universe (the grand unification era) is not yet known and must be searched within the framework of the conditions in the early universe (Perkins 2005).

SUMMARY AND OUTSTANDING QUESTIONS

The events during the two minutes of the big bang nucleosynthesis have shaped the 13.8-billion-year history of our universe (Table 6.1). In fact, our very existence today is a direct consequence of the events happening within those two minutes from the beginning of the universe. Table 6.1 summarizes the timeline during the nucleosynthesis epoch up to the time when the first atoms were formed. We know physics of the nucleosynthesis fairly well and can make accurate predictions of the events during that time. There are a number of parameters that governed synthesis of light elements when the universe was a few minutes old. These include the half-life of neutrons, the expansion rate of the universe, rate of proton-neutron interaction, the beryllium bottleneck, absence of stable elements with atomic numbers 5 and 8 ratio of the number density of baryons (protons and neutrons) to photons. If any of these were slightly different from what they were, we would not be here today.

Big bang nucleosynthesis is a step-by-step process, making heavier nuclei by combining simpler elements. Fusion of a proton and neutron results in a deuterium nucleus (^2H) with a binding energy of 2.2 MeV (the energy required to dissociate a deuterium nucleus). Deuterium is fragile, and given the high energy of photons at that time (about two minutes after the big bang), it was destroyed soon after it was formed (deuterium bottleneck). This delayed formation of the elements heavier than deuterium. Therefore, the binding energy of the deuterium nucleus played an important role in the timescale of the synthesis of elements following deuterium. The time of the synthesis of deuterium is close to the lifetime of neutrons (840 seconds). This means that by the time deuterium is synthesized, a large fraction of neutrons are already decayed to protons, leading to an abundance of protons over neutrons. This has direct implications on the abundance of deuterium and helium.

Table 6.1. Timeline and characteristics of the universe during and after the nucleosynthesis

Epoch	Time after big bang	Temperature	Description
Big bang nucleosynthesis	10 to 10^3 sec	10^{11} to 10^9 K	Protons and neutrons are bound into atomic nuclei.
Photon	10 to 10^{13} sec (~380,000 years)	10^9 to 10^3 K	The universe contains electrons, nuclei, and photons. Temperature is too high for electrons to combine the nuclei.

The rate of the expansion of the universe was responsible for the drop in its density and temperature that, in turn, govern the rate of reactions taking place between the light elements. This led to the synthesis of the elements during the first few minutes of the evolution of the universe. There was indeed competition between the rate of the expansion of the universe and the rate with which particles combined to form nuclei of heavier elements that fixed the abundance of these elements. As the rate of expansion dropped with time, reactions became more efficient, leading to the formation of heavier elements. The final product is the ^4He nucleus, which is very stable (with a binding energy of 28.3 MeV). There are no stable elements with atomic numbers 5 and 8. Therefore, the process of big bang nucleosynthesis stopped at ^4He (although small traces of ^5Li are produced that are immediately destroyed to other elements). Also, there is no way to bridge the gap between elements with atomic number 8 and heavier nuclei. These heavier elements are made in the dense center of stars through the "triple-alpha" reaction ($^4He + {}^4He + {}^4He \rightarrow {}^{12}C$) (see Chapter 14). This is the reason that only light elements are made through the big bang nucleosynthesis. All the elements heavier then ^7Li are made in stars billions of years later.

The above discussion shows that if any of these processes had happened differently, the consequences would have been dramatic for our present universe. There are, however, many unanswered questions, among them: What fixed the baryon-to-photon ratio in the universe? Why this ratio is so small? How could the neutrino background be detected?

REVIEW QUESTIONS

1. Explain how the competition between the rate of weak interactions and the rate of expansion of the universe affects the abundance of light elements.
2. What significant event happened at 0.1 seconds after the big bang?
3. Explain the events that led to the neutrino background.
4. How did the rate of neutron decay affect formation of light nuclei in the early universe? Also, what role did the ratio of neutron-to-proton play in the relative abundance of light elements today?
5. Explain the deuterium bottleneck and its effect on the formation of light elements.
6. Explain steps that led to the formation of ^4He.
7. Why couldn't ^4He turn to heavier elements by attracting a proton or neutron?
8. When did the big bang nucleosynthesis end and why?
9. Predict the expected fraction of hydrogen and helium in the universe based on big bang nucleosynthesis.
10. What role does the baryon-to-photon ratio in the universe play in constraining the abundance of light elements?
11. Explain how the density of the universe is related to the ^4He and ^3He abundance?

CHAPTER 6 REFERENCES

Alpher, R., H. Bethe, and G. Gamow. 1948. "The Origin of Chemical Elements." *Physical Review* 73 (7): 803–4. doi:10.1103/PhysRev.73.803.

Perkins, D. 2005. *Particle Astrophysics*. Oxford, UK: Oxford University Press.

Weiss, A. 2006. "Equilibrium and Change: The Physics behind Big Bang Nucleosynthesis." *Einstein Online* 2: 1018.

CHAPTER LEARNING OBJECTIVES

This chapter will cover:

- The origin of the first atoms
- Matter and radiation in the early universe
- Cosmic microwave background radiation
- The dark ages
- How the universe became transparent—the journey out of darkness

"Nothing in life is to be feared, it is only to be understood. Now is the time to understand more, so that we may fear less."

- MARIE CURIE

"The laws of nature are constructed in such a way as to make the universe as interesting as possible"

- FREEMAN DYSON

We learned in chapter 6 that the nuclei of light elements (deuterium, helium, and lithium) were all generated through the process of big bang nucleosynthesis. However, the intense radiation, extreme temperature, and high density at that time forbid electrons to join the existing nuclei to form neutral atoms. At this time, matter was in the form of mostly hydrogen and helium nuclei (with positive charges) and negatively charged electrons that scattered photons. As the universe expanded and cooled, stable atoms formed through a process called *recombination* (electrons joining the nuclei forming neutral atoms) that took place 280,000 years after the big bang. At this time, matter (in the form of hydrogen and helium atoms) coexisted with radiation. At some point, matter and radiation decoupled and evolved independently. The radiation formed the cosmic background photons we observe today, and the matter formed stars, galaxies, planets, and eventually us.

The neutral atoms made the universe opaque, bouncing photons around and stopping them from escaping. This epoch is known as the *dark ages,* which lasted for almost 600 million years and ended by formation of the first generation of stars and galaxies. The high-energy radiation from these primordial stars and galaxies ionized the atoms in a process called *reionization*, which led to the photons being able to escape without being impeded by matter, making the universe transparent. This is the reason we can see to great depths in the universe. During this time, the universe went through a number of transitions that affected its subsequent evolution.

This chapter presents an account of the universe before, during, and after the dark ages. This then leads to the study of the origin of atoms and the subsequent decoupling of matter from the cosmic background radiation that fills the universe today. Different physical processes involving the interaction between matter and radiation will be studied. The process of re-ionization that made the universe transparent will then be discussed.

FORMATION OF THE FIRST ATOMS

At the end of the primordial nucleosynthesis, completed around ten minutes after the big bang, 75 percent of the baryonic matter in the universe (protons and neutrons) ended up in hydrogen while 25 percent in helium nuclei, with a very small trace of heavier elements. At this point, the temperature of the universe was sufficiently low to allow formation of elements heavier than beryllium and lithium but still high enough to forbid the positively charged nuclei and negatively charged electrons to combine and form stable atoms (Figure 7.1). As soon as the electrons joined the hydrogen and helium nuclei, they were ejected by collision with high-energy photons (fig 7.1, top panel). This process continued until 240,000 years after the big bang, when the temperature of the universe was 3,740 degrees Kelvin. Because of the expansion of the universe, the temperature (and energy of photons) dropped sufficiently so that the negatively charged electrons could join the positively charged nuclei (of mostly hydrogen and helium) through the recombination process to form the first stable atoms (figure 7.1). This epoch in the history of the universe is called the *recombination* (Box 7.1), where the first atoms were formed and photons were continuously scattered by the neutral atoms and the free electrons and hence not able to travel unimpeded (figure 7.1- bottom panel; Box 7.1). The rate at which photons scatter depends on the number of free electrons at the time of recombination (figure 7.1- top panel). Therefore, as the number of free electrons decreases due to the recombination process, so does the rate of photon scattering. At this time, the rate of photon scattering competed with the rate of expansion of the universe (the Hubble constant). The rate of scattering of photons rapidly decreased (proportional to the volume $1/a^3$, where a is the factor proportional to the radius of the universe) as the universe expanded until the time between subsequent scattering of photons became longer than the age of the universe at that epoch (a Hubble time, corresponding to H^{-1}, where H is the rate of the expansion of the universe; the Hubble time almost corresponds to the age of the universe). At this point, photons ceased to interact with free electrons and therefore decoupled (figure 7.1- lower panel). This epoch is called the *decoupling*. As a result of this, photons could travel independently

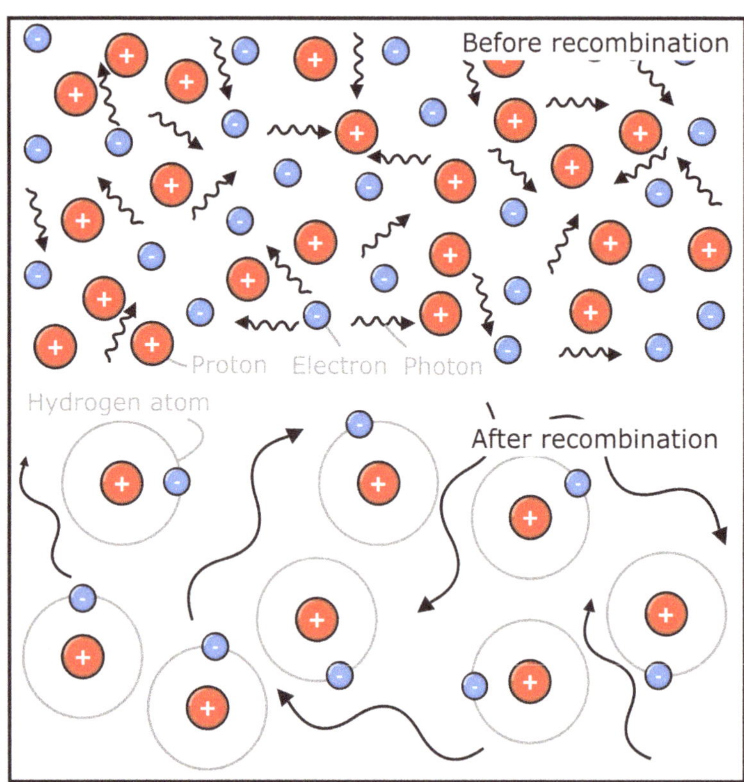

Figure 7.1. The process of scattering of photons by electrons and hydrogen atom nuclei (protons) before recombination (upper panel)-(see Box 7.1). At this epoch the temperature of the universe was so high that it would eject electrons from atoms as soon as they recombined (upper panel). Due to the expansion of the universe, the energy of the photons decreased and they were no longer able to remove electrons from the atoms. This corresponds to the epoch of recombination around 240,000 years after the big bang (lower panel).

BOX 7.1: INTERACTION BETWEEN MATTER AND RADIATION

Matter and radiation interact in the following ways.

Ionization: This is the removal of one or more electrons from atoms. As electrons have a negative electric charge, by losing them, the atoms become positively charged. The result is called *ion*. Ionization is expressed as H^+ (hydrogen atom losing one electron—singly ionized) or He^{++} (helium atom losing two electrons—doubly ionized). The ionization process takes place by collision of atoms or when electrons associated with atoms are hit and ejected by photons.

Recombination: This is the opposite of ionization, when an electron joins an ion and produces a neutral atom. In the early universe, when the temperature decreased (due to the expansion of the universe) and photons lost their energy, they could no longer remove electrons from the atoms, and hence stable and neutral atoms form. The rate of recombination depends on the density of the electrons and the nuclei in the medium as well as their speed (temperature).

Absorption: This is when a photon hits an electron in an atom and transfers all its energy so that the electron moves to a higher (quantized) energy level. In this case the photon has to have the same energy as the difference between the energy levels in the atom.

Scattering: This is when photons collide with free moving nuclei or electrons and lose energy. This reduces the energy of the photons or transfers the energy to and from the electrons and the nuclei. The scattering of photons by electrons or free nuclei took place before the recombination (about 240,000 years after the big bang). This corresponds to the radius of an imaginary surface, called *surface of last scattering*.

without interacting or being impeded by the matter, and hence the universe became transparent. The process of decoupling happened in a short time and was completed by around 380,000 years after the big bang, when the temperature of the universe was 3,000 degrees Kelvin (see figure 6.1). The time a photon went through its last scattering by an electron is called the *last scattering time,* which is very close to the decoupling time (Bennett et al. 2010).

COSMIC BACKGROUND RADIATION

Before the decoupling epoch, matter and radiation in the universe were in a state of thermal equilibrium, meaning that there was a uniform temperature throughout. As a result, the fluctuations in the temperature distribution of the radiation at the time of the decoupling (380,000 years after the big bang) directly reflects the matter distribution at that time. After the formation of neutral atoms, light was no longer scattered by free electrons and therefore could pass through the existing matter, making the space transparent (figure 7.1, bottom panel). This is the thermal remnant from the big bang, shifted to longer (microwave) wavelengths due the expansion of space. The remnant radiation has been detected today and is known as the *cosmic microwave background*, providing the strongest observational evidence for the big bang. The radiation has a blackbody spectrum (figure 7.2), implying that it has come to thermal equilibrium with the matter in the universe, and is uniformly distributed, indicating that it is from a cosmological origin. Over the cosmic time (since it decoupled), its temperature has reduced from 3,000 degrees Kelvin to 2.754 degrees Kelvin, observed today.

The cosmic microwave background fills the universe with microwave photons. However, it also shows distinct variations in temperature caused by fluctuations in matter distribution in the universe around the time of the

The cosmic microwave background radiation was discovered in 1968 by Arno Penzias and Robert Wilson. To understand the origin of the cosmic microwave background, and whether it indeed comes from the beginning of the universe, one needs to measure its spectrum and the degree of homogeneity in its temperature distribution. Due to the stretching of the wavelength of primordial radiation caused by the expansion of space, the observations need to be performed in the sub-mm wavelengths, requiring satellite probes. This is because sub-mm wavelengths are absorbed by water vapor in Earth's atmosphere.

Over the past twenty years, three different satellites were launched, aiming to measure physical characteristics of the cosmic microwave background. These were Cosmic Background Explorer (COBE), Wilkinson Microwave Anisotropy Probe (WMAP), and the Planck. To avoid background noise created by the sun and Earth, the satellites are launched to Lagrangian 2 orbit—about 1.5 million km from Earth. At that orbit, the gravity of the Earth, the sun, and the moon cancel each other out, and the satellite is in a "noise free" environment. These probes all surveyed the entire sky and produced maps of the temperature distribution. The improved resolution of the maps is shown in figure 7.3 (from COBE to Planck). These experiments confirm that the temperature is uniform to 1 in 10^{-5} units with a resolution better than 0.2 deg (Smoot 2006).

Observations of the cosmic microwave background have found the fluctuations in its temperature distribution, which grew to produce today's structures in the universe: provided the most accurate estimate for the age of the universe to be 13.8 billion years; constrained the curvature of space to be flat within 0.4 percent; confirmed that ordinary matter only contains 4.6 percent of the universe; showed that the universe contains 24 percent dark matter and 71.4 percent dark energy; and measured the amplitude of fluctuation of density in the universe, leading to the formation of the first galaxies.

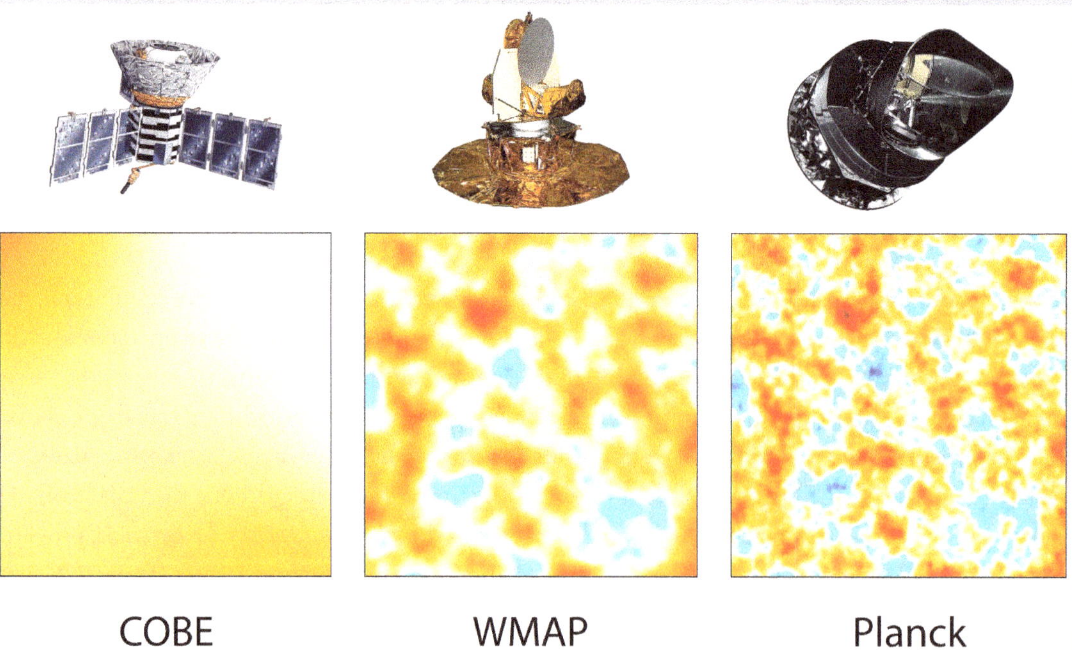

Figure 7.3. Temperature distribution of the cosmic microwave background photons at the time of the decoupling. This reflects the matter distribution when the universe was 380,000 years old. The red and blue points correspond to hot (high-density) and cold (low-density) regions respectively.

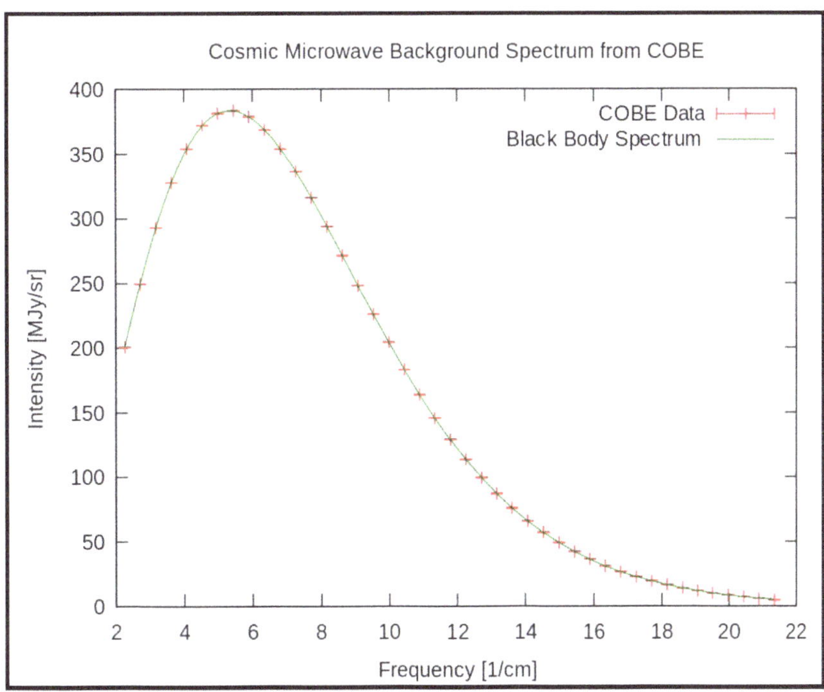

Figure 7.2. Spectrum of the cosmic microwave background showing a perfect match to a blackbody spectrum at a temperature 2.754 degrees Kelvin. This was observed by the Cosmic Background Explorer (COBE) satellite (see Box 7.2 for details about the cosmic microwave background).

decoupling (figure 7.3 and Box 7.2). These clumps of baryonic matter provided seeds for the formation of stars and galaxies. They collapsed under the force of their gravity, forming the structures (from clusters of galaxies to planets) we see in the universe today (Wilson 1978).

THE DARK AGES

The period after the formation of the first atoms and before the formation of the first stars is referred to as the *dark ages*. At this time the universe was devoid of any source of light and was mainly dominated by dark matter (which eventually collapsed to form stars and galaxies, Chapter 8). The stars and galaxies did not yet exist, and the only photons were those from the background radiation that decoupled from the matter around 300,000 years after the big bang. During this epoch, neutral matter (atoms) created an opaque wall, blocking light from passing through. Therefore, no information could be received from the dark ages, with the evolution of the universe significantly slowing down during this time. The dark ages started around 300,000 years after the big bang and ended when the first generation of stars and galaxies started to form around one billion years after the big bang (figure 7.4). The light generated by these objects increased the photon background in the universe by producing intense flux of ultraviolet photons. The high-energy light ionized the atoms (removed the electrons, Box 7.1) and hence made the universe transparent to light. This is referred to as the *reionization* (Figure 7.4 and next section). The emergence of the universe from the dark ages was gradual and took over 600 million years until the reionization was completed (figure 7.4; Bennett et al. 2010) around one billion years after the big bang.

What is the Reionization Era?

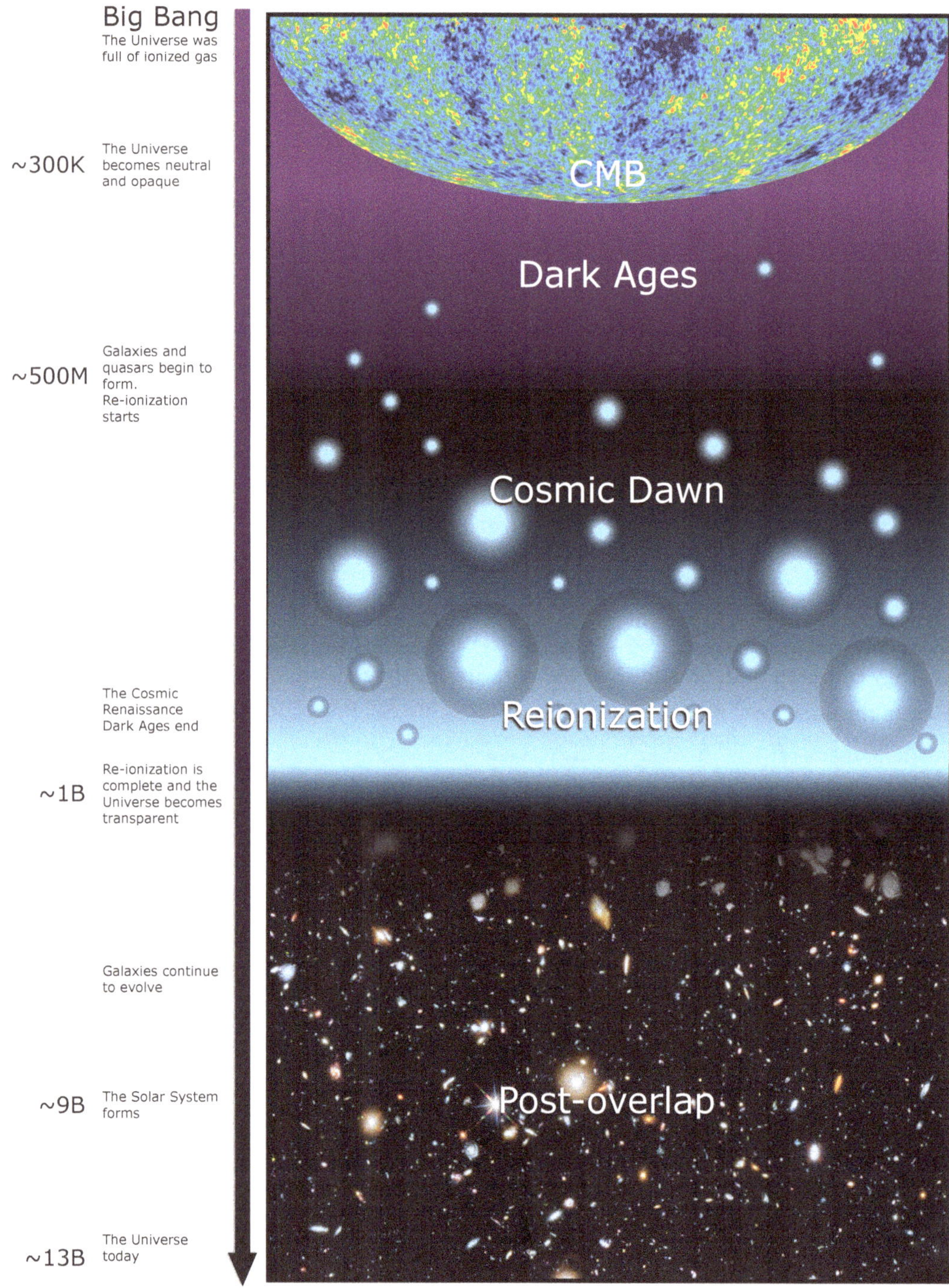

Figure 7.4. Different stages in the history of the universe. The dark ages were followed by the "cosmic dawn" after which the first generation of stars and galaxies formed and ionized neutral matter in the universe.

REIONIZATION OF THE UNIVERSE

The temperature fluctuations seen on the cosmic microwave background maps reflect the matter distribution in the universe at the time of the decoupling (figures 7.3 and 7.4). This indicates inhomogeneity (clumpiness) in the matter distribution. Under their gravity, these clumps collapsed, and with high density and temperature at their core, fusion of lighter elements (hydrogen and helium) took place, leading to formation of the *first generation* of stars and galaxies. These produced high-energy ultraviolet radiation that in turn ionized the surrounding neutral matter. Through this process neutral atoms were hit by radiation and were ionized (Box 7.1). As a result, light could freely move through the matter, making the universe transparent. It is not clear when the reionization was completed, but observations of the most distant galaxies indicate that this happened around 1 billion years after the big bang (figure 7.4; Bennett et al. 2010).

The reionization is studied by observing the light from distant and bright objects (that is, quasars, population III stars, and primordial galaxies). Along the line of sight, the light hits clumps of matter formed in the intergalactic medium and is absorbed in certain wavelengths by specific chemical elements in these clumps. Therefore, by studying the spectra of these objects, we see absorption features that indicate chemical composition of the clouds in the intergalactic medium. Shorter wavelengths are more susceptible to absorption, and therefore most of the light at these wavelengths is absorbed. The reionization is the reason we can see distant parts of the universe today. The reionization epoch lies at the edge of the observable universe, and today, using the deepest images of the most distant regions of the universe, we have been able to detect the first generation of stars and galaxies at that epoch (Hester et al. 2014).

SUMMARY AND OUTSTANDING QUESTIONS

There are a number of important phase transitions that shaped the history of our universe (figure 7.4). These are caused mainly because of the expansion of the universe and drop in its temperature, making the conditions suitable for formation of the first atoms through the recombination process around 240,000 years after the big bang. At about the same time, due to the expansion of the universe, the rate of the scattering of photons by the matter (electrons and nuclei of hydrogen and helium) was significantly reduced to the point that they were no longer scattered (due to increased time between two subsequent collisions of photons with electrons and protons). This led to the *decoupling* of matter and radiation. The matter was mostly in the form of baryonic dark matter while the radiation formed the cosmic background, the only source of light in the universe at that time. When the recombination process completed (all the atoms were formed), the universe went through a period of inactivity and very slow evolution. At this time no light sources were present, and the neutral atoms blocked the diffuse light from escaping. This is the so-called *dark ages*. During this period the gravity of dark matter led to the collapse of the structures, attracting more matter and growing in size. This period ended gradually by the formation of the first generation of stars and galaxies. The intensity of the high energy ultraviolet light generated by the stars ionized the matter in the universe again. This is called the *reionization* epoch. Today, with the power of our ground-based and space telescopes, we can detect the first generation of galaxies at the edge of the reionization epoch, around 1 billion years after the big bang. The transition from recombination into the dark ages and out was gradual and happened through many millions of years. Table 7.1 lists the timeline of the universe during this time.

Table 7.1. Timeline of the universe during recombination, dark ages, and reionization

Epoch	Time after big bang	Temperature	Description
Recombination	380,000 years	3,000 K	Electrons and atomic nuclei become bound, forming neutral atoms. The photons are no longer in thermal equilibrium with matter. They decouple from matter to form cosmic background radiation.
Dark ages	380,000 to 150×10^6 years	4,000–100 K	This is the time between recombination and formation of the first stars and galaxies. During this time the only source of radiation was the cosmic background photons.
Reionization	150×10^6 to 10^9 years	60–19 K	The first generation of stars and galaxies were formed in this epoch. The light from these objects ionized neutral hydrogen. The earliest population III stars formed at this time.

There are a number of outstanding questions still waiting to be answered. It is not yet clear what the first stars—*population III stars*—looked like. These were expected to be very massive stars (around ten to one hundred times more massive than our sun) and hence short lived (with an age of about a few million years). As a result, not many of these stars have been discovered yet. It is likely that the population III stars were the first sites of formation of heavy elements we see around us today. Also, it is not clear how dark matter affected formation of the first galaxies. This is complicated by our lack of knowledge about the nature of dark matter. Finally, it is not clear exactly when the reionization epoch ended and what is the nature of galaxies at the reionization epoch. These questions, plus many more, are so fundamental in understanding galaxies and the universe that a significant amount of observation time on the largest telescopes on the ground and in space has been dedicated to this aim, complemented with detailed simulations. These provide some of the most fascinating topics for future studies.

REVIEW QUESTIONS

1. What delayed formation of the first atoms?
2. Explain the process of recombination.
3. What were the physical conditions in the universe that led to the decoupling of matter and radiation?
4. Explain the epoch of last scattering.
5. How could one map the density fluctuations in the early universe using cosmic background radiation?
6. What does a blackbody spectrum for the cosmic background radiation mean?
7. Name different probes that observed the cosmic background radiation.
8. How did the epoch of dark ages start and end?
9. Explain the reionization of the universe.
10. How could one probe the medium between galaxies?

CHAPTER 7 REFERENCES

Bennett, J., M. Donahue, N. Schneider, M. Voit. 2010. *The Cosmic Perspective*. 6th ed. Boston: Pearson/ Addison Wesley.

Hester, J., B. Smith, G. Blumenthal, L. Kay, and H. Voss, H. 2014. *21st Century Astronomy*. 3rd ed. New York: Norton.

Smoot, G.F. 2006. "Cosmic Microwave Background Radiation Anisotropies: Their Discovery and Utilization" (lecture). Nobel Foundation. http://www.nobelprize.org/nobel_prizes/physics/laureates/2006/smoot-lecture.html.

Wilson, R. 1978. "The Cosmic Microwave Background Radiation" (lecture). Nobel Foundation. https://www.nobelprize.org/nobel_prizes/physics/laureates/1978/wilson-lecture.html.

FIGURE CREDITS

- Fig. 7.2: Source: https://en.wikipedia.org/wiki/File:Cmbr.svg.
- Fig. 7.3: Source: https://en.wikipedia.org/wiki/File:PIA16874-CobeWmapPlanckComparison-20130321.jpg.
- Fig. 7.4a: Source: https://en.wikipedia.org/wiki/File:Ilc_9yr_moll4096.png.
- Fig. 7.4b: Source: https://commons.wikimedia.org/wiki/File:NASA-HS201427a-HubbleUltraDeepField2014-20140603.jpg.

THE ORIGIN OF STRUCTURE IN THE UNIVERSE

CHAPTER LEARNING OBJECTIVES

This chapter will cover:

- The development of structures in the universe
- Different structures in the universe
- The origin of density fluctuations in the early universe
- The first generation of galaxies and stars
- Population III stars

The universe is filled with structures at different scales—from small-scale structures like planets and stars, to intermediate structures like galaxies, to large-scale structures like clusters and superclusters of galaxies. The origin of these structures and their evolution throughout the age of the universe are among the most outstanding questions in modern cosmology. For example, for many years there was a long-standing debate as to whether the clusters formed first, fragmented to galaxies and then stars, or the galaxies formed first and then were grouped together to form clusters and larger structures. Whatever the first structures, their origin lies within the inhomogeneity over and above the smooth distribution of matter we observe at large cosmological scales. These fluctuations in matter distribution then grew under the force of gravity, leading to the formation of structures in today's universe (stars, galaxies, and clusters of galaxies, Box 8.1). The scale of these structures depends on the size of the initial fluctuations. It is clear that gravity, resulting from dark matter, played a significant role in creating the structures we observe in the universe today.

The process of structure formation in the universe started after the matter and radiation were decoupled and continued during the dark ages but before the universe was reionized, less than 1 billion years after the big bang. It is believed that at the time of the reionization, many of the smaller-scale structures (stars and galaxies) were in place. Therefore, study of the process of structure formation will elucidate the nature of the first stars and galaxies that were responsible for the reionization of the universe as well as help explain their formation process.

"The greater danger for most of us lies not in setting our aim too high and falling short; but in setting our aim too low, and achieving our mark."

- MICHELANGELO

"It takes considerable knowledge just to realize the extent of your own ignorance."

- THOMAS SOWELL

This chapter examines the origin of the structure in the universe and its evolution with cosmic time. It investigates the origin of the density fluctuations that led to structure formation and growth, forming the first generation of stars and galaxies.

PRIMORDIAL STRUCTURE FORMATION

Matter and radiation decoupled around 380,000 years after the big bang, after which matter was free to follow its own fate. As a result, under the force of gravity, matter (both ordinary and dark matter) collapsed and formed dense and massive structures (figure 8.1). The prevailing theory today for the formation of structure in the universe is the *cold dark matter* scenario. The name "cold" is conceived as the velocity of the particles constituting cold dark matter is much less than the speed of light (unlike hot dark matter, which moves very fast). Therefore, fluctuations governed by cold dark matter can potentially grow to massive and dense structures (Box 8.1). This then produced sheets and filaments, collectively known as the *cosmic web*. The first stars and galaxies were formed within these filaments (figure 8.2; Bennett et al. 2007).

At the time of the initial collapse, matter and radiation were mixed. This led to interplay between gravity (which pulls things together) and radiation pressure caused by the radiation produced through the fusion of light elements at the core of the collapsing clouds (which pushes things out). This resulted in a pattern of oscillation

BOX 8.1: STRUCTURES IN THE UNIVERSE

Depending on their scale, there are different types of structures in the universe, including:

Stars: These are small structures that are being formed throughout the age of the universe. Small-scale primordial fluctuations are thought to be responsible for the formation of the first generation of stars—called *population III stars*. Stars are also formed more recently by collapse of gas clouds. They produce their light by the fusion process, converting light to heavier elements. Our sun is a moderate age star (5 billion years old) with an average mass.

Galaxies: On average, galaxies contain 10^{11} stars of different types, with their observed light being the integrated light of all those stars. Some types of galaxies also host sites of current star formation activity. Apart from stars, galaxies also contain gas and dust as well as a significant amount of dark matter. Galaxies are small islands in the universe, moving away from one another by the expansion of space. Some galaxies actively form stars, and some have no such activity. Our own Galaxy is called the *Milky Way*. It has a spiral shape with our sun being in one of its spiral arms. The mass of an average galaxy is about 10^{11} M_{sun} (where M_{sun} is the mass of our sun).

Galaxy clusters: These are larger-scale structures and contain many galaxies—from a few tens to millions, depending on the richness of the cluster. Most of the galaxies in the universe reside in groups or clusters. The high-density environment of clusters (accommodating many galaxies within a small volume) causes collision between galaxies and hence influences the shape and evolution of galaxies residing in them. Clusters have a mass of about 10^{15} M_{sun}.

Superclusters: These are the largest structures in the universe and are formed by many individual clusters coming together. These are likely to have been produced through initial density perturbations. They have sizes of about 100 Mpc. Our Milky Way galaxy is in the local super cluster.

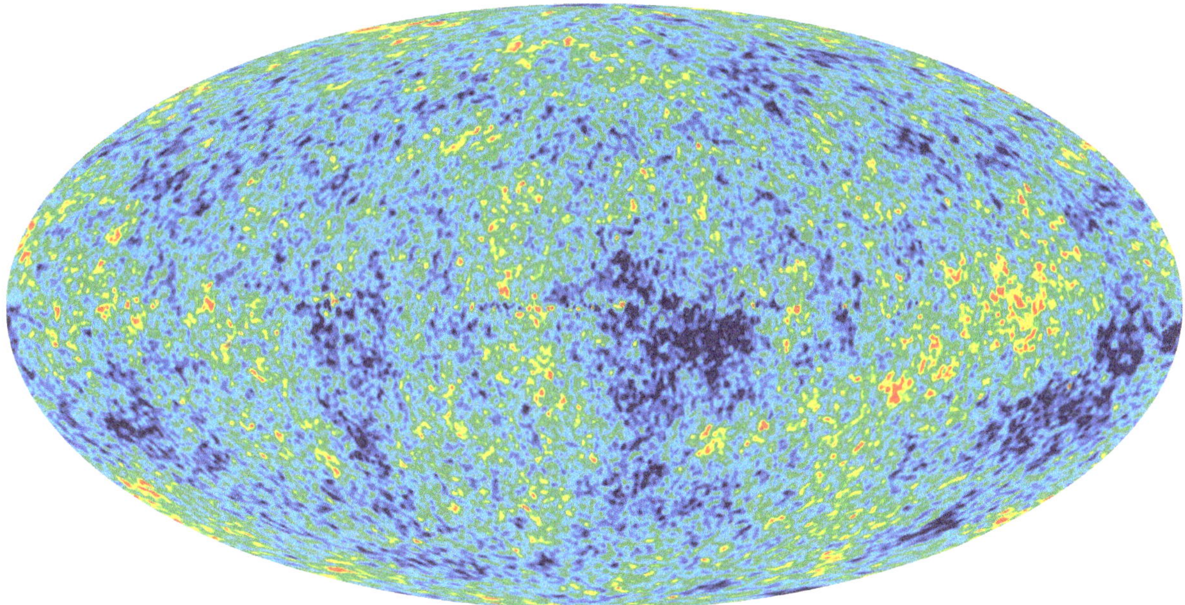

Figure 8.1. Distribution of the temperature of the cosmic microwave background. The imprints of density enhancements at the time of the decoupling are reflected in the temperature distribution of the cosmic background radiation. Red and blue regions have respectively high and low temperatures respectively. This reveals distribution of matter in the universe around 380,000 years after the big bang.

in the form of sound waves (propagation of density fluctuations throughout matter). Since gravity is a result of both dark and ordinary matters but radiation pressure is only caused by ordinary matter (as dark matter particles do not interact with photons), the shape of these oscillations reveal the ratio of the ordinary to dark matter. Furthermore, as dark matter is not coupled to photons, any concentration of them will grow very fast to produce dense and massive structures. The initial nonuniformities in the matter distribution in the universe were reflected in the temperature distribution of the cosmic background radiation (figure 8.1). This gives clues about the size and extent of these fluctuations that led to the present structures in the universe (Hester et al. 2014). Figure 8.1 shows temperature distribution of the cosmic background radiation at the time of the decoupling, with blue and red/yellow regions corresponding to low and high matter densities respectively (low and high temperatures).

THE ORIGIN OF INITIAL DENSITY NONUNIFORMITIES

In order to start the process of structure formation, somehow initial perturbations (nonuniformities) need to be introduced to the uniform matter distribution. These provide the initial seeds that grow to produce the structures we observe today (figure 8.2). Before the matter and radiation decoupled around 380,000 years after the big bang, they interacted and came to thermal equilibrium. Therefore, the nonuniformities in the matter distribution at that time are reflected in the temperature distribution of the cosmic background radiation today (figure 8.1; Bennett et al. 2007).

The question then is, how did these initial nonuniformities come about? Physicists believe that this happened by a process called *vacuum fluctuation*. According to the laws of quantum mechanics, there is a temporary change in the amount of energy in space, leading to the creation of "virtual particles" in empty space in extremely short times. The physics behind this prediction is the *uncertainty principle* (Box 8.2). This implies that in small scales, the energy field at any point in space (even in a vacuum, defined as the lowest state of energy) is always fluctuating

BOX 8.2: THE UNCERTAINTY PRINCIPLE

According to the uncertainty principle, there is a limit to the accuracy with which one could measure energy and time for any particle. If the error in energy is ΔE and the error in time is Δt, they follow the equation

$$\Delta E \cdot \Delta t \sim h / 4\pi$$

where h is the Planck's constant. This implies that it is impossible to make a precise measurement of energy and time simultaneously. The formulation may seem to violate the law of conservation of energy. However, this is preserved by the creation of virtual particles and antiparticles (energy pockets) for extremely short times. These in turn are responsible for the initial seeds in the nonuniformities in the matter distribution from which structures were eventually formed.

(jumping from one value to other, Box 8.2). These generate quantum ripples that grow into fluctuations, which created the seeds responsible for today's structures in the universe. These fluctuations are characterized by their wavelength, which correspond to their size (large- or small-scale fluctuations). During the inflation era, due to the rapid expansion of the universe, the wavelengths (fluctuations) increased by a factor of 10^{30}, becoming close to the size of the structures observed today (stars and galaxies). Such quantum fluctuations are hence the origin of the seeds that led to the structures in the universe before gravity took over.

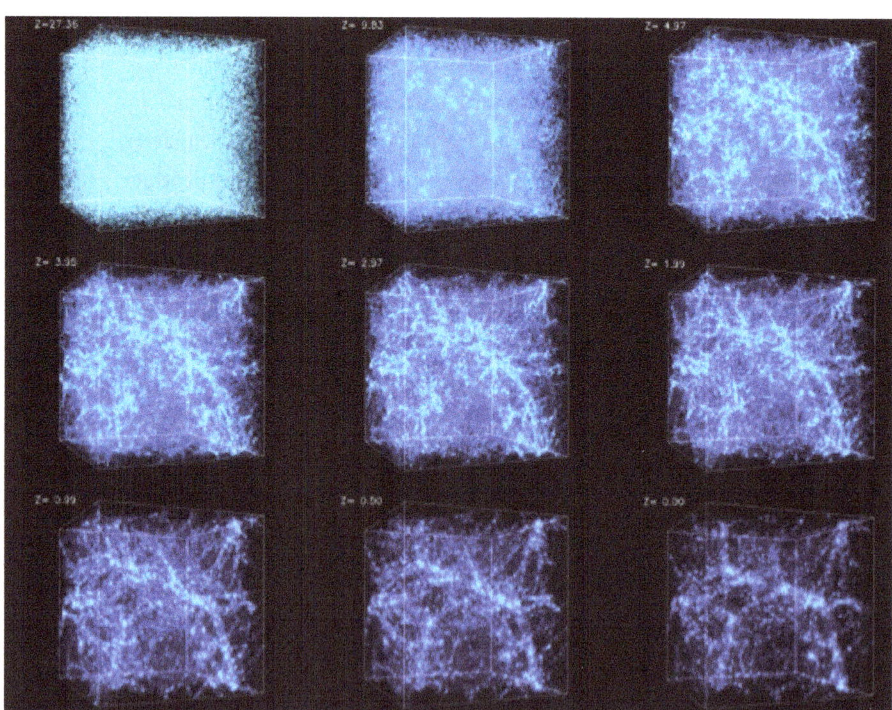

Figure 8.2. The growth of structure with the cosmic time. Initially (at early cosmic time), matter distribution in the universe was uniform (top panels). Small ripples led to small structures that grew over time to form bigger structures (middle panels). The large mass of these structures attracted more matter, increasing the degree of inhomogeneity in the present universe (bottom panels).

Figure source Simulations were performed at the National Center for Supercomputer Applications by Andrey Kravtsov (University of Chicago) and Anatoly Klypin (New Mexico State University). Visualizations by Andrey Kravtsov.

THE FIRST GENERATION OF GALAXIES

Following the big bang, the universe was hot, homogeneous, and smooth. Small density fluctuations of size one part in one hundred thousand (10^{-5}) grew under the force of gravity, and as they cooled, they formed dense bound systems within which gas molecules were made (figure 8.2). At this point, the gas and dark matter were attracted to regions of higher density, forming *dark matter halos*. These represent seeds for the first generation of galaxies. The halos collapsed under the force of gravity, forming protogalaxies. After this, the hydrogen and helium within the halos fused to form the first generation of stars. Over the time, the halos merged to form larger and more massive structures in the form of galaxies. The formation of galaxies started from small and low mass systems (that is, dwarf galaxies), then merged to produce larger galaxies and eventually very massive systems like our Milky Way galaxy. In this scenario, formation of structure is a bottom-up process, with first small galaxies forming and then merging, resulting in larger galaxies and finally clusters and superclusters of galaxies. The process eventually produces the *cosmic web* (Figure 8.3), within which galaxies are forming through collapse of dense regions (Bennett et al. 2007). Simulations of the early structures in the universe show the cosmic web consisting of filaments containing galaxies. Galaxy clusters form in the intersection of these filaments in the cosmic web (figure 8.3).

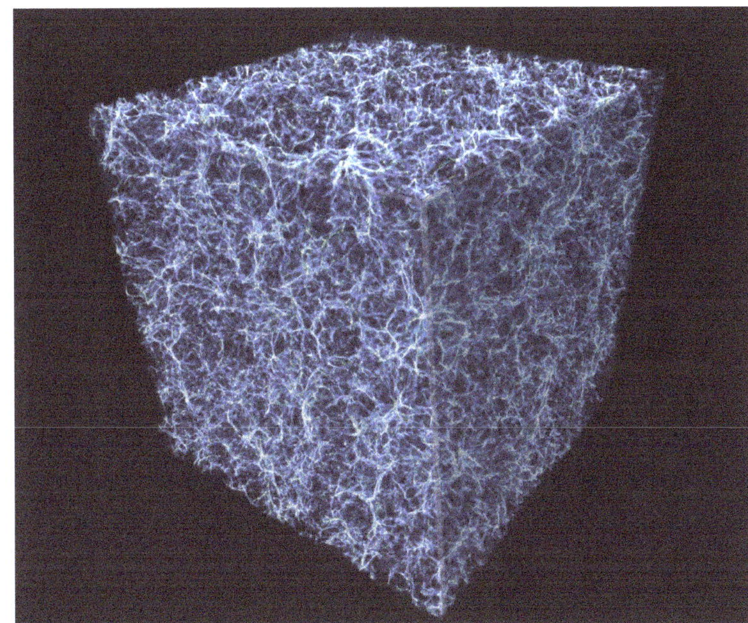

Figure 8.3. Simulation of the cosmic web containing galaxies. Clusters are believed to have formed through interaction of the filamentary structures in the cosmic web.

THE FIRST GENERATION OF STARS

The first stars were formed from the pristine gas, left over after the big bang nucleosynthesis. These stars are responsible for the creation of the first generation of *metals* in the universe. In astrophysics, metals are referred to as the elements heavier than hydrogen and helium. Therefore, carbon, oxygen, and nitrogen are all referred to as metals. Astronomers classify the stars into two populations: population I stars, which are young, metal rich (contain heavy elements), and reside in the disks of galaxies; and population II stars, which are old, metal poor (contain light elements), and mostly live in the central bulges and halos of galaxies. The reason that young population I stars are more metal rich than the older population II stars is because they were formed later from a gas that had had the time to produce the metals through nuclear reactions. Therefore, the first generation of stars that were formed after the dark ages (population III stars) were extremely metal poor since the gas from which they were formed had not had enough time to produce the metals.

Figure 8.4. Typical image of a star-formation site. Population III stars are bright in ultraviolet radiation (because they are young and emit energetic radiation) with very low metallicity.

Due to the collapse of the initial matter clumps, the density and temperature at their center increased and reached a level that could ignite the fusion process, combining two or more light elements to produce heavier elements and, in the process, a lot of energy in the form of short wavelength (high energy) ultra-violet light. This signals the birth of *population III* stars (figure 8.4). They are formed from hydrogen and helium (and small amounts of lithium and beryllium). These stars initiate the process of metal enrichment, producing the heavy elements (metals) that eventually end up in population II stars. *Population III* stars are very massive, with a mass of the order of sixty to three hundred times the mass of the sun. Massive stars are short lived. Therefore, it is expected that *population III* stars have short lifetimes (on the order of a few million years). This is the reason there are not a large number of them left around today. These stars were responsible for producing the enriched gas that was recycled, leading to the formation of next generation of metal-rich stars.

SUMMARY AND OUTSTANDING QUESTIONS

Structures of various sizes exist in today's universe. This is an observational fact, as we can see them (figure 8.5). The question now is how these structures were formed to begin with and how they were developed throughout the life of our universe. The seeds of these fluctuations are imprinted in the temperature distribution of the cosmic background radiation (figure 8.1). These led to the collapse of clumps of matter under their gravity (figure 8.2). As they collapsed, the density in their center increased, resulting in the fusion of hydrogen to helium nuclei, producing lots of energy and radiation that generated an outward pressure to counter balance the collapse (figure 8.2). This led to the formation of first generation of stars (population III stars). Through this process, a significant amount of high-energy photons were generated (in the form of ultraviolet radiation), and some escaped, reionizing the surrounding space (figure 8.4). This happened around the reionization epoch, less

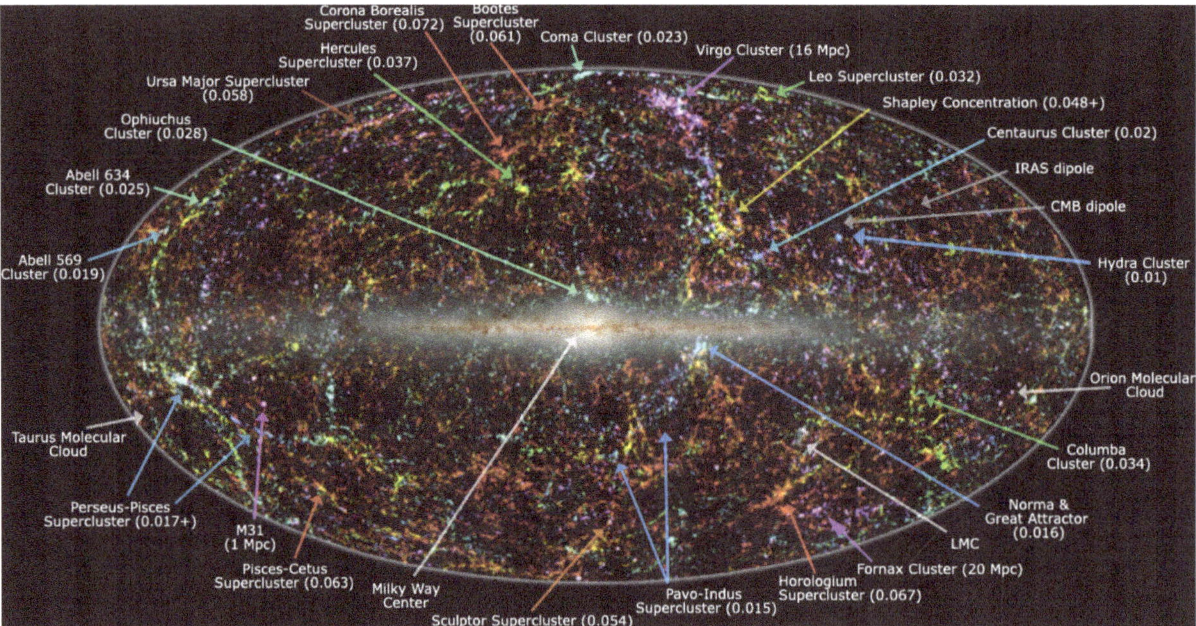

Figure 8.5. Shows structures in the nearby universe. More than 1.5 million galaxies are present in this diagram, shown as colored points. The galaxies are color coded depending on their redshift (distance) from us. The numbers in parentheses indicate the redshift of each galaxy or cluster of galaxies. The image is taken from 2 Micron All Sky Survey data showing all the sky in infrared wavelengths.

than 1 billion years after the big bang. The size of these fluctuations is equivalent to temperature variations in the cosmic background radiation.

We are now getting close to developing a consistent picture for structure formation in the universe. Studying the observed structures in the present universe (figure 8.5), astronomers have been able to generate simulations to understand the growth of these systems. There are, however, details that still need to be carefully examined. For example, what is the nature of dark matter halos that host the first generation of galaxies? What fraction of the high-energy photons generated this way can escape from the system to the intergalactic medium? Exactly when did the first generation of galaxies form? And what is the nature of metal-poor but luminous and massive population III stars?

A significant amount of theoretical and simulation work is in progress to study the dark matter halos and the emergence of structures within them. This involves formulation of fundamental physical principles and often nonlinear effects (when events do not happen in isolation, with one event affecting another one and so on). There is also much observational and theoretical effort to detect and study the first generation of galaxies around 1 billion years after the big bang. The most distant systems are found to have formed around 500 million years after the big bang, almost at the edge of the observable universe. Meanwhile, the search for population III stars is ongoing. This is hard, as the high mass and short lifetime of these stars indicate that they are very rare. Nevertheless, with increased sensitivity of our detectors and new instruments, reliable candidates for population III stars are being discovered.

REVIEW QUESTIONS

1. Explain the cold dark matter scenario for the formation of structure.
2. What do astronomers mean by the *cosmic web*?
3. How do density fluctuations propagate through collapsing gas clouds?
4. What is responsible for the growth of density fluctuations in the universe?
5. Explain different known structures in the universe.
6. What is the origin of the nonuniformities in the initial matter distribution?
7. Explain the uncertainty principle and how this could lead to explaining the generation of virtual particles.
8. What are dark matter halos?
9. How did small galaxies in dark matter halos end up with today's giant galaxies?
10. Explain the characteristics of population III stars.

CHAPTER 8 REFERENCES

Bennett, J., M. Donahue, N. Schneider, and M. Voit. 2007. *The Cosmic Perspective*. 4th ed. Boston: Pearson/ Addison Wesley.

Hester, J., B. Smith, G. Blumenthal, L. Kay, and H. Voss. 2014. *21st Century Astronomy*. 3rd ed. New York: Norton.

FIGURE CREDITS

- Fig. 8.1: Source: https://en.wikipedia.org/wiki/File:Ilc_9yr_moll4096.png.

- Fig. 8.2: Andrey Kratsov and Anatoly Klypin / The Center for Cosmological Physics, "Formation of the Large-Scale Structure in the Universe," http://cosmicweb.uchicago.edu/filaments.html.

- Fig. 8.3: Source: https://en.wikipedia.org/wiki/File:Structure_of_the_Universe.jpg.

- Fig. 8.4: Source: https://en.wikipedia.org/wiki/File:Stellar_Fireworks_Finale.jpg.

- Fig. 8.5: Source: https://commons.wikimedia.org/wiki/File:2MASS_LSS_chart-NEW_Nasa.jpg.

THE PRESENT STATE OF THE UNIVERSE

CHAPTER LEARNING OBJECTIVES

This chapter will cover:
- The expansion of the universe
- The geometry of space
- The age of the universe
- The density of the universe
- The edge and center of the universe
- Olber's paradox

"It is far better to grasp the universe as it really is than to persist in delusion, however satisfying and reassuring"

- CARL SAGAN

"I'm astounded by people who want to 'know' the universe when it's hard enough to find your way around Chinatown"

- WOODY ALLEN

Extensive observations by space-borne and ground-based facilities have greatly enhanced our knowledge about the universe. We have discovered that gas clouds collapse to form stars of different masses and types. These in turn constitute islands of matter called galaxies that, on average, consist of one hundred thousand million (10^{11}) stars. Galaxies are then attracted to each other to form groups and clusters that form the largest structures in the universe. All these massive structures formed by the force of gravity and then evolved to the present time. The collective mass of these systems and the matter distributed between them affect the geometry of space-time as well as the dynamics of individual galaxies.

We can detect matter (in the form of stars and galaxies) because of the light emitted by them. This is found to only constitute 4 percent of the matter content of the universe, the rest being nonluminous matter, called *dark matter*, only detected through its gravity (Chapter 10). Although not observable, dark matter significantly contributes to the density of the universe and hence dynamics of galaxies because of its mass and the gravity generated by that mass. The matter density of the universe affects its geometry (by modifying space-time), its age and its future fate. Using the largest observatories and most sensitive detectors, we could measure the observable parameters in the present universe. These also allow us to reach the edge of the observable universe. Today, for the first time in the history of civilization, we could contrast theories with observations to have a deep understanding of the working of our universe.

This chapter is about the present state of the universe. It shows that we could perform precise measurement of physical parameters governing our universe. The chapter studies the expansion of the universe as well as its density and geometry of space-time and therefore, how we could determine future fate of the universe. The chapter demonstrates that cosmology is now becoming a precision science, one that like other disciplines can be expressed by experimental data.

MATTER AND THE UNIVERSE

One of the predictions of Einstein's general theory of relativity (the only theory for gravity we have today) is that mass causes space around it to curve. For example, the light passing by the sun is deflected. This is because the mass of the sun curves the space around it, and the light (or any object) moving in curved space is diverted from its main trajectory as if an external force is acting upon it. Depending on the amount of matter in the universe, space takes different shapes, with its geometry often described in terms of its matter-induced curvature—that is, zero (flat), positive (spherical), or negative (saddle-shape) curvature (figure 9.1).

The mass content of the universe is responsible for the geometry of space (or its curvature) that, in turn, determines if the universe will expand forever or its recent expansion would stop at some point in the future. For example, if space has negative curvature, the matter density of the universe (mass per unit volume) is insufficient to stop its expansion, and as a result, the current expansion will

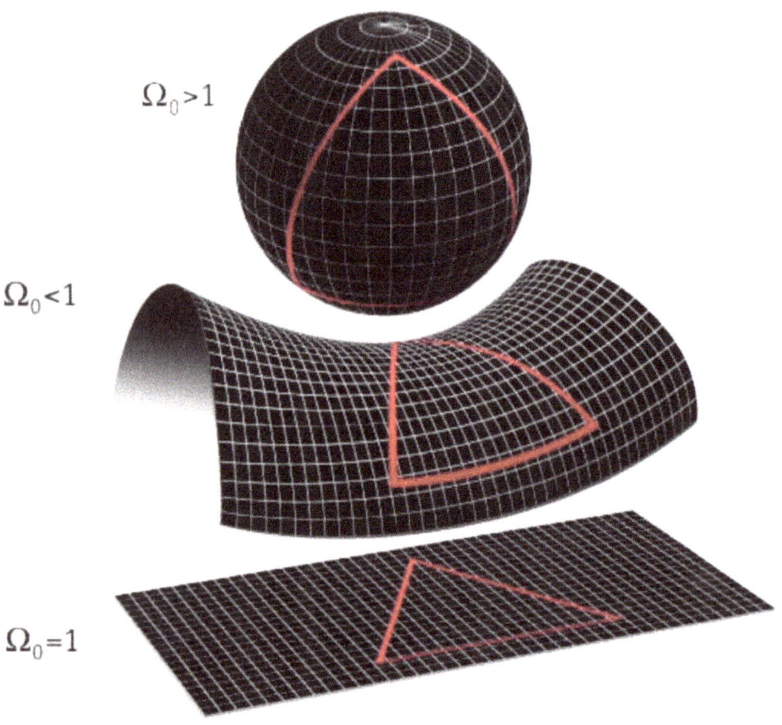

Figure 9.1. Shows the present geometry of the universe, fixed by the density parameter, Ω_0 (Box 9.2). The shape of the universe is determined by the density paramter if Ω_0 is greater than one (spherical), equal to one (flat), or smaller than one (hyperbolic).

BOX 9.1: THE COSMOLOGICAL PRINCIPLE

The cosmological principle states that matter distribution in the universe is homogeneous and isotropic. This means that at large scales, matter is uniformly distributed independent of the line-of-sight direction. This is the principle on which formulation of the mathematical equations for the universe is based. The cosmological principle is confirmed by observations of the homogeneity and isotropy of the cosmic microwave background radiation.

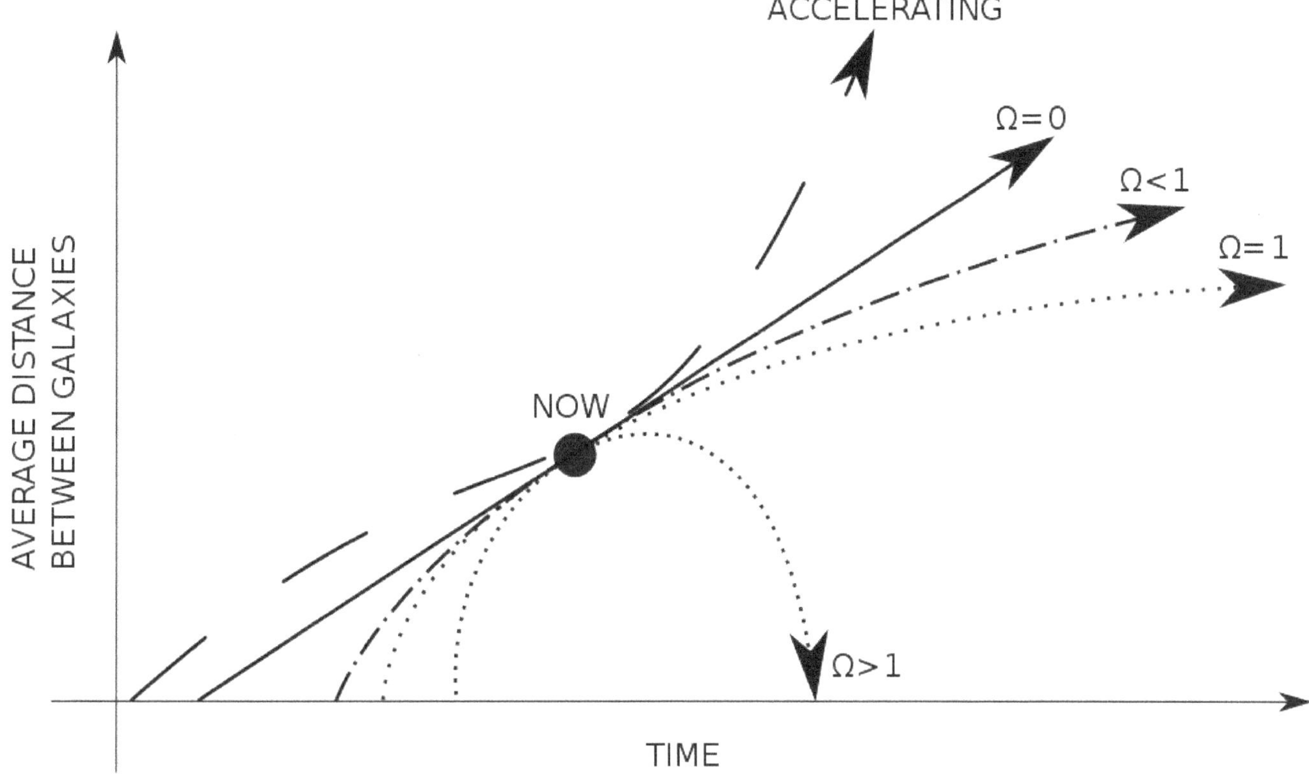

Figure 9.2. Shows the behavior of the universe as a function of cosmic time. A closed universe stops expanding and collapses on itself. An open universe will expand forever, with its radius increasing with time. A universe with the same density as the critical value (flat space) will expand uniformly for an infinite time. The Omega parameters (Box 9.2) corresponds to the matter density of the universe that fixes the geometry of the space and future fate of the universe.

continue forever—the universe is open (figure 9.1; middle panel). On the other hand, if it has positive curvature, the matter density is enough to stop its expansion, and at some point in the future, it would collapse on itself—the universe is closed (figure 9.1; top panel). Finally, if the space has zero curvature (figure 9.1, bottom panel), the matter density of the universe is just sufficient to stop its expansion but only after infinite time—the universe has no boundaries and will expand forever with uniform speed—the space is Euclidean. These scenarios and their effect on the fate of our universe are shown in figure 9.2. Recent theoretical predictions, based on the inflationary scenario, followed by various observations, indicate that the universe at cosmological scales is flat, with its density close to the closure value ($\Omega = 1$; Schneider and Arny 2015).

The formulation of the geometry of the universe is based on the consideration that, on large scales, the matter distribution in the universe is homogeneous and isotropic (does not depend on direction), known as the *cosmological principle* (Box 9.1).

THE DENSITY OF THE UNIVERSE

The density of matter and energy in the universe, defined as the amount of matter and energy per unit volume, determines its geometry, age, and future fate (whether it is open, closed, or flat). An important parameter here is the *critical density*, which is defined as the density required to close the universe (figure 9.2 and Box 9.2). If the observed

BOX 9.2: THE DENSITY PARAMETER

A fundamental parameter in cosmology is the "critical density." This is the average density (amount of matter per unit volume) needed to halt the expansion of the universe, ρ_c. The volume over which the density is defined must be large enough to be representative of the whole universe. Otherwise, this will be affected by local density inhomogeneity and would not express the cosmological value. The critical density is

$$\rho_c = 3\,H^2 / 8\,\pi\,G$$

where H is the Hubble constant and G is the constant of gravity. The critical density of the universe is 10^{-26} kg/m^3 or ten hydrogen atoms per cubic meter.

The density parameter, Ω, is the ratio of the real (observed) density of the universe, ρ, to the critical density and is the key parameter in determining the geometry of the universe. This is defined as

$$\Omega = \rho/\rho_c.$$

Total Ω is the sum of contributions from matter (Ω_M), radiation (Ω_R), and dark energy in the form of the cosmological constant (Ω_Λ):

$$\Omega = \Omega_M + \Omega_R + \Omega_\Lambda.$$

A flat, closed, or open universe respectively have $\Omega = 1$, $\Omega > 1$, and $\Omega < 1$ (figure 9.2).

matter and energy density of the universe is larger than the "critical density," the universe will have positive curvature and would hence be closed like a sphere. If the observed density is smaller than the critical density, the universe will have negative curvature (saddle shape), and the universe would be open. An observed value close to the "critical density" would indicate a flat and ever expanding universe (figure 9.2). To find the density parameters, astronomers measure the change in the speed of the expansion of the universe with cosmic time. In other words, they measure how fast or slow the universe has been expanding over the past many billions of years. This determines the change in the rate of expansion of the universe (technically this corresponds to the slope of the lines or tangent to the curves in figure 9.2) with cosmic time. The larger the density is, the higher is the decrease in its rate of expansion (Schneider and Arny 2015).

THE EXPANSION OF THE UNIVERSE

As we know by now, the universe is expanding, with galaxies moving away from one another. The recession velocity of galaxies increases with their distance from us. This means that the more distant galaxies are, the faster they move away from our own Galaxy. This velocity-distance relation is linear, called *Hubble's law* (figure 9.3), defined as

$$V = H \times D$$

where V is the velocity of each galaxy in kilometers per second (km/s), D is the distance to that galaxy in megaparsecs (Mpc), and H is the rate of expansion of the universe, the Hubble constant in km/sec/Mpc. The rate of expansion of the universe depends on its matter/energy density. Any deviation from linearity in this relationship (figure 9.3) is caused by the increase or decrease in the average expansion velocity of galaxies and hence total density of the universe (Kirshner 2003).

To measure the Hubble constant, one needs to estimate distances to galaxies far away from our own Galaxy to make sure the galaxy is taking part in the cosmic expansion (the noise introduced to the recession velocity of any given galaxy due to the gravity of galaxies close to it must be negligible compared to the "Hubble expansion velocity", the velocity with which that galaxy takes part in the cosmic expansion (figure 9.3)). This is accomplished by using objects that are bright enough to be seen at large distances and have similar and known intrinsic luminosities (called *standard candles*). The luminosity of an object at distance d diminishes with the inverse square of its distance from us. Therefore, by comparing the intrinsic and apparent luminosities of a standard candle inside a galaxy, one could estimate the distance to that object and hence to its host galaxy. The recession velocity of galaxies (also called redshift) is directly measured by performing spectroscopy. The relationship between the distance and velocity of galaxies with respect to us is called the *Hubble diagram*

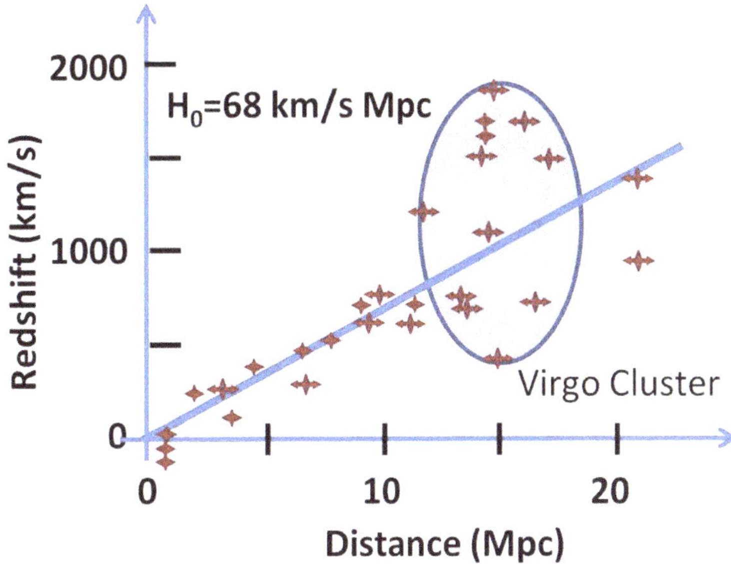

Figure 9.3. The velocity-distance relationship (Hubble diagram). The red symbols correspond to individual galaxies. The slope of this relationship, V/D, gives the Hubble constant. The deviation from the linear relationship appears when there is nonuniformity in the cosmological expansion. For example, a nearby galaxy cluster (a large number of galaxies at almost the same distance from us moving with almost the same velocity) could influence the velocity field, introducing nonlinearity. The nonlinearity could also be caused by the change in the geometry of space. In this case the observed non-linearity can be measured and used to constrain the geometry of space. I will discuss this in Chapter 10.

(figure 9.3), with the slope of this corresponding to the present value of the Hubble constant (Livio and Riess 2013). The latest estimate of the Hubble constant based on observations by the Hubble Space Telescope is 73.00 +/− 1.75 km/sec/Mpc.

THE AGE OF THE UNIVERSE

By measuring the rate of the expansion of the universe, astronomers calculate the time taken by a galaxy to travel to its present location from where it was formed in the early universe, assuming the galaxy has moved with constant speed. In other words, the quantity D/V, corresponding to the ratio of the distance of a galaxy from us (that is, the distance the galaxy has traveled from our Galaxy during the lifetime of the universe) to its velocity with respect to our Galaxy, gives an estimate of the age of the universe. This ratio is the inverse of the Hubble constant, and therefore the age of the universe can be estimated as $t_u = 1/H_0$, where t_u is the age of the universe and H_0 is the present value of the Hubble constant. This measurement depends on the mass density of the universe and only equals the age of the universe if the universe is flat (that is, the mass density of the universe is the same as its critical density or the universe has been expanding with uniform speed over its lifetime). A universe with a higher matter density is younger than one with a lower density (Kirshner 2003).

Clearly, the age of the universe must be older than the age of the oldest stars. These stars are believed to reside in *globular clusters*. These systems yield accurate age measurements, as will be discussed in chapter 13. This provides a lower limit to the age of the universe. Today the age of the universe measured from different independent methods has converged to 13.8 billion years.

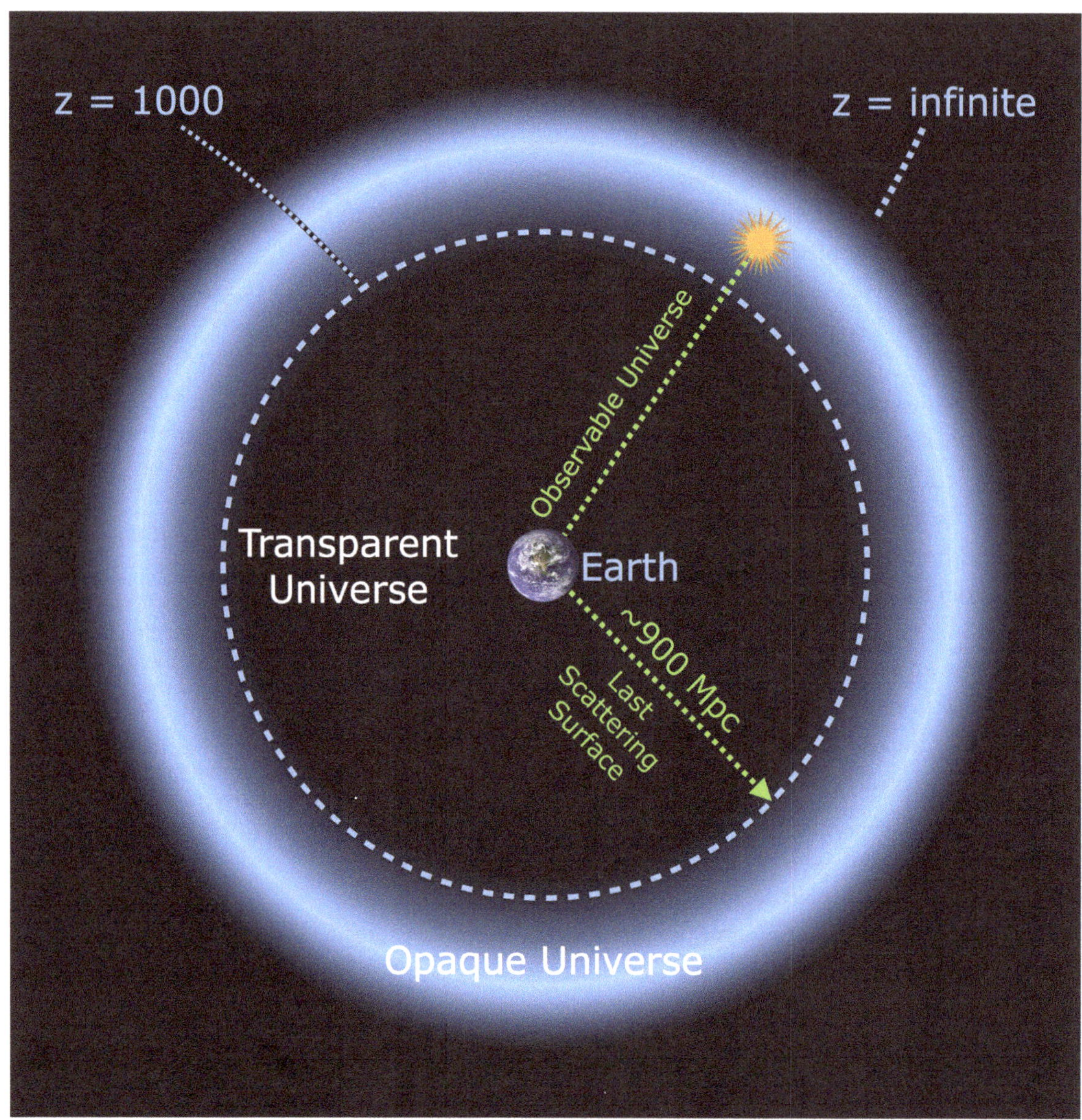

Figure 9.4. Shows the size of the "observable universe," the distance light travels during the age of the universe, the horizon distance. No signals can be received from the region beyond the horizon distance. Here, "z" denotes redshift. The last scattering surface is an imaginary surface on which cosmic back ground photons were last scattered by particles. It is the boundary between an opaque and transparent universe about the time the reionization happened (Chapter 7).

THE EDGE OF THE OBSERVABLE UNIVERSE

The longest distance one can see in the universe is called the *horizon distance* (Box 5.3). To recall, this is the distance light travels during the age of the universe, $d_h = c \cdot t_u$, where c is the speed of light and t_u is the age of the universe. The horizon distance extends to the edge of the observable universe (figure 9.4). Parts of the universe that are separated by a distance larger than the horizon distance could not communicate with each other, as one needs to either move with a speed faster than the speed of light or wait a time longer than the age of the universe—both impossible. These regions are called *causally disconnected* (Schneider and Arny 2015).

Where is the center of the universe? The short answer is that any point in the universe could be considered as its center. The big bang happened about 13.8 billion years ago. Therefore, if you look 13.8 billion light years away in any direction, you see the point where the universe started (or shortly after that when the dark ages ended). As a result, that point can be considered to be the point where the universe started (the center of the universe). Now, imagine an observer about 13.8 billion light years away, looking at our own Galaxy. The observer sees our Galaxy (or the space around our Galaxy) as it was at the beginning of the universe. To that hypothetical observer, we are where the big bang took place, at the center of the universe. Therefore, any point could be considered as the center of the universe.

OLBERS' PARADOX

In an infinite and unchanging universe, there would be an infinite number of stars. If the distribution of the stars were

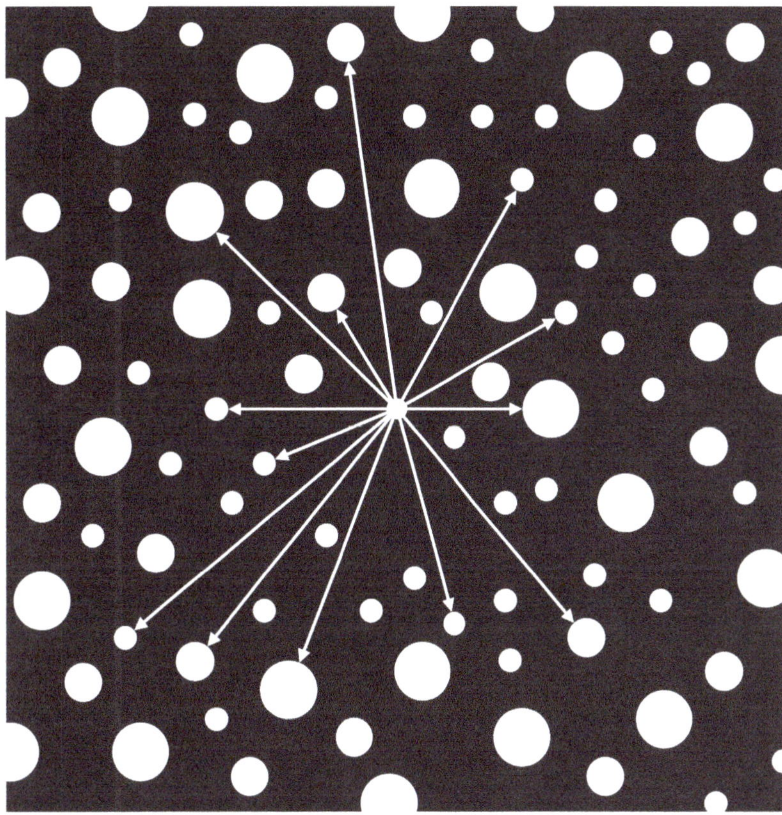

Figure 9.5. An infinite number of stars in the universe make our line of sight to encounter a star in whichever direction we look at. This leads to the conclusion that night sky is bright, contrary to observations. This is called the Olber's paradox.

uniform, we would then expect to see a star in every direction we look out. This means that every point in the sky would be bright, making the night sky illuminated (figure 9.5). Even if there is dust in the universe blocking the starlight, the conclusion still stands, as the intense starlight would heat the dust, making it glow or evaporate. This chain of argument leads to a paradox, as it contradicts with the observations that the night sky is dark. This was first noted by Johannes Kepler but is referred to as *Olbers' paradox* after German astronomer Heinrich Olbers (1758–1840).

There are two explanations to resolve this paradox. First, the universe has a finite age. Therefore, light from distant stars has not yet had time (within the age of the universe) to reach us. Second, as the universe expands, objects at larger distances move faster, and the light from them shifts more and more to the redder part of the spectrum and out of the visible part of the electromagnetic spectrum. Therefore, these objects cannot be seen at visible wavelengths (Schneider and Arny 2015).

SUMMARY AND OUTSTANDING QUESTIONS

The science of observational cosmology is a relatively young discipline, started with the discovery of the expansion of the universe by Edwin Hubble in 1929. For many decades there were serious controversies regarding the exact value of the rate of the expansion of the universe. In fact, one of the aims of constructing the Hubble Space Telescope was to measure this number by improving distance measurement to remote galaxies. Today astronomers have converged to a value of $H_0 = 73 +/- 1.75$ *km/sec/Mpc*, which means that the velocity between any pair of galaxies increases by *73 km/sec* for every Mpc increase in their relative distance. The inverse of the Hubble constant gives a lower limit to the age of the universe. Another independent constraint on the age of the universe is the age of the oldest stars. The latest estimate for the age of the universe is 13.8 billion years.

The expansion velocity of galaxies in the nearby universe is affected by the density enhancements (increase in the local density due to clustering of galaxies) attracting individual galaxies. This introduces noise to the estimated velocity of galaxies over and above the component of the velocity that takes part in the global expansion of the universe. This "noise" corresponds to the noncosmological component of velocities. Since the velocity of galaxies increases proportional to their distances from us (the Hubble law), such a noise in the velocity field is negligible for distant galaxies that move faster (see figure 9.3). Therefore, the degree of this excess velocity provides a measure of the local velocity field for galaxies from which astronomers estimate the matter density (including both luminous and dark matter) of the universe (given that the local density affects dynamics of galaxies).

What is clear from the above discussion is that cosmology is now a precise science. The outstanding question here is how present observations of the universe could help constrain fundamental laws of physics. Many of the parameters that directly relate to the physical properties of our universe can now be accurately measured. There are, however, uncertainties in such measurements. For example, the luminosity of a galaxy is reduced by the inverse square of its distance from us—the more distant the galaxy is, the less luminous it seems. Similarly, the presence of dust in galaxies also extinguishes their luminosity, making galaxies look as they are farther away. This is a serious source of uncertainty and astronomers sometimes confuse a nearby dusty with a truly distant galaxy. Indeed, it was the presence of dust that initially led Edwin Hubble to derive a value for the rate of the expansion of the universe eight times larger than its present value (as he failed to account for extinction due to dust).

What we know about our universe today is through the study of billions of galaxies we see, which are the basic constituents of the observable universe. Each of these galaxies, on average, consists of billions of stars, with the integrated light from these stars generating the observed luminosity of galaxies. Therefore, the study of the galaxies and their evolution, and using them as test particles for exploring physical properties of the universe, depends on the type and the luminosity of their constituent stars and to a large extent on the formation and evolution of the stars. For example, it is the ejected material from some of these stars or eventual demise of massive stars into supernovas (chapter 13) that is responsible for the dust in galaxies. Astronomers today have been very successful in confronting theoretical models for the formation and evolution of stars, galaxies, and the universe with the observations made by our telescopes. Future surveys of galaxies at different wavelengths, reaching the most distant parts of the universe will reveal new secrets about our universe and its evolution. This is one of the greatest achievements in science over the past decades.

REVIEW QUESTIONS

1. Describe the cosmological principle.
2. How does the matter content of the universe affect the geometry (shape) of space?
3. Explain the density parameter of the universe and how it is related to the fate of the universe.
4. What is Hubble's law? And what are the implications of deviation of this law from linearity?
5. What is the physical significance of the Hubble constant?
6. What is a standard candle?
7. How is the age of the universe measured?
8. Describe the horizon distance.
9. What do we mean by two regions in the universe being causally disconnected?

10. Explain Olbers' paradox and how it is resolved.

CHAPTER 9 REFERENCES

Kirshner, R.P. 2003. "Hubble's Diagram and Cosmic Expansion." *Proceedings of the National Academy of Sciences* 101 (1): 8–13. Bibcode:2003PNAS..101....8K. doi:10.1073/pnas.2536799100.

Livio, M., and A. Riess. 2013. "Measuring the Hubble Constant." *Physics Today* 66 (10): 41. Bibcode:2013PhT....66j..41L. doi:10.1063/PT.3.2148.

Schneider, S.E., and T.T. Arny. 2015. *Pathways to Astronomy*. 4th ed. New York: McGraw-Hill.

FIGURE CREDITS

- Fig. 9.2: Source: https://en.wikipedia.org/wiki/File:Universe.svg.
- Fig. 9.3: Copyright © Brews ohare (CC BY-SA 3.0) at https://en.wikipedia.org/wiki/File:Hubble_constant.JPG.

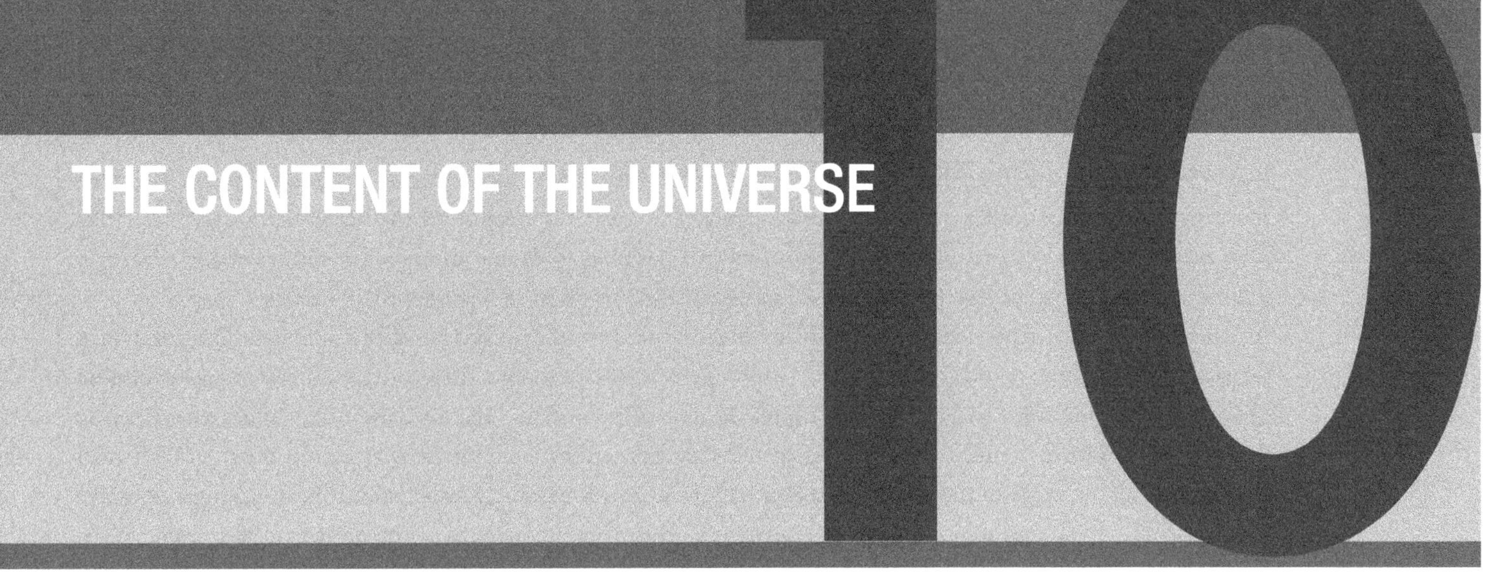

CHAPTER LEARNING OBJECTIVES

This chapter will cover:
- The evidence for dark matter in the universe
- Nature of dark matter
- The dark energy in the universe and its nature
- The future fate of our universe

The past history and future of the expansion of our universe depend on its content. If our current understanding is correct, the main cause of the gravitational field that governs the universe is *dark matter*. Dark matter does not emit light but directly affects the mass budget of the universe and hence plays a major role in formation of the structure (including stars and galaxies), speed of the expansion and the future destiny of the universe. Dark matter was first proposed in the 1930s by Fritz Zwicky (1898–1974) to explain the discrepancy he found between the mass of clusters of galaxies measured from the sum of the masses of individual (luminous) galaxies and from their dynamics (affected by both luminous and nonluminous matter). Zwicky found that dynamical mass always exceeded over the luminous (stellar) mass and postulated a kind of matter that only contributed through its gravity and not light—this was referred to as *dark matter*. In the 1960s this study was followed by Vera Rubin (1928–2016), who noticed that stars in the outskirt of the Andromeda galaxy move significantly faster than expected from that predicted by Newton's theory of gravity, indicating a stronger gravitational force than could be explained by the mass of the stars alone.

For nearly four decades, dark matter was assumed to constitute the main content of the universe. In 1998, two groups of astronomers led by Adam G. Riess and Saul Perlmutter independently showed that distances to galaxies measured from their expansion velocity (Hubble's law) were consistently smaller than their "real" distance measured using a distance indicator (that did not depend on a model for the dynamics of the universe). To explain this discrepancy and to compromise the

"We know there is gravity because apples fall from trees. We can observe gravity in daily life. If we could throw an apple to the edge of the universe, we would observe it accelerating"

- ADAM RIESS

"Our knowledge can only be finite, while our ignorance must necessarily be infinite"

- SIR KARL POPPER

two independently measured distances, they postulated the concept of an accelerating universe in which galaxies move away from one another with an accelerating rate. The entity responsible for this acceleration is referred to as *dark energy*. It produces a repulsive energy pushing galaxies away from one another. The nature of dark energy is not clear, but its existence has been confirmed independently by many independent techniques.

These two components combined constitute over 96 percent of the content of the universe. The rest, which forms the visible universe and all we observe, is only 4 percent of its content. Therefore, to understand the evolution of our universe, one needs to have detailed knowledge about its content—mainly dark matter and dark energy. So far, we have mainly concentrated on the 4 percent of the content of the universe that we can observe. How does this relate to the question of the origins? The content of the universe is solely responsible for everything we observe today, including galaxies, stars, planets, and hence life. Therefore, knowledge of the nature of 96 percent of the universe reveals how the universe arrived at its present state and how it is going to evolve in the future.

This chapter presents observational evidence for dark matter and dark energy and studies the nature of each of these components. It then investigates the future evolution and eventual destiny of our universe.

THE CONTENT OF THE UNIVERSE

Independent measurements have now confirmed relative contribution from different constituents of the universe. Although the nature of these components is not yet fully understood, astronomers, are confident about their relative fraction. The universe today consists of 72 percent dark energy, 23 percent dark matter, and 4.6 percent ordinary matter in the form of atoms (figure 10.1, left panel). As I mentioned in Chapter 9, in the distant past the universe was dominated by dark matter (which led to the formation of structure), photons (which constitute the microwave background radiation from which we can extract information about the universe at the time of the decoupling), atoms (which formed the stars and galaxies we see today), and neutrinos (which are still floating in

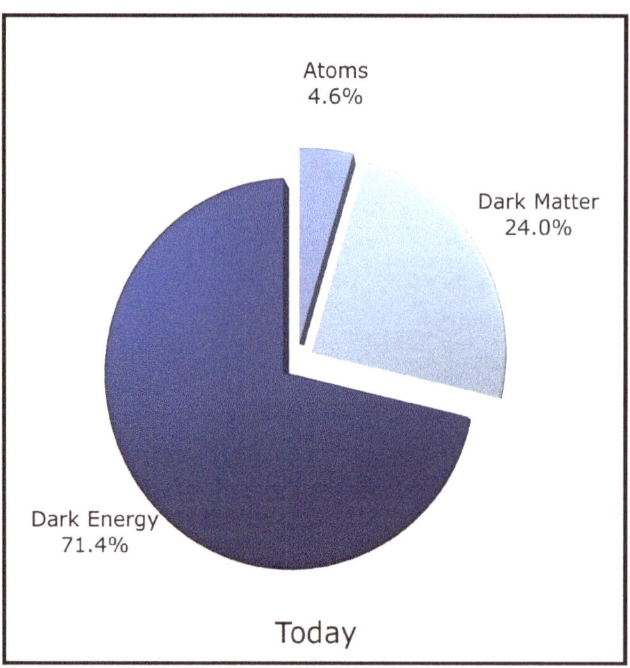

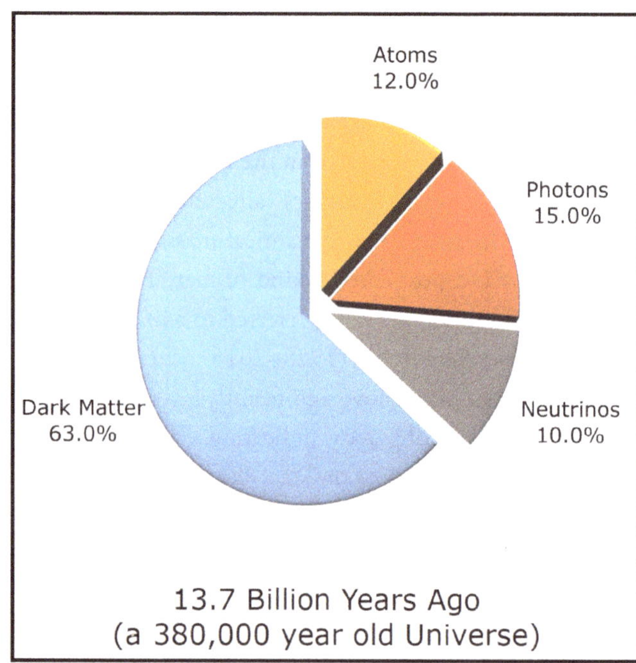

Figure 10.1. The relative contribution from different components to the content of the universe today (left panel) and at the time matter and radiation decoupled (380,000 years after the big bang) (right panel).

the universe)(Figure 10.1, right panel). However, dark energy has only recently (over the last 4 billion years) come into play to dominate the content of the universe (figure 10.1; Bennett et al. 2007).

DARK MATTER

Dark matter has been discovered on different scales, from our own Galaxy, the Milky Way, to external systems in groups and clusters of galaxies. It has also been found that the fraction of dark matter scales with the size of the structure within which it exists. Different independent observations have confirmed the existence of dark matter.

EVIDENCE FOR DARK MATTER

The most direct evidence for dark matter comes from the motion of stars and gas in the outer region of galaxies. The presence of dark matter is revealed through the *rotation curves* in gas-rich galaxies. This is the change in the rotational velocity of gas and stars in galaxies as a function of their distance from the center of the galaxy (figure 10.2). Studying motion of the material in the outer region of galaxies and using the laws of Newtonian dynamics, astronomers measure the total mass (due to luminous and dark matter) within any given radius from the center of the galaxy. Since this is based on dynamics of the gas, it provides a direct measure of the gravity and hence, gives an estimate of the amount of both dark and luminous matter. For most galaxies that contain gas, outer parts of the rotation curves are remarkably flat, significantly deviating from theoretical predictions based on the laws of Newtonian gravity that predict a rapid decrease in the rotational velocity with increasing radius (Figure 10.2). This indicates that there is extra (unseen) matter located in the outer parts of galaxies. For

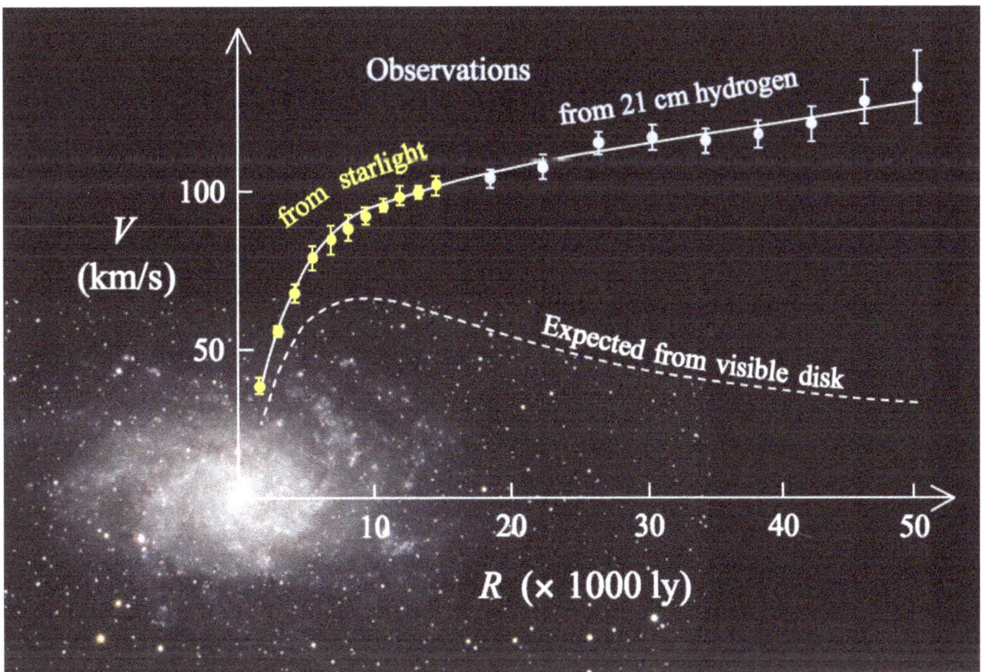

Figure 10.2. Changes in dynamics (velocity) of gas surrounding spiral galaxies as a function of distance from the center (rotation curve). The yellow symbols indicate the luminous part of the galaxy. Observations of the outer part of the galaxy (gray symbols) show an almost flat rotation curve when compared to predictions based on the expected relation from Newtonian dynamics (dashed line). The difference between the observed and predicted rotation curves is indicative of the presence of dark matter. The rotational velocities in the unseen (outer part) of the galaxy are measured by observing radio emission (at 21 cm) from neutral hydrogen produced by flip in the spin of the hydrogen atom.

galaxies that do not contain gas (elliptical galaxies), the presence of dark matter is confirmed through measurement of the motion of stars inside the galaxy (Schneider and Arny 2015).

The main evidence for dark matter in clusters comes from measurement of the motion of galaxies around the center of the cluster. In this technique the recession velocity of galaxies in clusters is measured (this is the component of the velocity that takes part in the expansion of the universe). Subtracting this from the velocity of individual galaxies will then give the component of the velocity of galaxy within the cluster. Astronomers use velocities of individual galaxies in clusters to find dynamical mass of the cluster (including both luminous and dark mass) using Newton's law. Comparing this to the sum of the masses of individual galaxies in the cluster, they found that clusters have significantly more mass than exists in luminous galaxies and hence contain a larger fraction of dark matter.

At yet larger scales, the presence of dark matter is confirmed through gravitational lensing technique. This is based on the concept that the space-time is distorted in the vicinity of massive objects (i.e. galaxy clusters). As a result, massive objects bend the light from background sources when passing close to them, as predicted by the general theory of relativity (figure 10.3). This distorts the image of the background sources while concentrating more light from the object on our telescope (the light that would otherwise have missed the Earth and the telescope) and hence making the object look brighter. In fact, the foreground object acts as a lens, focusing the light from the background source and magnifying its luminosity. For this reason, this is called "gravitational lens" technique. The mass of the foreground object could then be estimated by measuring how strongly it diverts the path of the light

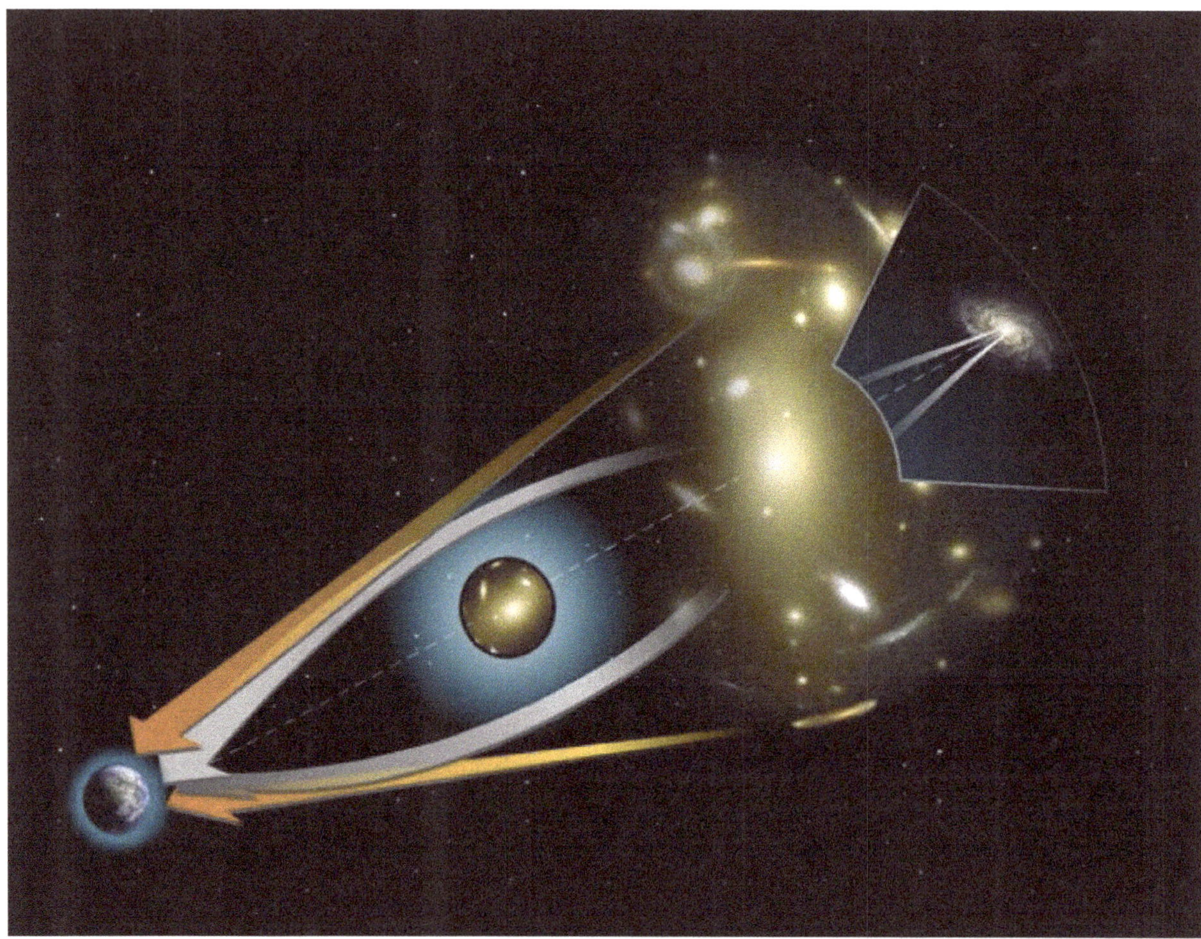

Figure 10.3. Light from background galaxies is bent by the gravitational field of foreground mass (galaxy cluster). The image of the galaxy is distorted and magnified. The degree of distortion depends on the mass of the intervening material.

passing by it from background sources. For very massive objects like galaxy clusters, this produces a distorted image of background sources (figure 10.3). Analysis of these images allows astronomers to measure the mass of the matter lying between the observer and the source behind the cluster independent of the relative ratios of dark to luminous matter. The gravitational lensing method not only allows measurement of the mass of clusters but any large scale structures between us and distant galaxies (Schneider and Arny 2015).

Dark matter is classified into two categories. The first is the ordinary dark matter that constitutes from protons and neutrons and is referred to as *baryonic matter*. The second is the extraordinary dark matter that is referred to as *nonbaryonic*.

Astronomers don't yet know the exact nature of dark matter. It is likely that a small fraction of it consists of ordinary matter that does not emit much radiation. These are mostly faint or dead stars or stars that failed to initiate nuclear fusion and hence have no light source, called brown dwarfs. Furthermore, if Jupiter-size planets exist in large numbers, they could constitute a large portion of ordinary dark matter. A strong candidate for dark matter is the MAssive Compact Halo Objects (MACHOs). These are dim, red stars that escape detection with our telescopes. They exist in large numbers in the halo of our Galaxy but not sufficiently large numbers to account for the entire dark matter (Bennett et al. 2007).

Most of the dark matter in galaxies and clusters is expected to be in the form of exotic particles. These particles do not have electric charge and hence cannot produce electromagnetic radiation. One candidate for this non-baryonic dark matter is neutrinos. However, due to their extreme speed and being weakly interacting (they only interact with matter through gravitational and weak forces), they escape through small structures in the universe. Another candidate for nonbaryonic dark matter is the Weakly Interacting Massive Particles (WIMPs). WIMPs could constitute a substantial mass of galaxies and galaxy clusters without emitting any electromagnetic radiation, since they rarely interact and exchange energy with other particles.

DARK ENERGY

Dark energy is a mysterious entity responsible for the observed acceleration of the expansion of the universe. Contrary to dark matter that is responsible for the attractive force bringing matter together, dark energy produces a repulsive force, driving galaxies away. Given that it dominates the present content of the universe, dark energy is in large part responsible for the future evolution and fate of our universe.

The repulsive force, called dark energy, was first discovered through measurement of distances to remote galaxies aiming to study the geometry of space-time (Box 10.1). Distances to the same galaxies were measured with two independent methods. One method was sensitive to the mass content of the universe and hence dynamics of galaxies, measuring the recession velocity of galaxies and using Hubble's law to find their distances, and the other by direct measurement using a distance indicator (and hence independent of the dynamics). If the two independently estimated distances to the same galaxy were different, this would mean that dynamics (expansion velocity) of galaxies does not follow a linear Hubble law, assuming that the "real" distances to galaxies were correct. As it turned out, the "real" distance was always larger than that measured from Hubble's law. To resolve this discrepancy required a revision of Hubble's law in that galaxies needed to move away faster than that predicted by the simple Hubble's law. In other words, galaxies needed to move away with an accelerating rate.

The distance indicators used in the discovery of dark energy were supernovae Type 1a (Chapter 13). These are the final products of the evolution of low mass stars when they explode. These are very good distance indicators for two reasons: (1) They are bright enough to be seen at large distances, and (2) they have almost the same intrinsic luminosity when their light reaches a maximum a few days after they explode (it normally attains that luminosity roughly after forty-five days). Comparing the "fixed" peak intrinsic luminosity (L) and the apparent luminosity (l) of the supernova that is directly measurable, astronomers estimate their distances (d), given that the luminosity decreases proportional to inverse-square of the distance ($l = L/d^2$).

Figure 10.4 demonstrates the effect of the repulsive force of dark energy. It shows the difference between the two independent distances as described above (vertical axis) as a function of the lookback time (from present to the past—horizontal axis). The models shown present a decelerating and accelerating universe as well as a scenario that is a combination of the two. The models, predicted based on Einstein's general theory of relativity, are compared with the observational data using supernovae as distance indicators (standard candles- see Box 10.1) (Figure 10.4; filled circles). The data best agree with a model for the universe that was decelerating in the distant past (due to the gravity from dark matter) and is accelerating in more recent time (due to dark energy; figure 10.4). The best model fit to the data indicates relative fractions of 73 percent, 23 percent, and 4 percent in the form of dark energy, dark matter, and luminous matter respectively (Bennett et al. 2007). The horizontal axis in Figure 10.4 (redshift) is also a measure of the age of the universe from $z = 0$ (present) to $z = 2$ (10 billion years ago). The redshift (cosmic time) where the slope of the best-fit model line (Figure 10.4; blue line) changes, corresponds to the time when the dark energy took over from dark matter (the universe transited from being dark matter dominated to dark energy dominated). This is at the redshift ~0.4 corresponding to a time 4.2 billion years ago (for comparison this is 400 million years after the earth was formed).

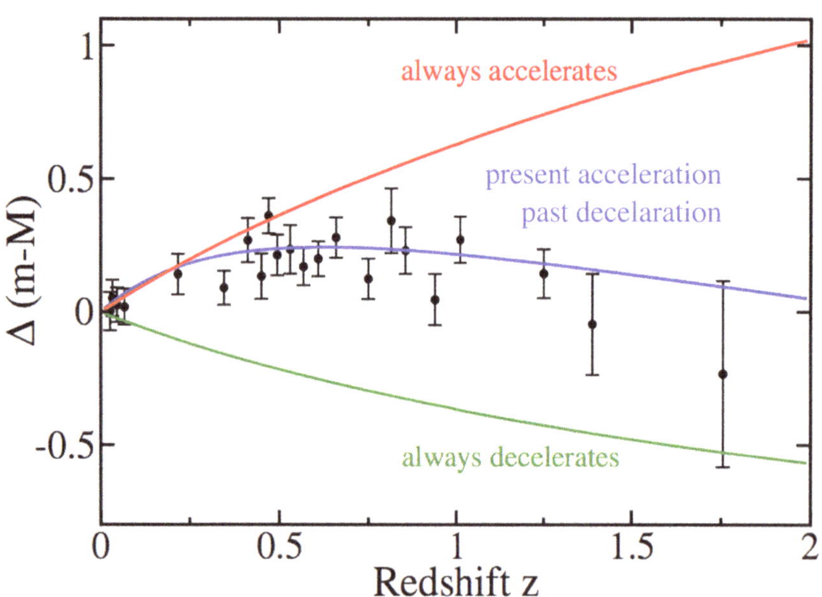

Figure 10.4. The difference between the "real" distances to galaxies and their dynamical distances, from Hubble's law (vertical axis), as a function of cosmic time- redshift-(horizontal axis- from the present on the left to the past on the right). The lines are different model predictions, and points are the data. Observations agree with a combination of decelerating (in the distant past) and accelerating (in the more recent past) universe.

NATURE OF DARK ENERGY

Although dark energy dominates the content of our universe, its nature is shrouded in mystery. It is caused by negative pressure, pushing the universe apart. In its simplest form, dark energy could be explained in the form of *vacuum energy*

BOX 10.2: THE COSMOLOGICAL CONSTANT

Einstein introduced the cosmological constant to his field equation when formulating the general theory of relativity. His theories predicted a dynamic universe, either expanding or contracting. He added the cosmological constant term to revise the solution to allow for a static universe. However, interest on the *cosmological constant* faded after Alexander Friedmann in 1922 derived solutions to Einstein's equations that allowed for an expanding universe and when the expansion of the universe was discovered by Edwin Hubble in 1926.

The cosmological constant also corresponds to the vacuum energy, which is defined as the lowest energy configuration. On the other hand, the uncertainty principle doesn't allow a zero energy state even in vacuum, leading to the creation of virtual particles that affect dynamics of the universe. Discovery of dark energy with its repulsive force revived interest in the cosmological constant, which is a manifestation of such energy. It produces negative pressure that could be responsible for the acceleration of the expansion of the universe.

and a consequence of the *cosmological constant* (Box 10.2). In this case dark energy is uniformly distributed in space, and its strength does not change with cosmic time. Therefore, to test the nature of dark energy, one needs to measure the change in the rate of expansion of the universe with time (over the age of the universe). Here, the time history of the expansion contains detailed information about the nature and strength of dark energy. Furthermore, the repulsive force caused by the presence of dark energy will oppose formation of structures. This is because rapid expansion of the universe would smooth out any structures. Therefore, the time scale of structure formation in the universe directly relates to the strength of dark energy. A lot of work still needs to be done to probe the nature of dark energy, and many space and ground-based missions are underway to investigate this.

THE FATE OF THE UNIVERSE

The fate of the universe is governed by the competition between two opposing forces—the attractive force of dark matter and the repulsive force of dark energy. Given different combinations of these forces, there are four scenarios for the fate of the universe (figure 10.5), as described below:

Closed and collapsing universe: A universe dominated by dark matter, with the density higher than its critical value, will stop expanding sometime in the future due to the collective force of gravity that would slow down and eventually stop the expansion. The universe will then collapse on itself under the force of gravity, ending in a big crunch, a similar state the big bang it started from (figure 10.5; bottom-left).

Open and uniformly expanding universe: This is when the matter density of the universe is exactly the critical value (Box 9.2). In this case forces causing the attraction and expansion will cancel each other out, and the universe will expand forever (basically, it will stop expanding at infinite time). Such a universe has a flat geometry. (figure 10.5; top-right).

Open and forever expanding universe: When there is no (or negligible) dark energy in the universe and the matter density is below the critical value, there is not enough matter in the universe to slow down the expansion. In this scenario, the universe expands forever with little change in the rate of the expansion. (figure 10.5; top-left).

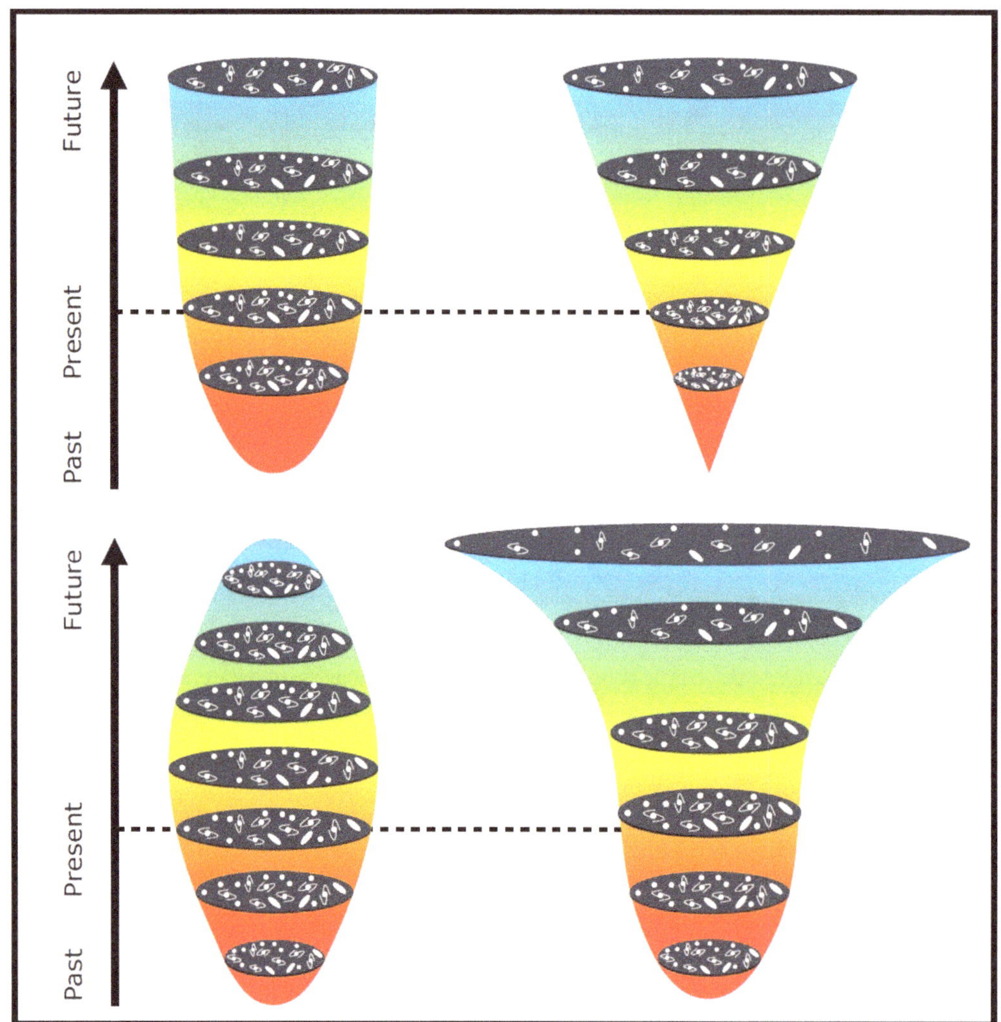

Figure 10.5. Shows different scenarios for the future fate of the universe. The observational data seem to support an accelerating universe, with the rate of expansion increasing (accelerating) with cosmic time.

Accelerating universe: If dark energy dominates the content of the universe, the repulsive force generated by it will accelerate the rate of expansion of the universe. The expansion rate will increase with time. Due to this force, all the structures in the universe will dissociate in distant future. (Figure 10.5; bottom-right).

Figure 10.5 shows that by studying the volume of space covered by galaxies and studying the change in the volume with time, one could constrain the geometry and future fate of the universe. Current observations are in agreement with an accelerating scenario for the universe. This implies an open universe, forever expanding. Recent data from cosmic background radiation missions (WMAP and Planck), also show that the universe is very nearly flat, with its matter density very close to the critical value, and is dominated by dark energy (Schneider and Arny 2015).

SUMMARY AND OUTSTANDING QUESTIONS

Since its discovery over half a century ago, astronomers have been studying the nature and properties of dark matter. It is easy to see why this is so important and why so many resources have been allocated to this study. Basically, dark matter is responsible for the formation of structures (at all scales) as well as for governing the expansion of the universe and its fate. Furthermore, it provides an exciting interface between cosmology and particle physics and how the physics of the very small could control the largest known structures. Some astronomers have even discussed the possibility of a modified Newtonian gravity to explain the excess gravity attributed to dark matter, although no concrete evidence is found to support this. The likely possibility is a mixture of ordinary (baryonic) dark matter as that constituting stars and planets and cold dark matter in the form of slow moving (cold) elementary particles that do not interact with matter or radiation. The cold dark matter scenario has been successful in explaining the observed structures. The structures grow from small to large systems—from low-mass dwarf galaxies to high-mass giant galaxies and clusters.

The discovery of dark energy is one of the greatest discoveries in contemporary physics. Dark energy is a more mysterious entity than dark matter with an opposite effect—pushing galaxies away from each other. It is analogous to negative pressure and theoretically has the same effect as the cosmological constant, a term in Einstein's equation of general relativity to counteract the expansion of the universe. Its true nature is not known yet. Observations confirm that the universe was dominated by dark matter over 10 billion years ago and went through a transition to become dark energy dominated around 4.2 billion years ago (figure 10.4) (for comparison, the age of Earth is 4.6 billion years). Since then, dark energy has governed the dynamics of the universe, with its strength increasing with cosmic time.

Today, 72 percent of the content of the universe is in the form of dark energy, 24 percent in dark matter, and only 4 percent in the form of ordinary matter, making everything around us. The relative fraction of dark matter and dark energy dictates the future evolution of our universe. Given the significant excess of dark energy, it is now believed that the universe will continue to expand with an accelerated rate. This implies that the current expansion of the universe will continue forever.

The nature of dark matter and dark energy pose some of the most outstanding challenges in physics, astronomy, and cosmology. What is the nature of dark matter and dark energy? And how do they affect the eventual fate of our universe? How is the strength of dark energy changing with cosmic time? Is the distribution of dark energy isotropic (independent of direction) in the universe? What is the role of the cosmological constant? These are among the most fundamental questions in cosmology today.

REVIEW QUESTIONS

1. What are the relative fractions of dark matter, dark energy, and ordinary matter in the universe?
2. Explain the evidence for dark matter.
3. What are the different candidates for dark matter?
4. Describe the methods for detecting dark matter.
5. What is the observational evidence for dark energy?
6. Explain the main characteristics of dark energy and how it may affect future fate of our universe.
7. What is a likely candidate for dark energy?
8. Briefly explain how distances to galaxies are measured.
9. How was the relative fractions of dark matter and dark energy in the universe measured?

10. At what epoch in the history of our universe did dark energy start to play a dominant role? How is this estimated?

CHAPTER 10 REFERENCES

Bennett, J., M. Donahue, N. Schneider, and M. Voit. 2007. *The Cosmic Perspective*. 4th ed. Boston: Pearson/Addison-Wesley.

Schneider, S.E., and T.T. Arny. 2015. *Pathways to Astronomy*. 4th ed. New York: McGraw-Hill.

FIGURE CREDITS

- Fig. 10.2: Source: https://en.wikipedia.org/wiki/File:M33_rotation_curve_HI.gif.
- Fig. 10.3: Source: http://hubblesite.org/newscenter/archive/releases/2000/07/image/c/.
- Fig. 10.4: Source: Turner and Huterer, 2007.

THE ORIGIN OF GALAXIES

CHAPTER LEARNING OBJECTIVES

This chapter will cover:
- Formation of galaxies
- Stellar population in galaxies
- Origin of different types of galaxies
- Evolution of galaxies with cosmic time

Galaxies are the basic constituents of the universe. They come in different forms and shapes, receding away from one another due to the stretching of space between them. A galaxy is defined as a collection of hundreds of millions of stars held together by the force of gravity. They are complex systems containing different types (populations) of stars with different masses, ages, and chemical enrichment histories, as well as gas and dust. The characteristics of the stars and fraction of gas and dust in galaxies determine their type. In the observable universe, there are around 10^{11} galaxies, each on average containing 10^{11} stars. Galaxies are used as mass particles to test existing models of the universe.

In the early 1900s the "spiral nebulae" were discovered as bright and cloud-like systems on the night sky. This initiated a long debate among astronomers concerning their nature—whether they were located within our Milky Way galaxy or were external objects outside our galaxy. This was the subject of the "great debate" between two astronomers, Harlow Shapley (1885–1972) and Heber D. Curtis (1872–1942). Shapley argued that the spiral nebulae were gas clouds in our galaxy that he proposed to be the "universe." Measuring distances to stars in our galaxy, he suggested that the sun was far away from the center of the galaxy (in his view, the center of the universe). Curtis, on the other hand, argued that the universe consisted of many galaxies, with ours being only one of them. He located the sun at the center of our galaxy. This debate was finally settled by Edwin Hubble in 1919. Using state-of-the-art telescopes and instruments, Hubble produced resolved photographs of one of these systems, the Andromeda nebula, and identified faint

"According to the standard model billions of years ago some little quantum fluctuation, perhaps a slightly lower density of matter, maybe right where we're sitting right now, caused our galaxy to start collapsing around here"

- SETH LLOYD

"Nature is the source of all true knowledge. She has her own logic, her own laws, she has no effect without cause nor invention without necessity."

- LEONARDO DA VINCI

stars in it. It then became clear around 1923 that the nebulae seen in night sky were in fact individual islands in the universe. This discovery changed humans' view about the universe. No longer was our Galaxy considered as the entire universe, but one of the billions of such systems. Today the story of the formation and evolution of galaxies in the universe and their interaction with the surrounding environment is among the most exciting areas of research. Astronomers use galaxies to measure the size and estimate the age of the universe as well as its evolution, while the medium between the stars (interstellar medium) in galaxies determines the elemental enrichment that is directly related to the origin and development of life.

This chapter studies the origin of galaxies and the different types they are divided into. It also discusses star formation activity in galaxies. The concept of the universe as a time machine will be discussed and used to study the evolution of galaxies. The deepest image of the universe ever seen at visible and infrared wavelengths will be presented and galaxies at the edge of the observable universe identified.

FORMATION OF GALAXIES

The observed maps of the cosmic background radiation reveal temperature fluctuations representing mass clumps on an otherwise homogeneous matter distribution at the time matter and radiation decoupled, around 380,000 years after the big bang (figure 8.1). The excess density in these regions gradually grew because of the gravity due to dark matter, attracting more mass. This inward collapse of matter into structures was competing with the outward force due to expansion of the universe. At some point, the attraction force of gravity overcame the expansion of space and the clumps started to collapse (figure 11.1). This led to the formation of protogalactic systems around one billion years after the big bang. These systems were formed from hydrogen and helium and were cooled through physical processes as they collapsed. Because of the very high density at the center of these systems, fusion took place (the process that combines the nuclei of light elements to produce heavier elements, releasing a lot of energy), generating light and forming the first generation of stars. At this stage, the stability of the gas cloud depended on the balance between the outward force of radiation produced by stars and the inward force of gravity due to dark matter (figure 11.1). These resemble massive population III stars that are identified by their low abundance of heavy elements (poor metal) but bright luminosities at high energy ultraviolet wavelengths. These stars started to produce heavy elements

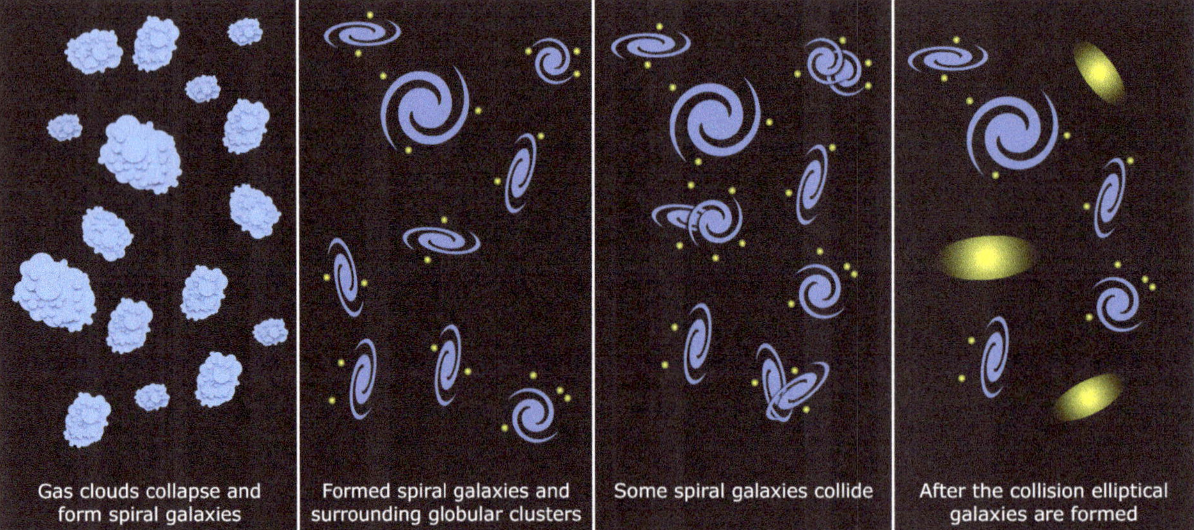

Figure 11.1. Formation of protogalaxies by collapse of initial gas clouds, initiating star formation at their center. Galaxies then collide and merge and lose gas to the intergalactic medium, forming more massive systems. The final shape of a galaxy depends on two initial parameters: the angular momentum and density of the primordial gas.

through fusion of lighter elements. Given their high mass the first generation of stars were short lived (of the order of a few million years) and soon exploded as supernovae, scattering the heavy elements in the gas clouds and hence enriching them (Chapter 13). This explosion produced an outward force, slowing down the speed of the collapse and allowing time for the formation of structure in these systems. This also created shock waves, condensing gas and initiating new star formation activities, this time from the "metal-enriched" material (Bennett et al. 2007). The end product depends on the mass of the initial gas cloud and its speed of rotation (angular momentum), as will be discussed in the following sections.

THE ORIGIN OF DIFFERENT TYPES OF GALAXIES

Galaxies are divided into two broad classes: spiral and elliptical. Spiral galaxies are relatively younger, have disks, contain gas, and are sites of star-formation activity. Elliptical galaxies are older, have no recognizable structures (no disks), have very little gas, and no ongoing star formation. These types have distinctly different formation histories and are represented by the Hubble fork (figure 11.2), where they are classified based on their observed morphologies (Chaisson and McMillan 2011).

At the protogalactic stage, gas clouds forming these galaxies had similar shapes and compositions (figure 11.1). However, they evolved into entirely different types depending on the speed with which the initial gas cloud rotated (angular momentum) and the density of the initial gas cloud. If the original cloud had a high angular momentum (that is, spin), it would rotate faster when it collapsed, due to the law of conservation of angular momentum, leading to formation of disks, with the resulting galaxy being a spiral. If the initial cloud did not have an angular momentum, the resulting galaxy would not develop a disk, and hence the product would be an elliptical galaxy (figure 11.2).

A protogalactic cloud with high density will collapse faster under its gravity, will radiate energy more efficiently, and hence will cool more rapidly. As the system loses thermal energy, it will continue to collapse (while not driven out by the internal radiation force), with its density increasing (because the volume decreases while

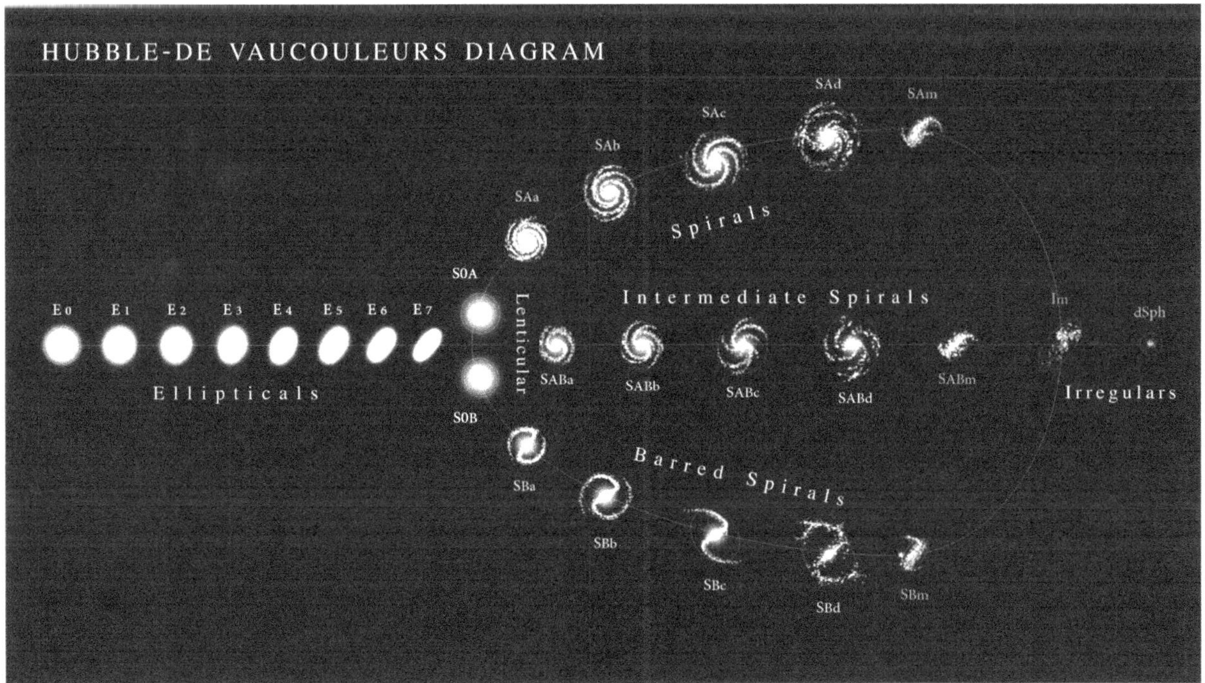

Figure 11.2. The Hubble fork, showing different types of galaxies. These have very different formation histories.

Figure 11.3. Image of the Andromeda galaxy (M31) with the bulge (dominated by young and old stars), the disk (site of star-formation activity), and halo (consisting of old stars) indicated.

the mass remains the same). This results in a more efficient star-formation activity as the gas in the system condenses. These clouds will not have time or have much gas left to form disks and therefore will end up as elliptical galaxies (figure 11.2). Similarly, a less dense system will have a slower rate of star formation and slower collapse rate, with a larger fraction of gas remaining and longer collapse time to allow formation of disks. This leads to formation of spiral galaxies. Through cosmic time, galaxies collide and merge, forming larger systems with little star-formation activity (figure 11.1). As galaxies merge, they often lose gas and destroy their disks, changing their morphology (Bennett et al. 2007).

Galaxies consist of three components: a bulge at the center, a disk in their outskirts, and a halo surrounding the galaxy (figure 11.3). It is the relative size of these components that fixes the shape and property of galaxies. For example, spiral galaxies have prominent bulge and disk while elliptical galaxies lack the disk component. There are also galaxies that show no clear morphologies, undergoing rapid star formation activity. Spiral galaxies continue to form stars until they run out of their gas. At that point they go through a phase of passive evolution, in which their stars evolve and age (Bennett et al. 2007).

THE ORIGIN OF STELLAR POPULATIONS IN GALAXIES

While disks of galaxies are rich in gas and are sites of ongoing star-formation activity (figure 11.3), their halos are mostly dominated by old stars, with bulges containing both young and old stars. The absence of dust and gas in halos indicate that no new stars are forming there, with the last stars formed around 10 billion years ago. As successive generation of stars form and die, they enrich the space between the stars by heavy chemical elements as a result of stellar nucleosynthesis (chapter 14). Observations have shown that halo stars contain little heavy elements compared to those in the disks or bulges of galaxies, indicating that these stars formed long ago when the gas in galaxies had not yet been enriched with metals. This indicates a time sequence for the stellar population in galaxies. The halo stars are old and poor in heavy elements and are the first stars formed in galaxies, while disk and bulge stars are younger and richer in heavy elements (since they were formed later from metal-enriched gas in galaxies). Astronomers refer to the young disk stars as *population I* and to old halo stars as *population II* (Schneider and Arny 2015).

THE UNIVERSE AS A TIME MACHINE

The light from distant galaxies takes many billions of years to reach us. Since the speed of light is finite, we therefore see galaxies as they were billions of years ago and not as they are today. A consequence of this is that when we look at the sky, essentially we look back in time. Therefore, if we find a galaxy that is about 13 billion light-years away (a light-year is the distance light travels in one year), we see it as it was 13 billion years ago. Given that the age of the

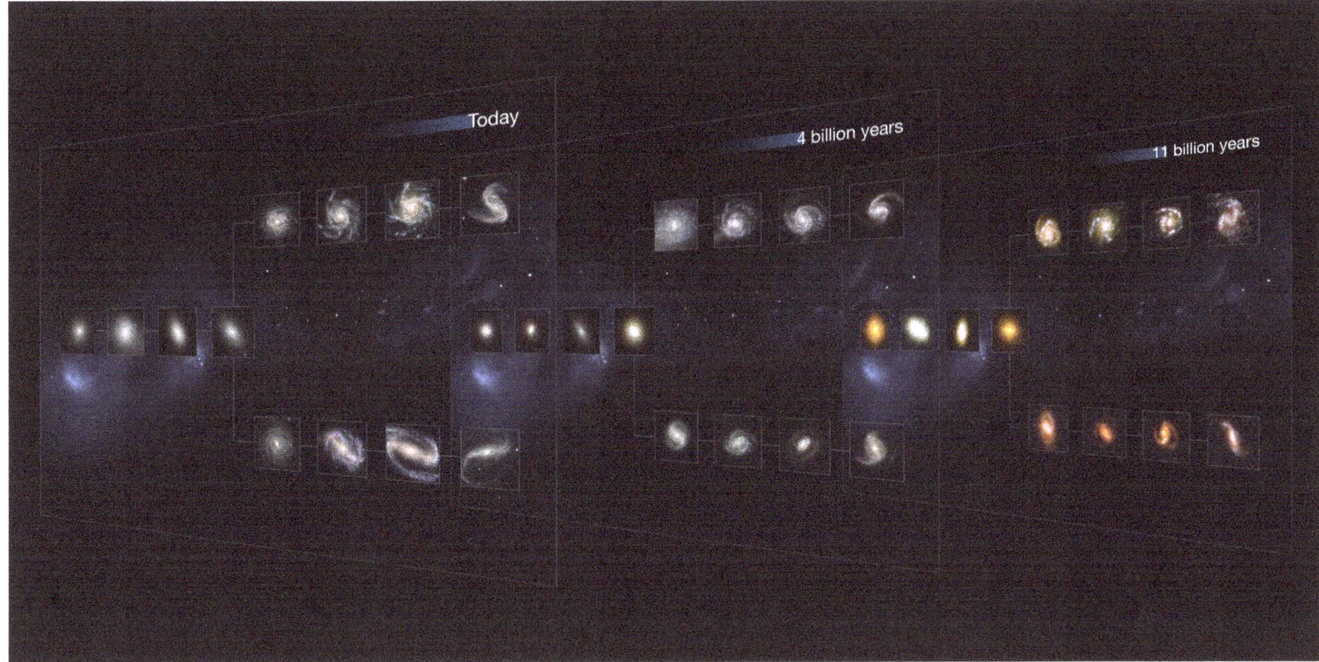

Figure 11.4. From left to right, galaxies as they were throughout cosmic history from today (left) to billions of years ago (right) and in between. By looking at distant regions in the universe, one looks back in time. Therefore, by studying properties of galaxies at different distances (time epochs), one can study the evolution of galaxies with cosmic time.

universe is 13.8 billion years, by looking at those galaxies, we look back at a time when the universe was 800 million years old. Therefore, we look back in time to when the first generation of galaxies formed (figure 11.4).

This discussion provides the incentive for astronomers to search for the most distant galaxies in the universe. By doing so, one could study galaxies as they form soon after the birth of the universe (~1 billion years) and hence understand the formation process and nature of the first generation of galaxies. Similarly, by identifying galaxies located at different distances from us, astronomers take snapshots at different times during the life of the universe. By comparing observations of galaxies at different distances (i.e., cosmic times), they study the evolution of galaxies through the cosmic time (figure 11.4; Chaisson and McMillan 2011).

Given the above concept, to observe the first generation of galaxies, one needs to peer as deeply as possible into space to reach the reionization time, when galaxies were assembled. Throughout the years, this quest has led to investment of a significant amount of telescope time to obtain deep images of the universe. An example of this is the Hubble Ultra Deep Field (HUDF), the deepest image ever seen of the universe in visible wavelengths (figure 11.5), taken by the Hubble Space Telescope (Box 11.1). Astronomers have extensively used the HUDF

BOX 11.1: THE HUBBLE ULTRA DEEP FIELD (HUDF) THE DEEPEST IMAGE OF THE UNIVERSE

The HUDF is the deepest view of the universe ever seen by humankind. It is the result of four hundred hours of observation with the Hubble Space Telescope over an area of 3×3 *arcmin²*—roughly corresponding the size of a coin. Figure 11.5 shows the HUDF with images of some of the most distant galaxies at the edge of the observable universe. By looking deeper into space and further back in time, one observes younger galaxies at the first stages of mass buildup when undergoing first bursts of star-formation activity.

Nasa/esa hubble space telescope

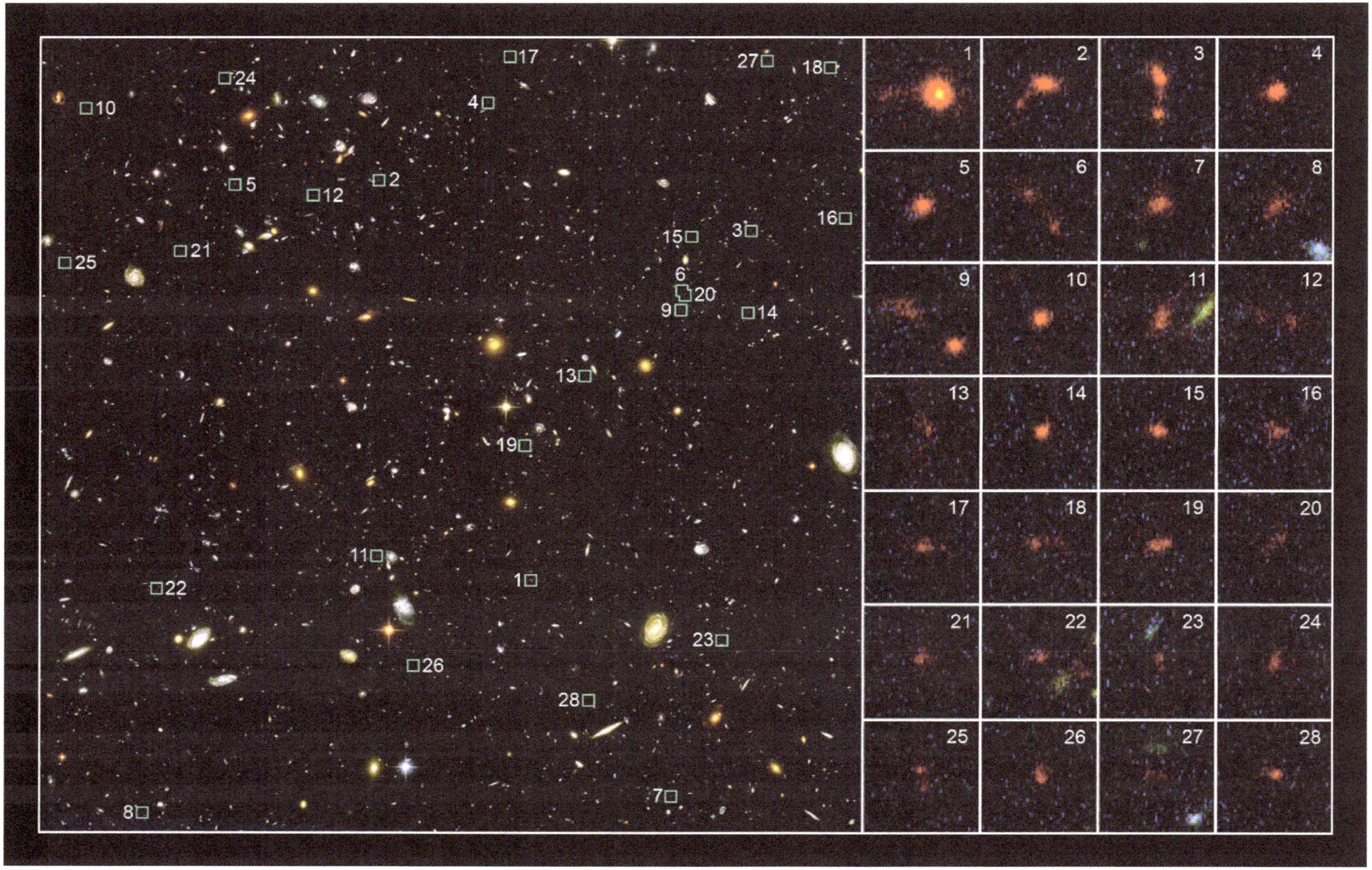

Figure 11.5. An image of the HUDF, the deepest image of the universe ever taken, at optical and infrared wavelengths. The galaxies here are at the edge of the observable universe—some around 12 billion to 13 billion light-years away. Examples of some of the most distant galaxies found in the universe (the first generation of galaxies formed around 12 billion years ago) are shown as postage stamps at the right of the figure.

to identify galaxies at the edge of the observable universe. Some galaxies in figure 11.5 were formed when the universe was 500 million years old (about 13 billion years ago) and the light we receive today on our telescopes left those galaxies well before our own Galaxy and the solar system were formed.

SUMMARY AND OUTSTANDING QUESTIONS

The process of formation of galaxies started as seeds of density fluctuations in the initial smooth distribution of matter. They then grew through the force of gravity, forming low-mass (dwarf) systems. How these developed to form the large massive galaxies we see today and how the shape and morphology of these galaxies were fixed are the subject of intensive research by astronomers. Once the initial gas clouds formed, they cooled down (through radiation by physical processes) and collapsed, forming spiral or elliptical galaxies (figure 11.1), depending on their angular momentum or mass density. A nonzero angular momentum results in the development of the disk and hence a spiral galaxy. This is the result of the law of conservation of angular momentum. A gas cloud with relatively high mass will collapse rapidly, there is not much time for development of the disk (as the system collapses, it is reduced in size. To keep the angular momentum conserved, it needs to rotate faster, resulting in the development of the disk), resulting in an elliptical galaxy. A lower gas density leads to slower collapse and longer collapse time, giving enough time for the disk to form. For high-density gas clouds, the star formation is more efficient, converting more gas to stars in shorter time and hence running out of gas sooner. The gas clouds collide and merge, forming more massive galaxies like our own Milky Way.

By identifying galaxies at different distances from us, astronomers look back in time to when the first generation of galaxies formed. By studying and comparing properties of galaxies at different distances from us, (different cosmic times), they then investigate the evolution of galaxies with cosmic time. The shape and type of galaxies change as a function of look back time. More distant galaxies (younger systems) often show signs of mergers and interaction, indicating that galaxy collision played a paramount role in the evolution of galaxies we see today. The main process contributing to the buildup of mass in galaxies is the star-formation activity. This is also responsible for enrichment of the interstellar medium inside galaxies and generation of heavy elements. The main episode of star-formation activity in the universe (prerequisites for life) took place around 10 billion years after the big bang.

Among the most outstanding questions in the study of galaxies include: What is the origin of the observed relationships between different properties of galaxies (star formation, metal content, and mass), and how do these relationships evolve with cosmic time? How do galaxies interact with their environment through exchange of gas and dust? What is the frequency of galaxy merger? And what are the most fundamental parameters governing formation and evolution of galaxies? It is recently discovered that there is a black hole at the center of every galaxy. It is not clear however, how the black holes resided in their host galaxy, and their role in the evolution of galaxies, star formation and the feedback process (retaining gas in galaxies and hence fueling the star formation process). These questions are the subject of intensive research using the latest telescopes and instruments. With the launch of new telescopes, astronomers will be able to peek deeper into the universe to reach more distant galaxies and obtain information that was not possible before. Many of the new generation of telescopes [eg. The James Webb Space Telescope (JWST) and Extremely Large Telescope (ELTs)] are over nine times more powerful than the most powerful telescopes currently available. These observatories will identify galaxies at the edge of the observable universe, allowing study of the nature of first generation of galaxies.

1. Describe what is known as the "great debate" that led to the acceptance of the view that galaxies are entities external to our own Milky Way Galaxy.
2. What do astronomers mean by protogalaxies? Explain their formation process.
3. Explain why the sites of star-formation activity in galaxies need to have low temperatures.
4. What are the main parameters influencing the formation of galaxies and why?
5. What is the origin of the morphology of galaxies we observe today?
6. Explain the Hubble fork concerning different types of galaxies.
7. What is the difference between the stellar populations in bulge, disk, and halo of galaxies?
8. Explain the process of mass build up in galaxies.
9. How do astronomers study the evolution of galaxies with cosmic time?
10. Why do astronomers need to obtain very deep images to study formation and evolution of galaxies?

CHAPTER 11 REFERENCES

Bennett, J., M. Donahue, N. Schneider, and M. Voit. 2007. *The Cosmic Perspective*. 4th ed. Boston: Pearson/Addison-Wesley.

Chaisson, E., and S. McMillan. 2011. *Astronomy Today*. New York: Pearson.

Schneider, S.E., and T.T. Arny. 2015. *Pathways to Astronomy*. 4th ed. New York: McGraw-Hill

FIGURE CREDITS

- Fig. 11.2: Copyright © Antonio Ciccolella (CC BY-SA 3.0) at https://en.wikipedia.org/wiki/File:Hubble-Vaucouleurs.png.
- Fig. 11.3a: Copyright © Adam Evans (CC by 2.0) at https://commons.wikimedia.org/wiki/File:Andromeda_Galaxy_%28with_h-alpha%29.jpg.
- Fig. 11.4: Source: https://en.wikipedia.org/wiki/File:The_Hubble_Sequence_throughout_the_Universe%27s_history.jpg.
- Fig. 11.5: Source: https://www.spacetelescope.org/images/heic0611a/.

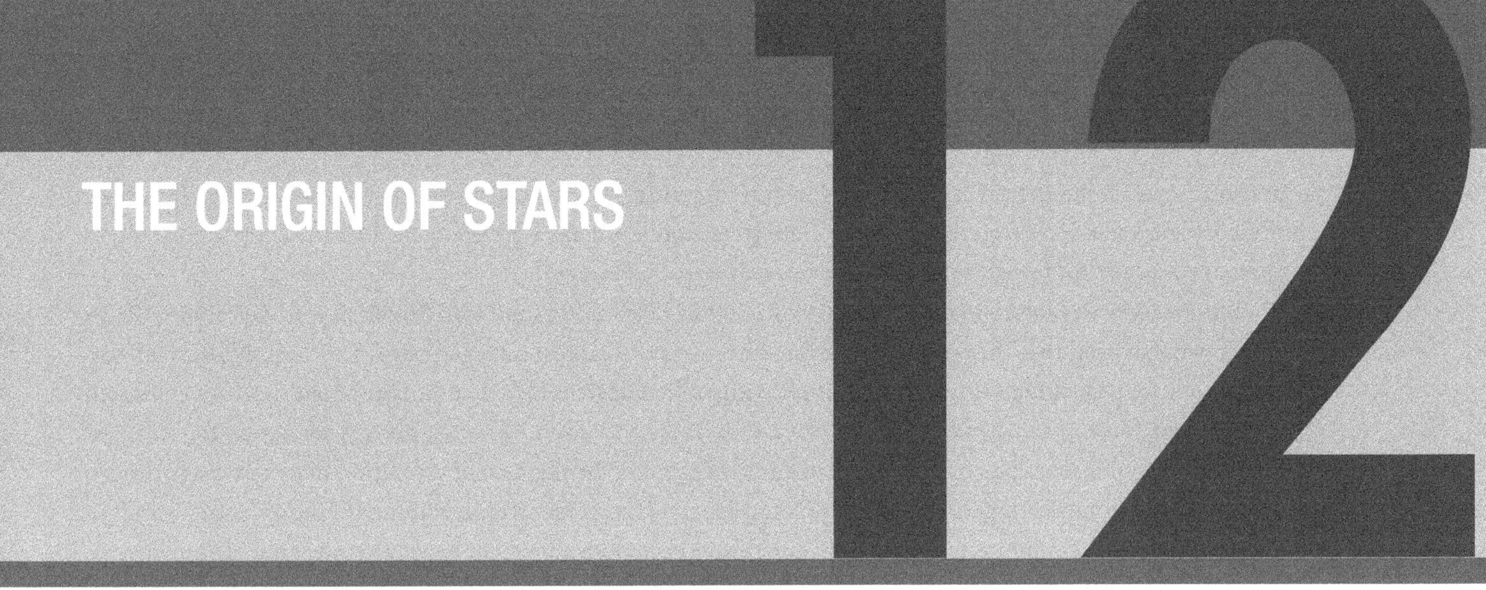

CHAPTER LEARNING OBJECTIVES

This chapter will cover:

- Steps toward formation of stars
- Why stars have the mass they have
- Early stages of the evolution of stars
- The main parameters governing the evolution of stars

Looking at the night sky with our naked eyes, we see an abundance of stars, all in our Galaxy. Although they have been there for millions and perhaps billions of years, they are not eternal. Some of these stars are just being born, some are actively using their fuel to produce the light and energy we see, and some are near the end of their life. In short, stars form and evolve for billions of years before they run out of fuel and die. Stars form in regions with a lot of gas within dark clouds in the spiral arms of galaxies. Once "protostars" are formed, they evolve into different types of stars, depending on their mass and the rate with which they convert their fuel into energy. The process of formation and evolution of stars takes billions of years. Therefore, to study the stars throughout their lifetime, astronomers search and identify different types of stars at different stages of their life. By putting these pieces together, they uncover secrets of stellar evolution.

Karl Schwarzschild (1873–1916) was the first to develop the theory of stellar evolution. He found that the distribution of matter in the sun could be determined by studying the exact dependence of the gas pressure on its temperature and density. He also discovered that the energy is transported from the core of a star to its surface by the process of *convection* (moving around as a result of temperature difference) or by direct streaming of energy. This work was further continued by Sir Arthur Eddington (1882–1944), who considered the effect of radiation pressure and showed that stars are mechanically stable only because of a combination of their mass and luminosity, fixed by the laws of physics. Eddington also discovered that as the gas clouds collapse, the temperature at their core increases, and as soon

"Even a fool knows you can't touch the stars, but it won't keep the wise from trying."

- HARRY ANDERSON

"A philosopher once asked, 'Are we human because we gaze at the stars, or do we gaze at them because we are human?' Pointless, really... 'Do the stars gaze back?' Now, that's a question."

- NEIL GAIMAN

as it reaches 20 million degrees Kelvin, they stop contracting. At that stage the system becomes stable, forming a *main sequence* star. However, the questions Eddington could not answer were: why does the contraction stop at this temperature? And what is the source of this temperature (energy)?

This problem was solved in the 1930s by Hans Bethe (1906–2005) and Carl Friedrich von Weizäcker (1912–2007), who showed that thermonuclear fusion, known as the carbon-nitrogen-oxygen cycle (CNO cycle), was responsible for generating a temperature as high as 20 million degrees Kelvin at the core of our sun. This generated the outward force of radiation pressure needed to balance the inward force of gravity. However, for stars less luminous than our sun, that constitutes the majority of stars on the sky, a nuclear reaction that converts hydrogen to helium (but not initiating the CNO cycle) could generate the required temperatures (of the order of 16 million degrees Kelvin). Today more complicated processes have been discovered that could generate even higher energies than the CNO cycle and explain the formation of even heavier elements.

In this chapter the origin of stars will be studied, from formation of a protostar to the more evolved systems. Physical properties of stars at different stages of their evolution will be investigated, and the question will be addressed of why only stars with certain mass or luminosity can exist and survive.

STEPS TOWARD FORMATION OF STARS

STEP 1. STAR-FORMATION MEDIUM

Stars are formed within cold and dense clouds of interstellar gas in galaxies. Our knowledge of the birth process of stars comes from the study of young stars and the medium within which they form. Astronomers refer to the gas and dust filling the space between stars as the *interstellar medium* (ISM). The first generation of stars formed from the material in the ISM that consisted of primordial hydrogen and helium (by the time the first generation of stars were formed, the ISM was not enriched and hence did not contain elements heavier than hydrogen and helium). Because of the low temperature (10 to 30 degrees Kelvin) and high density of these clouds, hydrogen atoms combine and from molecular hydrogen without being destroyed by high temperature or high intensity radiation. For this reason, these are called the *molecular clouds* (figure 12.1; Bennett et al. 2007). Apart from these gas clouds, there is also an abundance of dust in the ISM. Dust absorbs the light from the newborn stars, and as a result, these molecular clouds are often dark (figure 12.1). Therefore, to observe stars in molecular clouds, astronomers use observations in longer wavelengths (infrared or submillimeter) that are less affected by the absorption by dust. This is because the dust grains are smaller than the typical wavelength of the visible light and hence scatter the short wavelength blue light more efficiently than the long wavelength red and sub-millimeter light.

STEP 2. FORMATION OF PROTOSTARS

Stars form in cold gas clouds, as the gravity of the cloud makes it collapse without the outward force of radiation pressure that is produced by the fusion of hydrogen to helium, generating radiation and increasing the temperature. The collapse continues until the core of the cloud becomes dense enough to start nuclear fusion (combining light elements to produce heavier elements and releasing energy), forming *protostars* (Box 12.1). The radiation generated by the protostar produces an outward force due to radiation pressure. In order for a star to form, gravity of the collapsing gas cloud must be strong enough to overcome the outward force of radiation pressure (Box 12.1). Given the cold temperature of gas clouds before the fusion process, the radiation pressure inside stars is small. This, combined with the higher density at the core of these clouds, makes them appropriate sites for the formation of protostars. Stars could also form where two clouds collide, increasing the gas density in both clouds and hence triggering fusion and the star-formation process (Bennett et al. 2007).

BOX 12.1: FORMATION OF PROTOSTARS

The attractive force of gravity in gas clouds is resisted against the outward force of gas (radiation) pressure. This is indeed the reason why a cloud-hosting star formation needs to be cold to minimize the radiation pressure, allowing it to collapse under the force of gravity. The balance between these two forces is responsible for a star being stable. This is called gravitational equilibrium. For gravity to overpower thermal pressure, a molecular cloud needs to have a mass of at least one hundred times the mass of the sun.

STEP 3. FROM PROTOSTARS TO STARS

The protostars formed at the dense center of gas clouds will eventually grow to become stars, as a protostar is not yet able to support the fusion process due to lower temperature at its core.

As the molecular clouds collapse and their density increase, it becomes more difficult for thermal radiation (in the form of photons) to escape. The photons hit the molecules that impede them from leaving the medium. The thermal energy is therefore deposited back in the cloud. As the cloud further contracts, its core becomes denser and eventually reaches the stage where no photons could escape the medium. This produces an outward pressure, slowing down the contraction. Meanwhile, as the radius of the contracting cloud reduces, the force of gravity becomes more powerful, as it is inversely proportional to the square of the radius. As a result, the gas in the outer region

Figure 12.1. Cold gas clouds as sites of newly born protostars. The light produced by the star is absorbed and scattered in the cloud, depending on its gas and dust content. Long wavelength radiation is emitted from the cloud but shorter wavelength radiation (at ultra-violet or blue wavelengths) will scatter within the cloud. The radiation is also absorbed by dust in the clouds and re-emitted at lower energies and longer (redder) wavelengths.

experiences little inside pressure, falling onto the protostar under its gravity and increasing its mass. At this stage the main energy source of the protostar is from gravitational contraction (figure 12.2).

Due to the mass buildup by the protostar and the trapped radiation, the temperature at its core increases until it is high enough to initiate nuclear fusion (figure 12.2). The speed with which this happens depends on the mass of the star and the initial gas cloud. For a star the mass of our sun, this takes many millions of years (Hester et al. 2010).

STEP 4. HYDROGEN-BURNING PHASE

Part of the thermal energy produced by conversion of gravitational contraction escapes from the surface of the protostar, allowing further collapse of the system (figure 12.2). This results in trapping more thermal energy in the core and increasing its temperature. As the core temperature exceeds 10 million degrees Kelvin, hydrogen fusion starts, converting hydrogen to helium. The energy produced through this halts the contraction. At this stage a new star is born, a *hydrogen-burning* main sequence star (figure 12.3, Box 12.2). Therefore, a main sequence star is formed when the inward attractive gravitational force and outward force due to radiation pressure come into

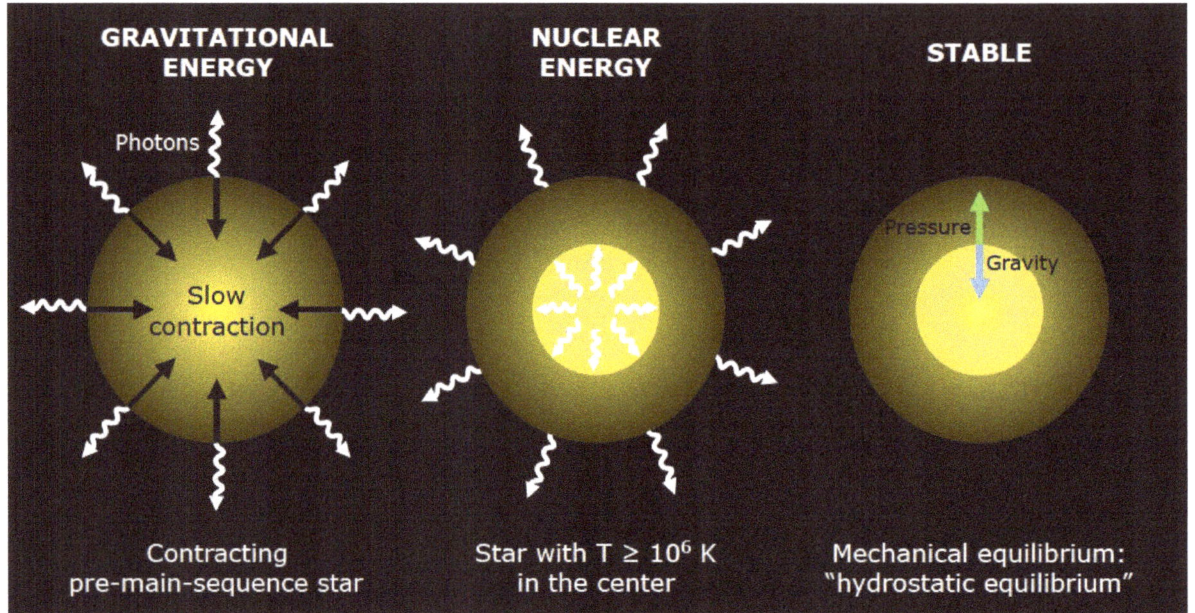

Figure 12.2. Formation of a protostar from the collapse of a gas cloud (left). Further collapse of the cloud causes the temperature to reach 10^6 degrees Kelvin, starting the fusion process (middle). The gravity and radiation pressure must balance for the star to be stable (right).

balance. The time needed for a protostar to evolve to a main sequence depends on the mass of the star (figure 12.3). More massive stars proceed faster, as they are subject to a larger gravity (due to their higher mass) and hence need to burn more of their fuel to generate more energy (radiation pressure) to balance the attractive force of gravity. It took around 30 million years for our sun to move from a protostar to the main sequence stage, while for stars significantly more massive than the sun, the same process could take as little as 1 million years (Bennett et al. 2007; Chaisson and McMillan 2011).

The main parameter governing the evolution of a star is its mass, which fixes the rate with which they evolve. Main sequence stars have different masses, with the most massive having the highest luminosity and surface temperature and largest radius. During their lifetime, the mass, luminosity, and surface temperature of main sequence stars are correlated (Bennett et al. 2007).

BOX 12.2: PHYSICS OF HYDROGEN-BURNING PROCESS

In the extreme density of the core of a collapsing gas cloud, helium is produced by fusion of hydrogen nuclei (protons) through the following process:

1. Two protons fuse to form deuterium (an isotope of hydrogen) containing one proton and one neutron (a proton here is converted into a neutron). This process happens 10^{98} times per second in our sun.
2. One of the deuterium nuclei fuse with a proton, producing helium-3 (two protons and one neutron).
3. Two helium-3 nuclei collide and produce the stable nucleus of helium-4 (two protons and two neutrons) and two free protons. (This sequence is shown in figure 14.1)

TEMPERATURE-LUMINOSITY RELATIONSHIP FOR STARS

Using the available observations at the time, Danish astronomer Ejnar Hertzsprung (1873–1967) and American astronomer Henry Norris Russell (1877–1957) found a well-defined correlation between the intrinsic luminosity and color (which is a measure of surface temperature) of stars. This relationship has become known as the *Hertzsprung-Russell* (H-R) diagram (figure 12.3). On this diagram the main sequence stars form a tight sequence based on their mass and age, called the *main sequence branch*. The evolution of stars past the main sequence branch and their entire history is well represented on the H-R diagram. The main sequence branch extends from the hot, luminous, blue stars in one side to the cool, dim, red stars in another part of the H-R diagram (figure 12.3). Stars move to the main sequence branch after their protostellar stage and spend all their hydrogen-burning phase on that branch. Other types of stars occupy different parts of the diagram, depending on their luminosity and temperature. The H-R diagram has played a major role in the study of the evolution of stars and in developing evolutionary models, predicting signatures of stars at different stages in their evolution (Hester et al. 2010; Bennett et al. 2007).

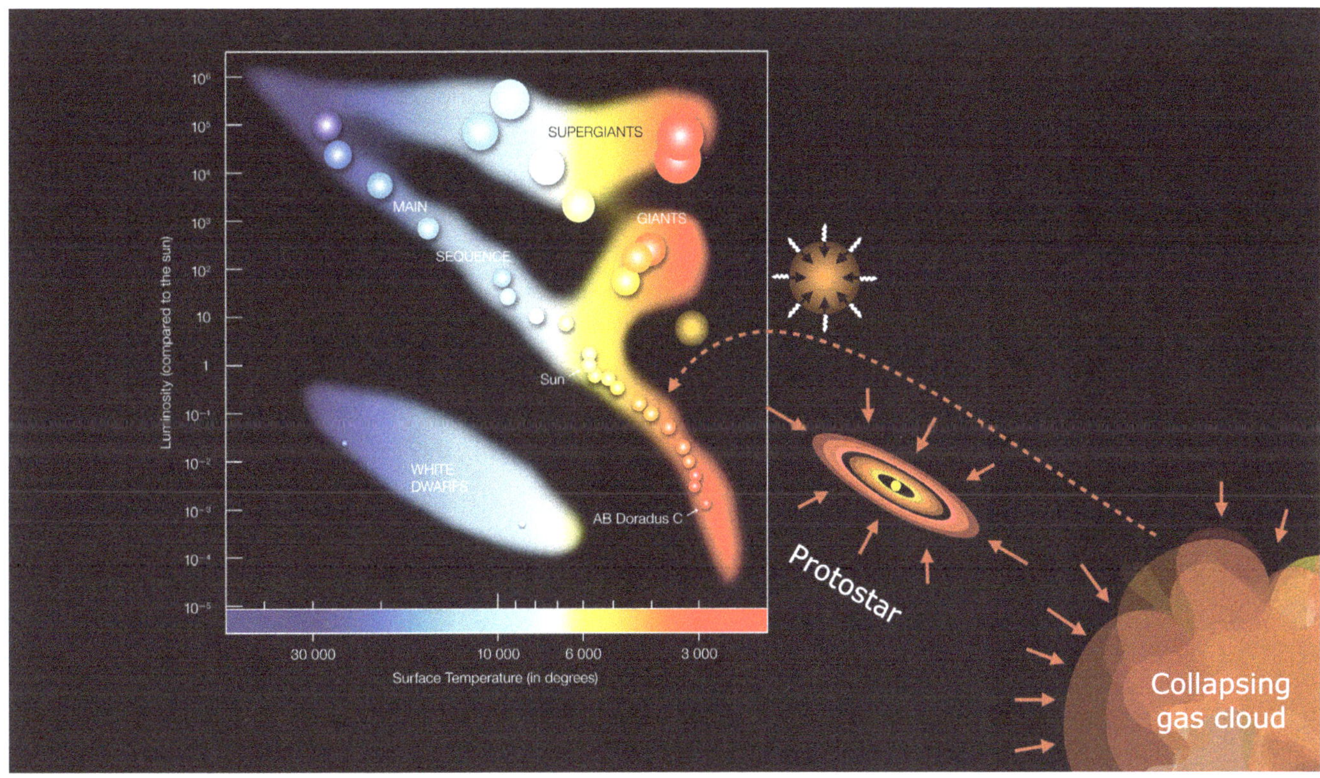

Figure 12.3. The H-R diagram showing the relationship between temperature and luminosity in stars. The band across the diagram is the main sequence branch. Location of stars on this branch depends on their mass and temperature. Protostars move to the main sequence branch as soon as they initiate the fusion process.

WHAT DETERMINES THE MASS OF STARS?

When the mass of the initial contracting cloud is too low, it cannot produce the required temperature at its core that is needed to initiate nuclear fusion. In this case the gravitational collapse would be counteracted not by radiation pressure (as previously described) but by degeneracy pressure that only depends on density and not

When a medium containing a mixture of elementary particles is squeezed, the particles come close to each other, producing a very dense system. However, the compression cannot be continued indefinitely. According to the laws of quantum mechanics (the exclusion principle), no two particles could have the same position, momentum, and spin (that is, all the atoms cannot lie on top of one another). Therefore, the compression must stop when the system gets to that density. The resistance against compression, produced by the extreme density at the core of stars is called *degeneracy pressure*.

The degeneracy pressure that stops gas clouds from collapsing is due to electrons being forbidden to share the same state. This is called *electron degeneracy pressure*. The same exact process could result from neutrons. However, such neutron degeneracy pressure takes over at much higher densities. Since neutrons have larger mass than electrons (about 1750 times more massive), at a speed close to the speed of light, they have a momentum 1,750 times that of electrons, meaning that their position can be this much more precise, allowing them to occupy a much smaller volume of space. This is the process that causes neutron stars to collapse and become black holes (Chapter 13).

temperature (Box 12.3). The degeneracy pressure does not allow formation of stars with masses less than 0.08 M_{sun} (nearly eighty times the mass of the Jupiter). Such stars cannot generate the core mass density required to initiate fusion of hydrogen. This is due to their small mass and the electron degeneracy pressure (Box 12.3). Therefore, they never produce their own luminosity and, as a result, no star with a mass less than this could be formed.

Similarly, there is a limit to how massive a star could be. As the mass of a star increases, it will collapse faster, and the fusion process at its core becomes more rapid, generating a lot of radiation. It is predicted that stars with a mass larger than 150 M_{sun} generate so much energy in the form of radiation pressure that override the force of the gravity, pushing their outer layer out into space. This halts formation of the star. High mass stars are short-lived because they need to consume more of their fuel (hydrogen) to generate the energy needed to counterbalance the force of gravity. As a result, they run out of fuel sooner. Given this, there are very few high mass stars around (Bennett et al 2007; Chaisson and McMillan 2011).

SUMMARY AND OUTSTANDING QUESTIONS

Active sites of star formation in galaxies are associated with gas reservoirs. Before the first generation of stars formed and enriched the ISM, the gas was pristine, only containing hydrogen and helium. As gas clouds collapse, the density at their core increases and eventually reaches a level sufficient for fusion to take place, forming protostars. Due to the small temperature at their core, protostars cannot support the fusion process. The gravitational energy responsible for the collapse of the clouds is converted to thermal energy inside the cloud which in turn is trapped in the system due to an increase in the density of the cloud, caused by its collapse. The radiation pressure generated this way counters the attractive force of gravity and hence slows down the collapse of the cloud. The increase in the temperature at the core initiates the nuclear fusion and the hydrogen-burning phase of the evolution of a main sequence star.

The stars follow a luminosity-temperature relation, known as the Hertzsprung-Russell (H-R) diagram. The protostars are shifted to the main sequence branch of the H-R diagram once they start their hydrogen-burning phase. The mass of stars is the main parameter controlling their evolution on the H-R diagram. Massive stars are short lived, as they are subjected to a stronger gravity; to retain the balance, they need to convert more of their fuel (mass) to generate energy and hence consume their mass sooner. Only stars within the mass range 0.08 M_{sun} and 150 M_{sun} could exist. This is limited by electron degeneracy (for the low-mass limit) and the need to generate significant radiation pressure to balance the extreme force of gravity (for high-mass limit).

Star formation is the most important single process responsible for the evolution of galaxies. Stars form in gas clouds in galaxies, and after passing a maximum activity, star-formation rate declines, with the galaxy becoming quiescent. Despite extensive study of the star-formation activity in galaxies, there are still a number of unanswered questions in need of further study. What is the number of stars formed within a mass range and what is this dependent on? What are the most fundamental parameters responsible for star formation? What was the star-formation rate for the first generation of galaxies? How does star formation build up the mass in the gas cloud and how does this change with time? What is the effect of dust on the star-formation process? Any study of the evolution of galaxies requires answers to the above questions. These are the subjects of significant current research.

REVIEW QUESTIONS

1. What is the physical process responsible for generating the high temperature at the core of protostars?
2. Explain the formation of dark molecular clouds that are sites of star-formation activity.
3. Why is a low temperature required in star-forming clouds to allow for the star-formation to proceed?
4. Why do astronomers use certain wavelengths to study the sites of star formation in molecular clouds?
5. Describe the steps that leads to the formation of protostars.
6. Explain the transition from protostars to main sequence stars.
7. Describe the Hertzsprung-Russell (H-R) diagram and its application in the study of stellar evolution.
8. How does the mass of a star control its life cycle?
9. Explain why stars could only have a mass within a certain range.
10. Explain the degeneracy pressure and its significance in the formation of stars.

CHAPTER 12 REFERENCES

Bennett, J., M. Donahue, N. Schneider, and M. Voit. 2007. *The Cosmic Perspective*. 4th ed. Boston: Pearson/Addison-Wesley.

Chaisson, E., and S. McMillan. 2011. *Astronomy Today*. New York: Pearson.

Hester, J., B. Smith, G. Blumenthal, L. Kay, and H. Voss. 2010. *21st Century Astronomy*. 3rd ed.New York: Norton.

FIGURE CREDIT

- Fig. 12.1: Source: https://en.wikipedia.org/wiki/File:Milky_Way_
  IR_Spitzer.jpg.

THE EVOLUTION AND DEATH OF STARS

CHAPTER LEARNING OBJECTIVES

This chapter will cover:

- The evolution of stars
- The death of stars and supernovae explosion
- Neutron stars
- Black holes
- Heavy elements in the space between stars
- The interstellar medium

One of the triumphs of modern astrophysics is the development of the theory of stellar evolution. Not only does this explain the origin of heavy elements (the chemical elements heavier than hydrogen and helium), it allows an understanding of how the elements (originally formed in stars) were distributed in the interstellar medium and their journey to Earth, as well as the processes that led stars to attain the characteristics they have. This requires a study of stars at different stages of their life. The main physical parameter responsible for the speed with which a star evolves and for the end product is its mass. This is because both the temperature and age of stars depend on their mass, which is particularly important during later stages of the stellar evolution. Low-mass stars evolve slowly while high-mass stars evolve rapidly with very different end products. Therefore, the mass governs the evolution of stars away from the main sequence branch and the speed of their evolution. The subject of stellar evolution is very complex, with many different steps involved. Here, my aim is not to explain the stellar evolution but to concentrate on the stages that are important in our quest for understanding the origins. Since the evolution and life of stars is dependent on their mass, I study the evolution of the low and high mass stars separately. The formation of stars and initial stages of their evolution followed by their journey to the main sequence branch were discussed in chapter 12.

This chapter continues the story by following the evolution of stars away from the main sequence branch. A step-by-step study of the evolution of stars will be presented. This is needed to understand the origin of heavy elements in the universe (chapter 14).

"The sun with all those planets revolving around it and dependent on it can still ripen a bunch of grapes as if it had nothing else in the universe to do."

- GALILEO GALILEI

"A hundred thousand million Stars make one Galaxy; A hundred thousand million Galaxies make one Universe. The figures may not be very trustworthy, but I think they give a correct impression"

- SIR ARTHUR EDDINGTON

LIFE STORY OF LOW-MASS STARS

The main source of energy for the main sequence stars is provided by *proton-proton* fusion (since the nucleus of hydrogen atom consists only of a single proton) converting hydrogen to helium and releasing energy (this is the same as hydrogen burning discussed in the previous chapter, Box 13.1). As the hydrogen reservoir of a main sequence star is used, helium builds up at its core. Due to the higher density and temperature at the core of the star, the helium build up is more efficient there. This continues until all the hydrogen at the core of the star is used. Since hydrogen is depleted at the core, the hydrogen burning shell moves to upper layers in the star. At this point the star leaves the main sequence branch (figure 13.1). For a star the mass of our sun, it takes a total of 10 billion years to get to this stage (our sun arrived at the main sequence branch about 5 billion years ago and will leave it in 5 billion years) (Box 13.1).

The stability of a star depends on the balance between the inward force of gravity due to its mass and the outward radiation pressure caused by the internal fusion process. As hydrogen burning at the core slows down, the internal radiation pressure diminishes and the core starts to contract. The collapse of the core accelerates as all the hydrogen at the core is consumed, turning it to a helium core (figure 13.1). The subsequent collapse of now the helium core produces gravitational energy, increasing the temperature at the core. The temperature needed for helium to fuse into heavier elements is around 10^8 degrees Kelvin, compared to 10^6 to 10^7 degrees Kelvin needed for hydrogen. As soon as the hydrogen-burning temperature is reached, the hydrogen in the outer shell starts to fuse. This causes an increase in the outer region size of the star while the inner helium shell continues to collapse. As a result, the star becomes larger in radius—about the size of the Mercury's orbit around

BOX 13.1: SOURCE OF A STAR'S ENERGY

In the core of stars such as our Sun, the extreme temperature and density transforms four hydrogen atoms into a helium atom. We know that one helium atom has a little less mass than the four hydrogen atoms combined. This means that in converting hydrogen to helium through the fusion process, roughly 0.007 of the original mass (the mass of four hydrogen atoms) has disappeared. In fact they have not disappeared but were converted to energy through Einstein's formula: $E = mc^2$, where m is mass converted to energy and c is the speed of light. Therefore in each fusion process, the equivalent of 0.007 of mass is released as energy. This is the source of the energy from our Sun.

Using this argument, the total energy generated by our Sun through the fusion process is:

$E = 0.007 \, M_{sun} \, c^2$

where M_{sun} is the mass of the Sun (2×10^{30} kg). Only 10% of the mass of the Sun is located at its core where it is dense enough and has a high enough temperature to enable a nuclear reaction. Therefore, the energy generated by the Sun through the fusion of hydrogen to helium is:

$E = 0.007 \times 0.1 \, M_{sun} \, c^2$

This gives the energy due to fusion of 1.3×10^{44} jules. Given the observed rate at which the Sun is producing energy, 3.8×10^{26} watts, and dividing this by the energy produced by hydrogen fusion as estimated above, gives a time period of 10 billion years for the Sun to convert all of its hydrogen to helium. This is the time a star like the Sun will spend on the main sequence branch.

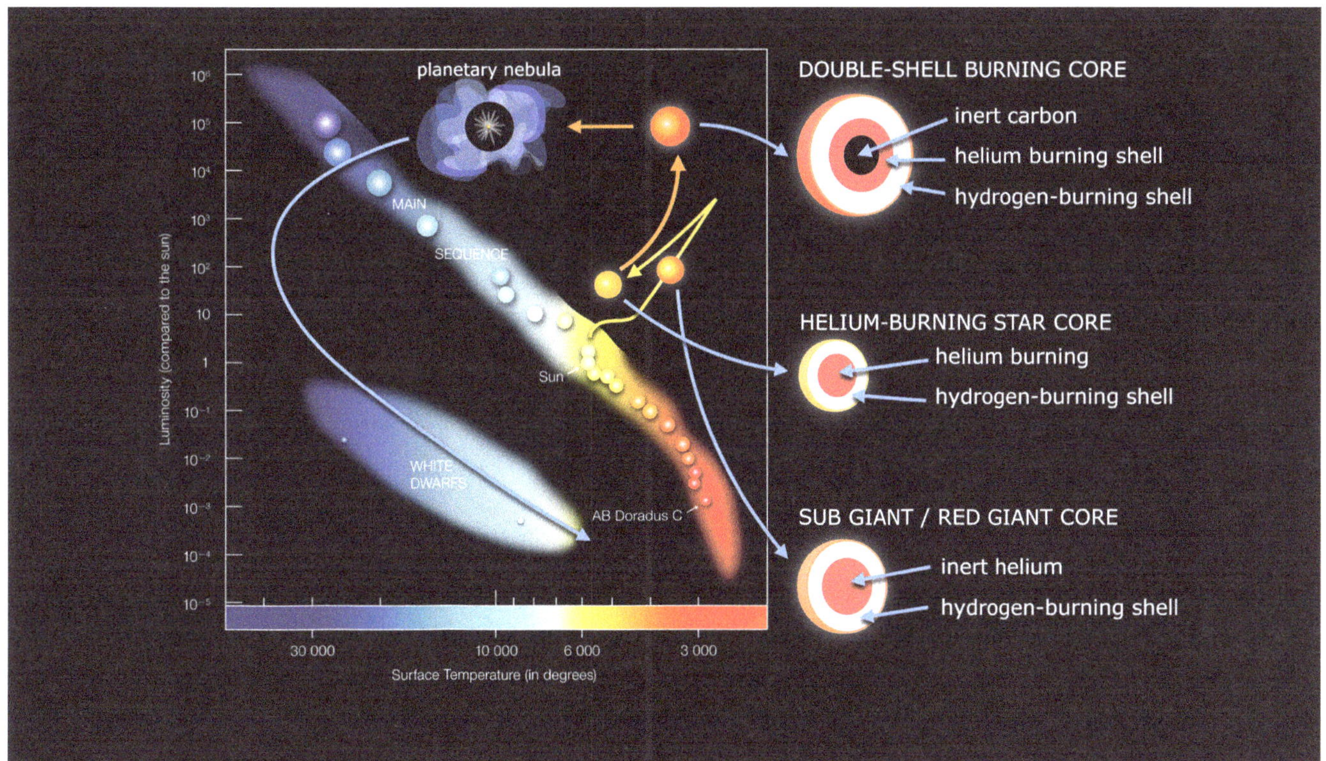

Figure 13.1. The H-R diagram showing the evolution of low-mass stars. They start from the low-mass end of the main sequence branch. After using the hydrogen at their core, they leave this branch and go through a red giant phase, where the outer radius of the star increases while its helium core shrinks under its gravity. The star eventually ejects its outer envelop as planetary nebulae with its core becoming a white dwarf star. It ends up as a white dwarf.

the sun. Around one hundred years after the star leaves the main sequence branch, it arrives at the *red giant phase* of its life (figure 13.1; Bennett et al. 2007). By the time our sun gets to this stage (around 5 billion years from now), it will swallow its closest planets (Box 13.1).

As the temperature in the contracting helium shell reaches 10^8 degrees Kelvin, the helium starts to fuse into ^8B (beryllium-8; a highly unstable nucleus). The density at the core is so high at this time that before beryllium decays, it will hit another ^4He (helium-4) and turns to ^{12}C (carbon-12). This results in a carbon-rich inner core, with helium depleting at the core of the star with a nonburning carbon core replacing it. Again, the carbon core shrinks under its gravity, with the outer hydrogen and helium layers burning and expanding similar to what happened during the helium-burning phase (figure 13.2). The end result is a second swollen red giant star. The temperature at this stage is much higher than before, with the star's radius and luminosity increasing more than what it did during the helium-burning phase (Bennett et al. 2007).

THE DEATH OF A LOW-MASS STAR

In order for the carbon core to start fusion to yet heavier elements, the temperature at the core should reach $\sim 6 \times 10^8$ degrees Kelvin. Low-mass stars are not able to attain this temperature. As the carbon core collapses, the core density and temperature become extremely high but stops short of the temperature needed to ignite carbon. Such large temperatures are only attainable for high-mass stars. As the carbon core shrinks and increases its density, the outer layers of hydrogen and helium burn faster. As a result, the envelope expands and cools,

Figure 13.2. Schematic picture showing the increase in the size of a low-mass star as a result of fuel burning at the core and shifting of the hydrogen- and helium-burning layers to larger radii. Eventually, the outer envelope of the star is ejected, forming a planetary nebula. Different colors in the planetary nebula indicate sites of different chemical elements.

reaching a radius three hundred times that of the sun (figure 13.2). At this point the core exhausts all its fuel, and the system contracts and heats up. This generates high-energy ultraviolet radiation that would ionize the envelope. The combined effect of the outward push due to internal radiation and reduced gravity in the outer region causes the outer envelope to be ejected, moving away with a speed of a few tens of kilometers. This ionized cloud is called the *planetary nebula* (figure 13.2; Bennett et al. 2007; Chaisson and McMillan 2011).

As the envelope in the form of planetary nebula moves away, the carbon core of the star becomes transparent. Due to continued contraction, this is now the size of Earth with a mass substantially smaller than the mass of the sun. Because of its small mass, it cannot support fusion and has its own "stored" heat, giving it a white surface. The star at this stage of its life is called *white dwarf* (Box 13.2). The evolutionary history of a low-mass star is summarized in figure 13.3.

LIFE STORY OF HIGH-MASS STARS

Stars with larger mass generate more heat due to their larger surface gravity, speeding up the process of nuclear fusion. As a result, they run out of their fuel in shorter time and hence leave the main sequence branch sooner than their lower-mass counterparts. All their subsequent evolution is governed by their mass. The stellar evolution process up to the time when they leave the main sequence branch is almost the same for low- and high-mass stars, with the difference being in the details (figure 13.3). High-mass stars reach temperatures in excess of 6×10^8

These are the end products of the evolution of low-mass stars. They do not produce energy themselves but shine their stored energy. They are the size of Earth and about half the mass of the sun. When a white dwarf is in a binary system, it accretes matter from its companion (which could be a main sequence star). This increases the mass of the white dwarf and hence its gravity. Theoretically, it is shown that the maximum mass a white dwarf could have is 1.4 M_{sun}, called *Chandrasekhar mass*. Exceeding this mass, electron degeneracy can no longer resist the force of gravity and the star collapses. As a result, its core temperature increases, reaching a point needed for carbon fusion, causing the star to explode as supernova—this is called carbon detonated supernova or type Ia supernova.

degrees Kelvin at the core. At this temperature, they are able to synthesize carbon, oxygen, and heavier elements. As synthesis of new elements are depleted to upper shells of the star, the core contracts and heats up leading to fusion of heavier (and new) elements. This continues until iron is produced. At this point stars contain different chemical elements, from the heaviest (iron) at the core to lighter elements close to their surface (figure 13.4).

Iron nucleus is the most stable among all the elements and cannot be fused by any element to release energy. When light nuclei fuse, the mass per particle decreases, and energy is released due to the law of conservation of mass/energy. Iron nucleus has the smallest mass per nuclear particle and hence is the most stable. As a result, iron cannot be combined with other elements to generate energy, and therefore the fusion process in high-mass stars stops with iron. Iron build up extinguishes the energy production at the core of massive stars and hence substantially reduces the internal radiation pressure that was balancing against the gravity. As a result, the star goes through a catastrophic collapse. Because of this, the temperature at the core of the star increases to 10^{10} degrees Kelvin (mainly by conversion of gravitational to thermal energy), enough to disintegrate elements, breaking them into their constituent elementary particles, protons and neutrons. The photo-disintegration process consumes some of the energy in the core and hence reduces the core temperature, further diminishing the outward pressure and accelerating the contraction. At this stage, the core only consists of the elementary particles, resulting in the protons and

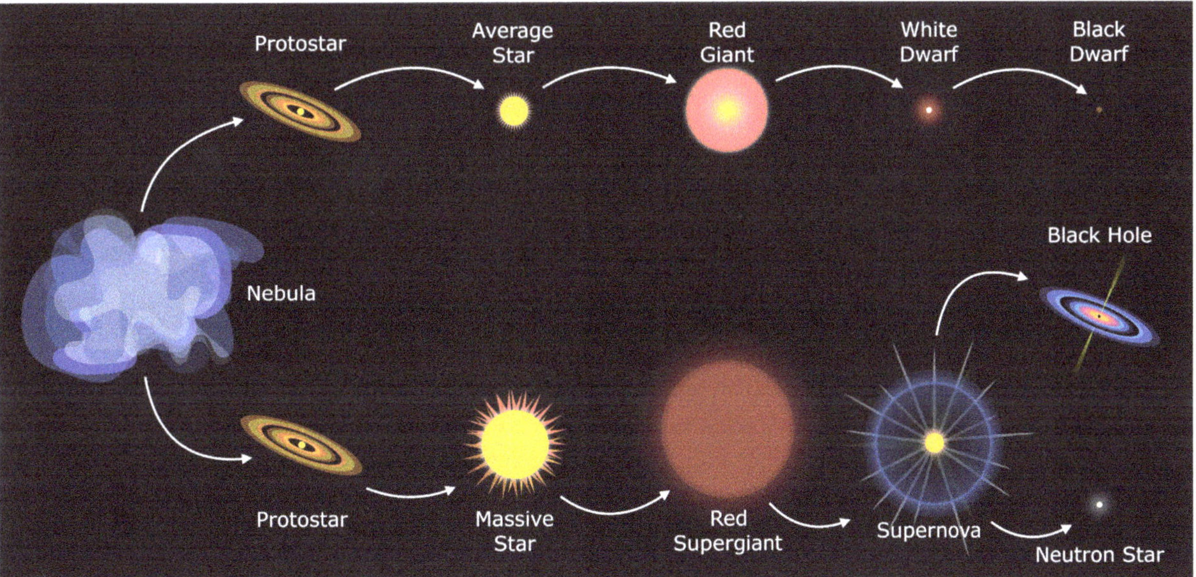

Figure 13.3. Shows different stages in the evolution of a low-mass star (top) and massive star (bottom).

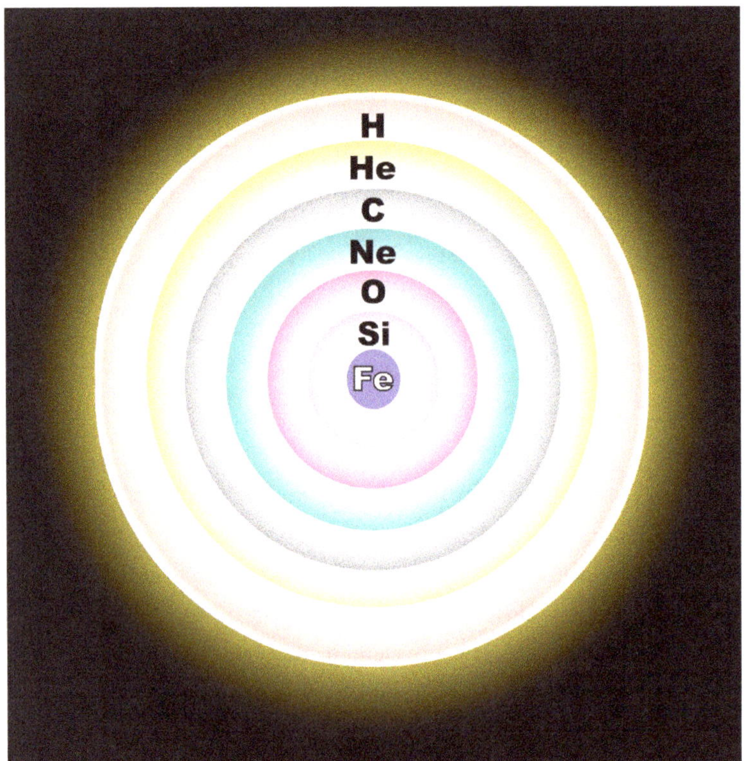

Figure 13.4. Different layers of elements produced during the evolution of massive stars. Heavier elements are at the core, while lighter elements are closer to surface of the star.

electrons crushing together, producing neutrons and a neutrino (Box 13.3). Neutrinos are neutral particles and do not interact with matter, and therefore escape from the medium, further decreasing the energy and causing the star to implode (figure 13.3; Bennett et al. 2007).

DEATH OF MASSIVE STARS

Due to the extreme density at the core of a massive star at advanced stages of evolution (10^{15} kg/m^3), neutrons are densely packed. The destiny of the star from this point on is controlled by neutron degeneracy pressure. This is similar to electron degeneracy pressure for low mass stars but, in this case, neutrons refuse to be in the same quantum state. As a result, further collapse is halted and the system rebounds back, causing an enormous shock wave through the star's outer layers of heavy elements. This leads to a huge explosion, resulting in a *core collapse supernova (*also called *Type II supernovae)* which is the end of life of a massive star (figure 13.3). The core of the star is not affected by the shock wave and the explosion and remains intact in the form of closely packed neutrons, called a *neutron star*. This is much denser than white dwarf stars. This is the remnant of a massive star after its outer envelope is blown away by a supernova explosion (Box 13.3) (Bennett et al. 2007; Chaisson and McMillan 2011).

Having a diameter of 10-25 Km with an extremely large mass, neutron stars have a huge density. For comparison, one tea-spoon full of a neutron star material would weigh around 10 million tons. If the mass of a neutron star becomes 3 times the mass of our Sun, then nothing, even neutron degeneracy pressure could halt the collapse of the star. At that point the star collapses and a black hole will be formed. Black hole resembles a point source with its size defined as the radius beyond which the escape velocity exceeds the speed of light (Box 13.4).

This is the final stage of the evolution of a massive star. These are very small stars (about 20 km across) consisting of densely packed neutrons. Their mass exceeds the mass of the sun, giving them a density of 10^{17}kg/m^3. At this density the system will not collapse, due to neutron degeneracy pressure. Neutron stars spin fast, due to the law of conservation of angular momentum (as their size shrink, they spin faster to keep angular momentum constant). They have strong magnetic field. Due to the collapse of the star, magnetic field lines are squeezed, making the field density stronger.

BOX 13.4: BLACK HOLES

If the mass of a neutron star exceeds three times the mass of the sun, neutron degeneracy that was supporting the star against the force of its gravity breaks down and the star collapses. Once neutron degeneracy is broken, no other force can resist gravity. At this point gravitational force becomes so strong that even light cannot escape from the star. This is called a *black hole*. There is a radius around the center of a black hole beyond which no information can escape–so-called *event horizon*. At the center of black holes, there is a singularity where space and time lose their separate identity and become one entity.

SUMMARY AND OUTSTANDING QUESTIONS

There are two basic principles underlying stellar evolution. First, stars evolve by fusion of light elements, resulting in the production of energy (in the form of light and heat) and heavier elements. The rate at which this process takes place depends on the mass of the star. Second, stars are in equilibrium, due to the balance between the inward force of surface gravity and the outward force of radiation pressure. When this balance is disturbed, the system either collapses or expands outward. The surface gravity of a star is controlled by its mass. Therefore, a star with large mass tends to collapse more rapidly. To counteract the force of gravity, more radiation must be produced to push the system outward. To generate high radiation pressure required to balance the inward force of gravity, more material must be processed resulting in a faster consumption of the fuel and hence a shorter life for the star. For example, it takes about 10 billion years for a star the mass of our sun to consume its hydrogen to helium (Box 13.1). Stars more massive than the sun require much less time to complete this process. Therefore, the mass of a star determines its destiny. Figure 13.3 shows different steps in the evolution of low-mass and massive stars.

After the hydrogen-burning phase in a low-mass star, the core collapses while its outer envelope expands out, still burning hydrogen. As a result, the density and temperature at the helium core of the star increases, and after reaching the *helium-burning* phase (a temperature of 10^8 degrees Kelvin), another fusion takes place, forming the carbon (chapter 14). Because of the small mass of the star, the energy and temperature at its core are not sufficient to convert carbon to heavier elements. This results in a diminished radiation pressure and rapid collapse of the star under its gravity. At this point the outer region of the star is ejected (in the form of a planetary nebulae), as there is little gravity affecting the surface of the star, while the carbon core of the star continues to contract, forming a star with the size of Earth and mass substantially less than the sun, called white dwarf star. These stars cannot initiate fusion process. The theoretical mass of a white dwarf cannot exceed 1.4 times the mass of our sun. If it exceeds the fusion process, electron degeneracy that was halting the star from collapse under the force of its gravity breaks down and the star explodes in supernovae Type Ia (these are objects undergoing very violent explosion producing extremely large luminosities). Through this explosion, heavy elements produced in the stars are distributed in the interstellar medium, leaving a white dwarf behind (figure 13.3).

The evolution of a massive star is similar to that of the low-mass star until the carbon core is formed. Due to the high mass of the star, the carbon core collapses, with the core temperature increasing until it reaches the value needed for carbon fusion. At this point elements heavier than carbon are produced through the fusion process. I will discuss this in the next chapter. The process continues until iron is formed at the core through fusion of lighter elements. This is one of the most stable nuclei and, as a result, does not take part in the fusion process. Therefore, the process of fusion stops at this point, and the star goes through a rapid collapse phase because of the absence of the outward pressure. The high density at the core causes the elements to disintegrate into their constituent elementary particles, protons, and electrons, which would then combine resulting neutrons. Because of neutron degeneracy at these high densities, the star explodes in a supernova (This is a core-collapse supernova that is different from that resulted from the evolution of low-mass stars). This leaves behind a neutron star. If the mass of the initial star exceeds three times the mass of our sun, the neutron degeneracy breaks and the system collapses into a black hole.

The theory of stellar evolution is among the most successful theories in astrophysics. It has been verified on many different occasions by detailed observations. However, there are still a number of ambiguities in details. For example, the physics of supernova explosion is not well understood. It is not yet clear if core collapse supernovae are all produced in galaxies undergoing star-formation activity; that is, spiral galaxies, as they are the result of evolution of massive (young) stars; or they also appear in non-star-forming systems; that is, elliptical galaxies. Also, the physics of black holes, the singularity at their center and the violent processes ending the life of stars are not fully understood.

REVIEW QUESTIONS

1. Explain the evolution of main sequence stars to the red giant phase.
2. Explain the processes involved in the synthesis of carbon.
3. What was the heaviest element formed as a result of the evolution of low-mass stars?
4. What is the event horizon in a black hole?
5. How is a planetary nebula formed? What are the constituents of a planetary nebula?
6. Explain the characteristics of a white dwarf star.
7. In the final stages of the evolution of massive stars, the core temperature rises to 10^{10} degrees Kelvin. Explain the consequence of this increase in temperature on the evolution of the star.
8. Iron is the heaviest element produced as a result of the evolution of massive stars. Explain why elements heavier than iron could not be produced in the star.
9. How does neutron degeneracy lead to explosion of *core collapse* supernova?
10. Explain the physical characteristics of a neutron star.

CHAPTER 13 REFERENCES

Bennett, J., M. Donahue, N. Schneider, and M. Voit. 2007. *The Cosmic Perspective*. 4th ed. Boston: Pearson/ Addison-Wesley.

Chaisson, E., and S. McMillan. 2011. *Astronomy Today*. New York: Pearson.

FIGURE CREDITS

- Fig. 13.2: Copyright © European Southern Observatory (ESO)/S. Steinhöfel (CC by 4.0) at https://en.wikipedia.org/wiki/File:The_ life_of_Sun-like_stars.jpg.
- Fig. 13.4: Copyright © Rursus (CC BY-SA 3.0) at https:// en.wikipedia.org/wiki/File:Evolved_star_fusion_shells.svg.

THE ORIGIN OF HEAVY ELEMENTS

CHAPTER LEARNING OBJECTIVES

This chapter will cover:

- Steps toward synthesis of heavy elements
- Parameters affecting the production and abundance of heavy elements
- The origin of the heaviest of the elements
- Why some elements are more abundant than others
- Enrichment of the interstellar medium

It is now well accepted that light elements like hydrogen and a significant fraction of helium are primordial and were all formed over the first ten minutes from the big bang. Also, after formation of stars and during the course of stellar evolution, hydrogen has continuously been converted to helium, producing the energy and the heat in stars. The question now is: how were all the heavier elements in the universe, responsible for everything around us, formed? In Chapter 6, I discussed formation of light elements during the first few minutes after the birth of the universe. I also explained why heavier elements could not be synthesized at that time. Elements heavier than hydrogen and helium were formed under completely different circumstances than the light elements in the early universe. We now have detailed knowledge as how these elements were formed. In astrophysical literature, elements heavier than helium are called *metals*, with the fraction (by mass) of metal to hydrogen called *metallicity*.

The metals were formed by nuclear fusion in the center of stars through a process called *stellar nucleosynthesis* (as opposed to big bang nucleosynthesis). The core of stars is the only place in the universe with high enough temperature and density to allow this process to take place. However, even the core of stars do not have a high enough temperature and density to synthesize elements heavier than iron. These are predicted to have formed at the last stages of stellar evolution in the core of very massive stars or through a violent explosion that signifies the death of stars with masses higher than our sun.

"In the visible world, the Milky Way is a tiny fragment; within this fragment, the solar system is an infinitesimal speck, and of this speck our planet is a microscopic dot. On this dot, tiny lumps of impure carbon and water, of complicated structure, with somewhat unusual physical and chemical properties, crawl about for a few years, until they are dissolved again into the elements of which they are compounded."

- BERTRAND RUSSELL

"The nitrogen in our DNA, the calcium in our teeth, the iron in our blood, the carbon in our apple pies were made in the interiors of collapsing stars. We are made of starstuff"

- CARL SAGAN

A successful theory for the synthesis of heavy elements should be able to explain their formation through known physical principles, as well as being able to predict their abundance. In a classic paper in 1957, E.M. Burbidge, G.R. Burbidge, W.A. Fowler, and F. Hoyle proposed new processes that lead to production of heavy elements in stars. In this paper, for the first time, they demonstrated that the atomic nuclei from lithium to uranium were formed in stars. Subsequent progress in nuclear physics confirmed many of the proposed reactions responsible for production of heavy elements, which were later tested through measurement of elemental abundance in stars.

Using the information about stellar evolution in the previous chapter, this chapter presents steps toward synthesis of heavy elements in stars. Processes leading to the synthesis of different elements will be discussed as well as the mechanism that is responsible for the enrichment of the interstellar medium. The chapter studies the physical conditions required for production of heavy elements.

FORMATION OF HEAVY ELEMENTS

The first step in the stellar nucleosynthesis process is formation of helium nucleus from fusion of four hydrogen nuclei

$$4\,(^{1}H) \rightarrow {}^{4}He + 2e^{+} + 2\nu + energy$$

where ^{1}H and ^{4}He indicate the number of nucleons (protons and neutrons) in the nuclei of hydrogen and helium, with a hydrogen nucleus containing one proton and helium nucleus containing two protons and two neutrons (Box 14.1). Positrons (e^{+}) are electron antiparticles and, immediately after being produced, are annihilated with electrons, producing high-energy gamma rays. Neutrinos (ν) are particles with very low mass, carrying energy and moving with a speed close to light. They do not interact with matter and, as a result, escape from the medium. The steps toward production of ^{4}He in stars are shown is figure 14.1.

The energy produced through this and other nuclear reactions is the source of the heat and light from stars, including our sun (Box 13.1). As helium is synthesized through hydrogen burning in main sequence stars, the star runs out of fuel (hydrogen), and as a result, it contracts under its gravity (because there is no longer an outward force of radiation pressure) and heats up, reaching around 100 million degrees Kelvin at the core. At this temperature, helium nuclei could fuse together by overcoming their mutual electrostatic force. Therefore, three helium nuclei combine to form a carbon nucleus and significant amount of energy

$$3\,(^{4}He) \rightarrow {}^{12}C + energy.$$

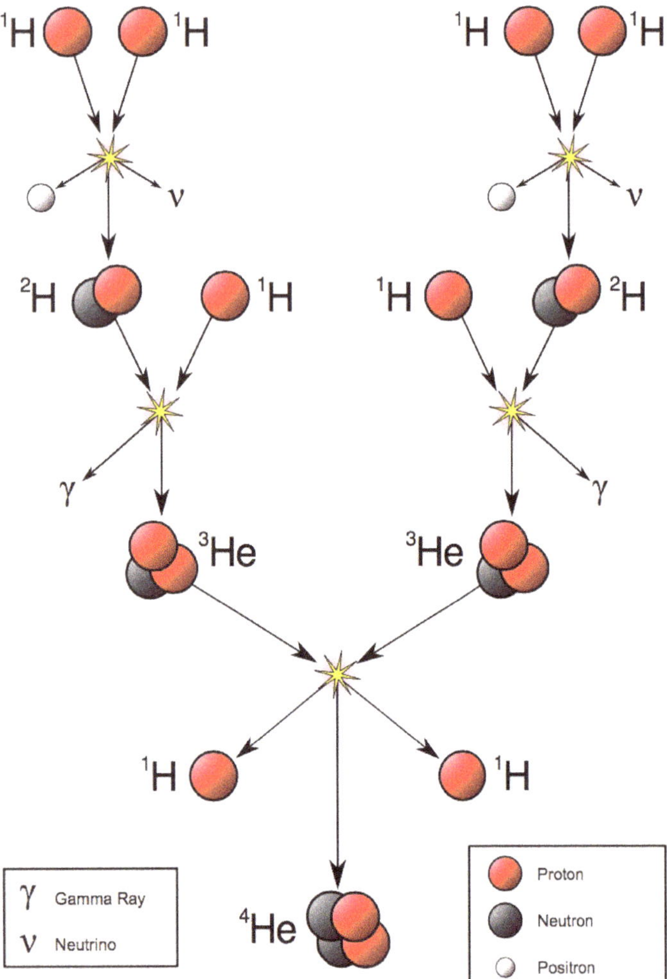

Figure 14.1. A step-by-step synthesis of ⁴He from hydrogen at the core of stars.

BOX 14.1: ATOMIC NUMBER AND MASS NUMBER

Chemical elements are recognized from their electronic configurations as represented by the number of their protons. Elements with the same number of protons have similar chemical properties. Those elements with the same number of protons and different number of neutrons are called *isotopes*. Isotopes with higher number of neutrons are denser.

Chemical elements are identified by their atomic number (the number of protons) and mass number (total number of protons and neutrons). An element with symbol X, atomic number Z, and mass number A is denoted by

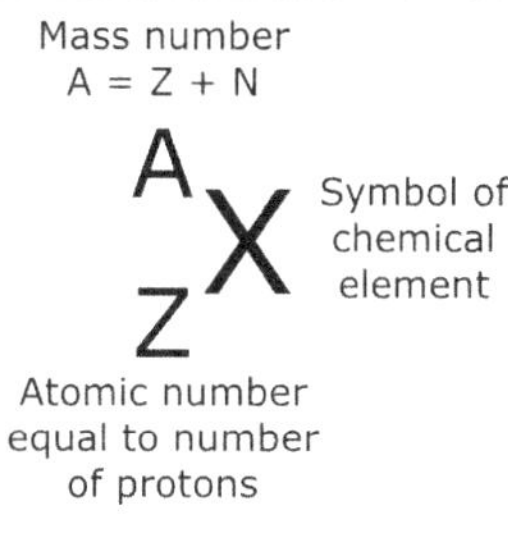

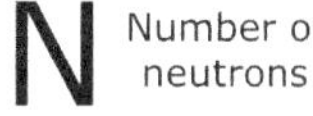

This process takes place in all stars regardless of their mass. The processes involving helium capture are called the *alpha process*. The above reaction, involving three helium nuclei, is called *triple-alpha process*. However, this process is extremely rare, given the likelihood of three helium nuclei colliding at the same time. What more frequently happens is the fusion of two ^4He nuclei producing ^8Be (beryllium-8). This element is unstable and will soon decay into two helium nuclei. As discussed in chapter 6, ^8Be is produced in the early universe, but since it is unstable, it immediately decayed and hence delayed formation of other elements beyond ^4He (i.e., *beryllium bottleneck*). The beryllium bottleneck is overcome at the center of stars where the density is excessively high. As soon as beryllium was produced and before it disintegrated, it immediately combined with another ^4He, producing ^{12}C:

$$^4He + {}^4He \rightarrow {}^8Br$$
$$^8Br + {}^4He \rightarrow {}^{12}C.$$

The next steps, leading to the production of elements heavier than carbon, could only take place at the core of massive stars where extreme conditions of high density and temperature prevail (figure 14.2a; Bennett et al. 2007).

When most of the fuel (helium) is consumed at the core of massive stars, the system ceases energy production and collapses under its gravity, increasing both the density and temperature at its core. When the temperature reaches around 600 million degrees Kelvin, carbon burning starts form magnesium through the reaction:

$$^{12}C + {}^{12}C \rightarrow {}^{24}Mg + energy.$$

Such temperatures can only be produced at the core of stars much more massive than our sun. In general, to fuse together elements heavier than carbon, due to the large number of protons in their nuclei causing huge electrostatic repulsive force, extreme temperatures and densities are needed to overcome this repulsive force. As a result, reactions like the above where two heavy elements are fused together are rare (figure 14.2b). For this reason

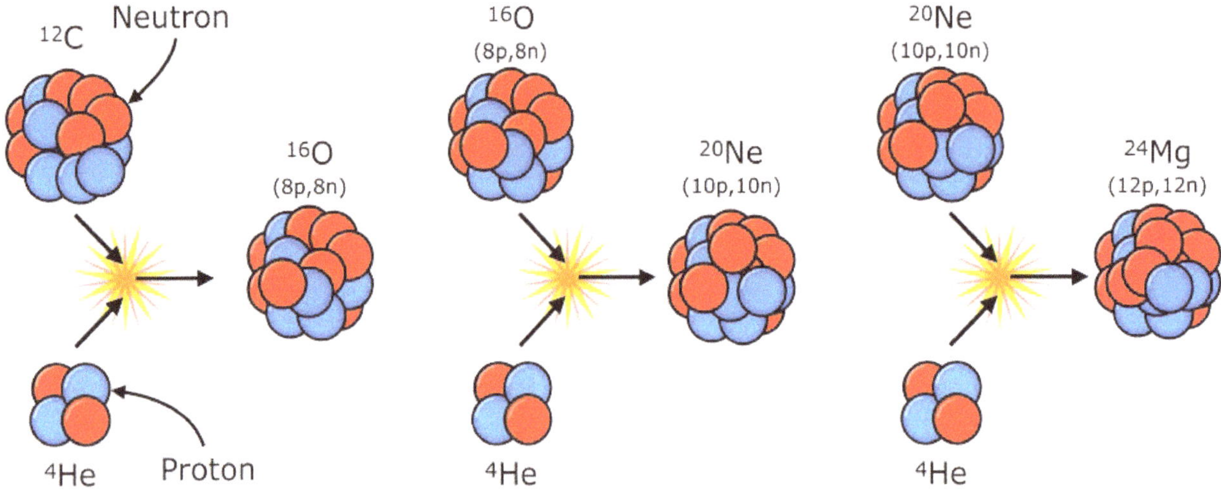

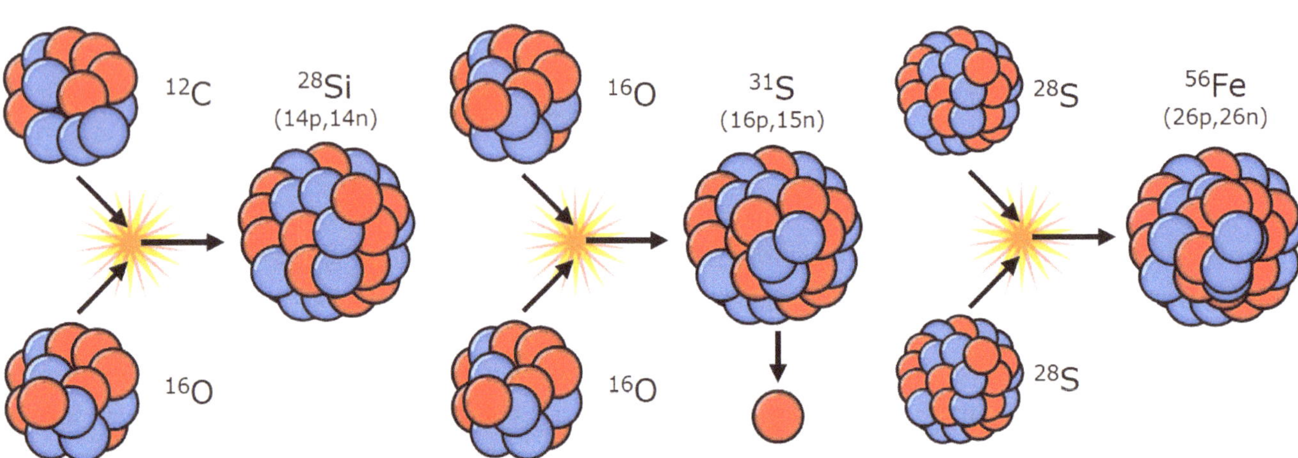

Figure 14.2a. (Top panel): The s-processes (meaning slow processes) are shown. Shows synthesis of heavier elements by capturing ^{4}He. The protons and neutrons are shown by blue and red colors respectively.

Figure 14.2b. (Bottom Panel): Formation of iron from fusion of heavy elements. Given that the abundances of heavy elements is lower and the likelihood of collision small, these are all r-processes (meaning rare-processes). The protons and neutrons are shown by blue and red colors respectively.

these are called rare processes (r-processes). However, processes in which a carbon nucleus captures a helium, forming oxygen, needs much lower temperature of about 200 million degrees Kelvin and therefore, are more likely to take place (figure 14.2a):

$$^{12}C + {}^4He \rightarrow {}^{16}O + energy.$$

Similarly, an oxygen nucleus (^{16}O) could capture a ^{4}He and produce ^{20}Ne:

$$^{16}O + {}^4He \rightarrow {}^{20}Ne + energy.$$

In summary, processes involving helium capture are more frequent compared to those that involve two heavy nuclei (figure 14.2a and 14.2b). These are called slow processes (s-processes). As a result, elements with mass numbers in multiples of four (the helium mass number; ^{12}C, ^{16}O, ^{20}Ne, ^{24}Mg, ^{28}Si) are more abundant and very stable (figure 14.3). Elements with intermediate mass numbers are formed when protons and neutrons are released from their parent nuclei and absorbed by others. For this reason, the abundance of these elements is not as high as those created by helium capture, the alpha elements.

By the time ^{28}Si is formed in the core of a star, the temperature inside the star is so high (around 3 billion degrees Kelvin) that it dissociates the heavier nuclei to their components, namely many helium nuclei. For example, the high-energy photons break ^{28}Si to seven ^{4}He nuclei. These in turn are captured by other nuclei, creating new elements through the alpha process, such as ^{56}Ni:

$^{28}Si + 7\,(^{4}He) \rightarrow {}^{56}Ni + energy.$

However, ^{56}Ni (Nickel) is radioactive and unstable, decaying to ^{56}Co (cobalt) and then into the most stable of all the elements, iron (^{56}Fe). This is the end result of the alpha process, ending in iron at the core of stars (figure 14.2b). The binding energy at the nucleus of iron (the energy needed to bind twenty-six protons and thirty neutrons) is larger than any other element. For this reason, iron is very stable, needing excessively large amount of energy to convert to yet other heavier elements. These energies do not exist even at the core of the most massive stars (Bennett et al. 2007; Schneider and Arny 2015).

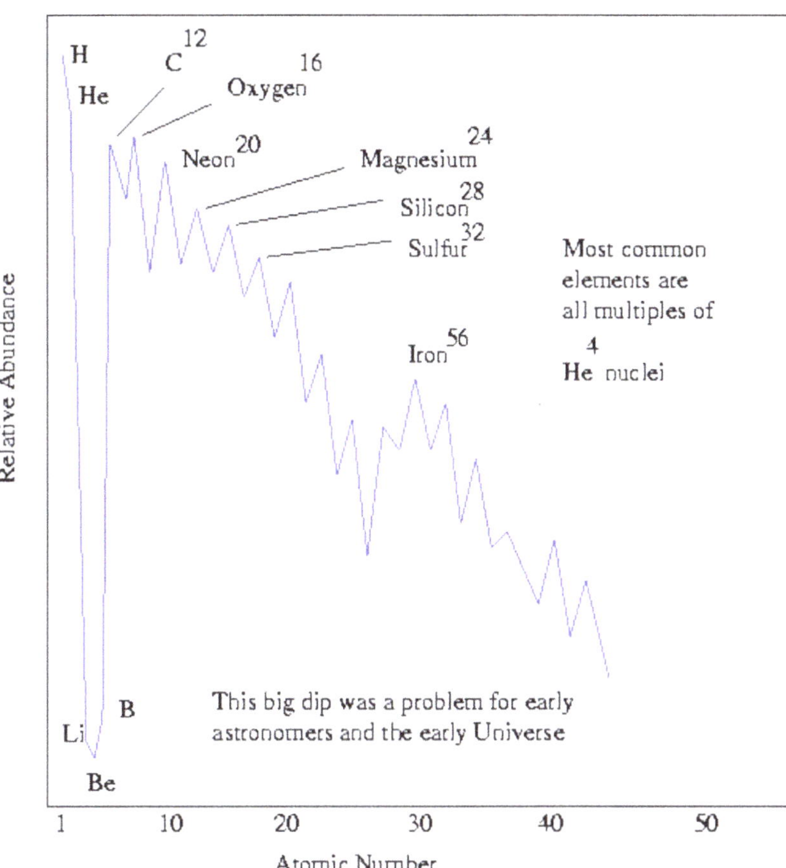

Figure 14.3. Elements with mass numbers (the number of protons and neutrons) in multiples of four are formed by capturing a ^{4}He. These are the most abundant elements because there is a higher likelihood of a ^{4}He capture by heavy elements to synthesize heavier elements than two heavy elements fusing together.

To synthesize heavy elements, one needs to have a constant supply of ^{4}He. There are two ways to produce this. One is by fusion of four hydrogen nuclei (figure 14.1) in low- or high-mass stars. The other dominant process occurs only at the core of massive stars (stars with masses in excess of 1.3 times the mass of our sun), through the Carbon-Nitrogen-Oxygen (CNO) cycle (figure 14.4). In this process a helium nucleus, two positrons, and two electron neutrinos are produced through a chain of reactions using carbon, nitrogen, and oxygen as catalysts (Schneider and Arny 2015).

ELEMENTS HEAVIER THAN IRON

In order to form elements heavier than iron, a different process other than helium capture must be invoked. At the core of evolved stars, nuclear reactions take place, producing neutrons as by-products. Neutrons do not have electric charge and can pass through atomic nuclei unimpeded by the repulsive force due to protons. The heavy elements therefore initiate a process called *neutron capture*, in which one neutron is added to their nucleus, producing a heavier isotope of the same element. For example, through the neutron capture process ^{56}Fe (iron-56)

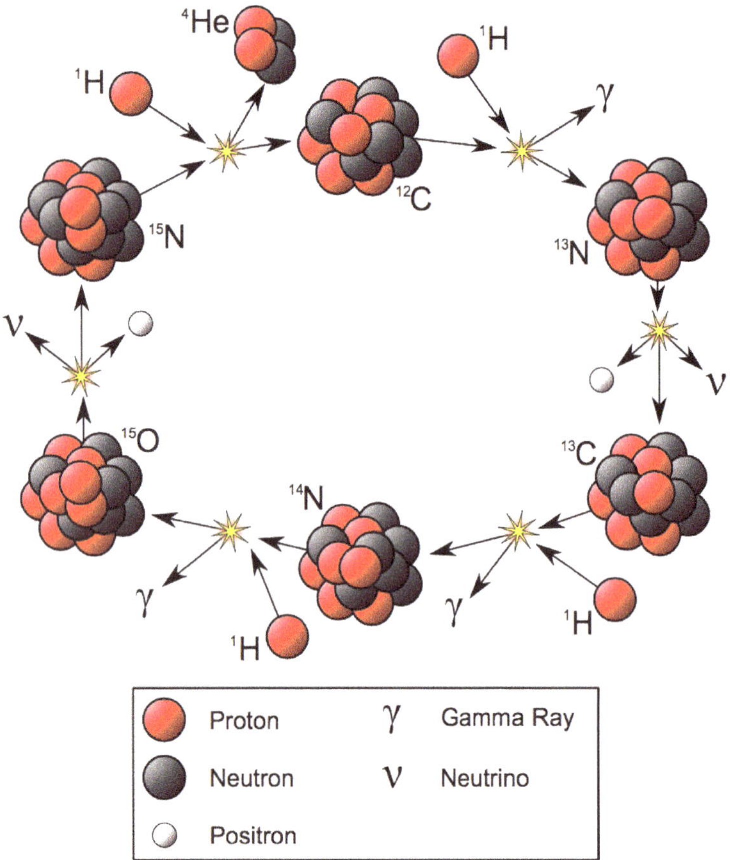

Figure 14.4. Example of a Carbon-Nitrogen-Oxygen (CNO) cycle, producing helium. This process is temperature sensitive and is dominant in stars with masses more than 1.3 times that of our sun.

nucleus can capture a neutron to become ^{57}Fe (iron-57) isotope. The neutron capture process again converts ^{57}Fe (iron-57) to ^{58}Fe (iron-58) and then ^{59}Fe (iron-59). Some of these isotopes are unstable, decaying to other new elements not formed before. It takes about one year for a neutron to be captured, giving enough time to the isotopes to decay into other elements before capturing another neutron. Because these processes take place slowly, they are called *s-processes*. Elements like copper, silver, and gold are produced through the s-processes (Box 14.2). The heaviest stable (nonradioactive) element produced through this process is ^{209}Bi (bismuth-209). (Box 14.3). Any element heavier than bismuth produced through neutron capture is radioactive, decaying to lighter elements and eventually to ^{209}Bi. A different physical process is responsible for the formation of the heaviest of the elements (Schneider and Arny 2015).

THE ORIGIN OF THE HEAVIEST OF THE ELEMENTS

The heaviest elements were formed during the violent supernova explosion, following the death of a star. These explosions are among the most powerful in the universe (after the big bang), breaking the nuclei of heavy elements and releasing significant number of neutrons. These neutrons are rapidly captured by the elements present during the supernova explosion. This happens so fast that the unstable elements do not have enough time to decay. Also, in many cases multiple neutrons are captured by the light elements leading to the formation of heavy elements. Therefore, the heaviest of the elements were formed during a very short time (about fifteen to thirty minutes) of extreme temperature during the explosion of supernovae. Because of the very short time during which the conditions are suitable for the formation of these elements, the heaviest of the elements are the least

BOX 14.2: EVIDENCE FOR STELLAR NUCLEOSYNTHESIS

The element technetium-99 is radioactive with a half-life of about two hundred thousand years. This means that almost all this element has long decayed, with no trace of it expected on Earth. However, astronomers doing spectroscopy have found evidence for it in red giant stars, indicating that it is produced at the core of these stars through neutron capture—the only known way this element could have been formed. This is taken as experimental evidence for the *s-process*.

abundant in the universe. In general, elements heavier than iron are about a billion times less abundant than hydrogen and helium. The heavy elements ^{238}U (uranium-238) and ^{242}Pu (plutonium-242) were formed through this process.

The conclusion from the above discussion is that different elements have different origins, depending on how heavy they are. Some elements are more abundant than others (those produced by capturing an alpha particle (^4He) with mass numbers that are multiples of four). Elements heavier than iron could only be synthesized through neutron capture at the core of massive stars or in supernova explosions. These are slow processes and therefore the elements produced through them are rare. The heaviest of the elements can only be synthesized by neutron capture in sites of supernova explosions. The origin of the elements is summarized in the periodic table in figure 14.5 (Bennett et al. 2007).

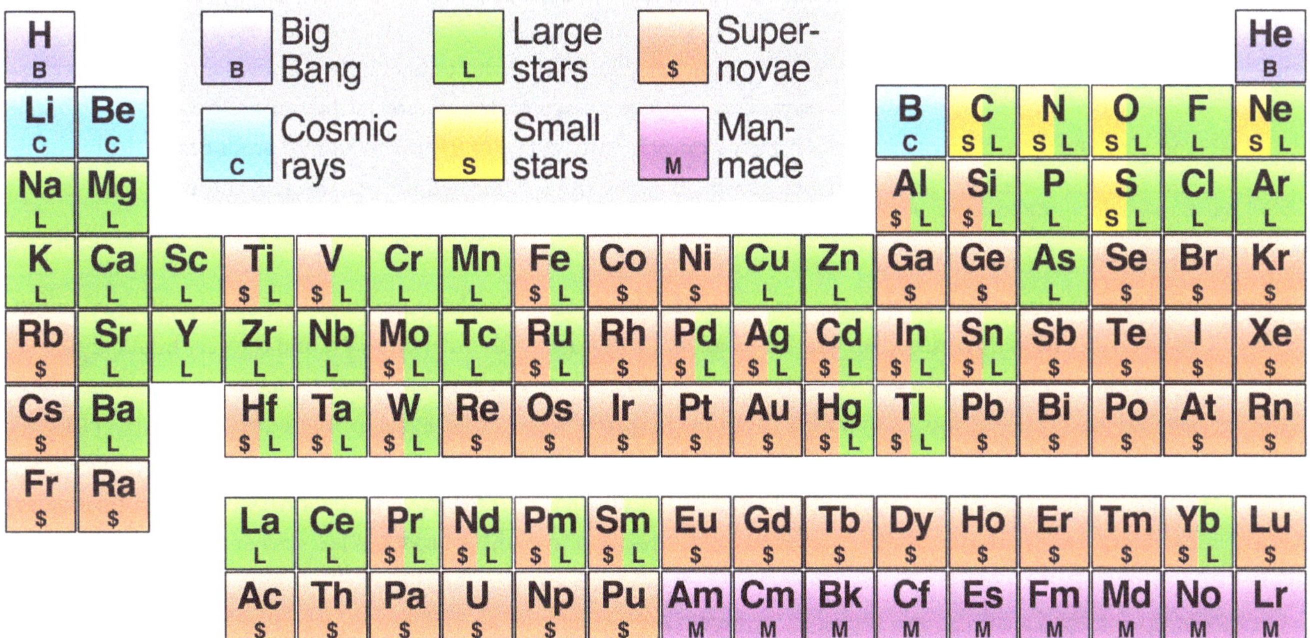

Figure 14.5. The periodic table with the origin of different elements indicated. Large- and small-mass stars have different evolutionary histories and therefore produce different elements.

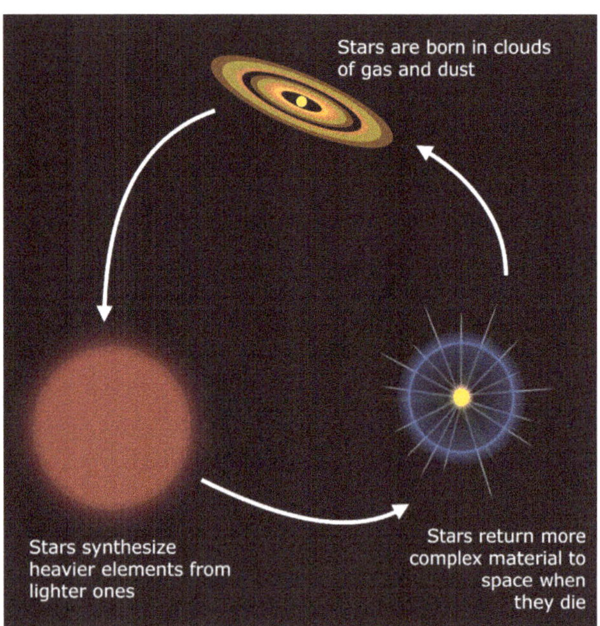

Figure 14.6. The cycle showing synthesis of heavy elements in stars and the enrichment of the interstellar medium when these elements are scattered in space because of supernova explosion.

ENRICHMENT OF THE INTERSTELLAR MEDIUM

New stars within a galaxy are born in the medium between the existing stars. These are formed from the primordial material (hydrogen and helium) generated at the big bang (first generation stars) or from the processed material (rich in heavy elements) produced through the stellar evolution (second generation stars) and thrown out to the interstellar medium by supernova explosion (figure 14.6). As a result, the second generation of stars that were formed from the processed material, are rich in heavy elements. These materials were produced in stars and were scattered in the space between the stars after the death of their host star through supernova explosions. The material that are distributed in the interstellar medium are responsible for the composition of the planets in the solar system, Earth, and the life it harbors. Therefore, each new generation of stars is more enriched in the heavy elements than the previous generation. The old stars (globular clusters) are less enriched in heavy elements than younger stars that are more recently formed from the processed material (Schneider and Arny 2015).

SUMMARY AND OUTSTANDING QUESTIONS

The interstellar medium in a galaxy contains large amounts of gas and dust that host sites of star-formation activity. The gas clouds collapse under their gravity, increasing the density and temperature at their core. Once the critical value of density and temperature is reached, fusion takes place, forming stars. In the center of stars four ^1H nuclei combine and form a ^4He. The interstellar gas clouds mostly consist of hydrogen atoms (produced at the big bang) and therefore, there are no shortage of hydrogen at the time. After helium was formed, through a step-by-step fusion process, heavier elements are formed. Two ^4He nuclei combine to form ^8Br (beryllium), and this fuses with another ^4He to form ^{12}C and so on, until ^{56}Fe is produced.

The elements produced through fusion of ^4He (and hence have mass numbers in multiples of four) are the most abundant as these processes take place fast and rapid due to abundance of ^4He and the very high density and temperature at the core of stars, needed to fuse them. These are called the *alpha processes*. The case of the beryllium is interesting as it was also formed soon after the big bang and since it is unstable, immediately decayed. However, at the core of stars due to higher density, before beryllium had time to decay, it was fused with another ^4He to produce ^{12}C. This is the most efficient way carbon can be produced. Other less abundant heavy elements that are not produced by fusion with ^4He are produced by neutron capture. These processes are slow and are therefore called s-processes. This process produces unstable isotopes that decay to other heavy elements (with the total number of protons and neutrons not multiples of four). The heaviest of the elements are formed during the supernova explosions that signify the death of stars. Depending on the mass of the progenitor star, different elements are cooked. The huge temperature at the time of supernova explosion breaks many heavy elements, releasing neutrons.

BOX 14.4: HOW MANY ELEMENTS ARE THERE IN NATURE?

There are about 115 different elements currently known. Out of this, 81 are stable, found on Earth, and constitute the basic matter content of the universe. There are 10 unstable (radioactive) elements on Earth. Due to steady decay, their abundance has diminished through time, and as a result, they are very rare today. There are a further 20 elements that have been artificially produced in laboratories. Four elements are either found in other stars but not on Earth or there is no experimental verification for their existence.

The neutrons will then be absorbed by other elements leading to the heaviest of the elements. These elements are very rare because the condition for their synthesis is only satisfied for a very short period of time (15–30 minutes) during the supernova explosion.

Given various processes that led to the formation of different elements, there are different sites responsible for their synthesis, ranging from the big bang (the lightest of the elements, H and ^4He), high-mass stars (elements up to iron), low-mass stars (elements lighter than carbon), and supernovas (the heaviest of the elements) (figure 14.5). This is the origin of different elements existing today on Earth, although some elements are only found to exist in space (Box 14.4).

It is not clear yet if there are still unknown elements waiting to be discovered or if the elements on Earth all arrived at the same time. Why some elements exist on in space and not on the Earth? Using large ground-based and space-borne observatories, astronomers will be able to look for heavy elements in more stars in our Galaxy and other galaxies to measure the metal abundance of these systems.

REVIEW QUESTIONS

1. Explain step-by-step process of formation of ^4He from ^1H.
2. What is an *alpha process*?
3. Describe different processes involved in production of ^{12}C. Explain which one is the more efficient method and why.
4. Where elements heavier than carbon were synthesized and how does this happen?
5. Why are elements with mass numbers (total number of protons and neutrons) in multiples of four more abundant and stable?
6. Explain the s-processes and the kind of elements they produce.
7. The high-energy photons at the core of massive stars break heavy elements to their constituents. How this process affects synthesis of heavy elements?
8. What is the CNO cycle? Explain its significance.
9. Explain formation of the elements heavier than iron.
10. Explain the process through which heavy elements are distributed in the interstellar medium.

CHAPTER 14 REFERENCES

Bennett, J., M. Donahue, N. Schneider, and M. Voit. 2007. *The Cosmic Perspective*. 4th ed. Boston: Pearson/Addison-Wesley.

Burbidge, E. M. Burbidge, G. R. Fowler, W. A. and Hoyle, F. 1957. "Synthesis of the Elements in Stars." Rev. Mod Phys. 29, 547.

Schneider, S.E., and T.T. Arny. 2015. *Pathways to Astronomy*. 4th ed. New York: McGraw–Hill.

FIGURE CREDIT

- Fig. 14.5: Based on information from https://en.wikipedia.org/wiki/File:Nucleosynthesis_periodic_table.svg.

THE ORIGIN OF THE PLANETARY SYSTEMS

CHAPTER LEARNING OBJECTIVES

This chapter will cover:

- The origin of the planetary systems
- Formation and characteristics of different types of planets
- The search for planetary systems beyond our solar system
- The habitable zone and conditions for life in other planets

In recent years enormous progress has been made in finding planets outside our solar system, thanks to advances in detector technology, the ever-increasing power of our telescopes, new space missions and development of new observing techniques. Due to the faintness of the planets, their small size, and dominance of planetary systems by light from their central star, it is very difficult to perform detailed study of planets external to our own, known as *exosolar* planets. However, this is very important, as the study of the nature and properties of other planetary systems will help us better understand our own solar system and its origin and to decipher the mystery of habitability. Given the above difficulties in studies of planetary systems, it is instructive to first understand the working of our own solar system.

For the theories of the origin of our solar system to be viable, they have to explain all its observed features. These theories also provide a starting point for studying exosolar planetary systems and their formation history. The first ever theory proposed for the origin of the solar system was by German philosopher Immanuel Kant (1724–1804) and French mathematician Pierre-Simon Laplace (1749–1827) in the 1700s. This is known as the *nebular theory*, which implies the solar system originated from an interstellar gas cloud. This remained as the only theory for the origin of the solar system until the first half of the twentieth century, when a competing scenario was developed, proposing that the planets in the solar system formed from debris due to close collision of the sun with another star. This is called the *close encounter theory*. This theory failed to explain some of the observed patterns in the solar system, while new observations of the planets in the solar system provided grounds

> "Many different planets are many different distances from their host star; we find ourselves at this distance because if we were closer or farther away, the temperature would be hotter or colder, eliminating liquid water, an essential ingredient for our survival"
>
> - BRIAN GREENE
>
> "The noblest pleasure is the joy of understanding."
>
> - LEONARDO DA VINCI

for revision of the nebular theory. As a result, a revised version of the nebular theory has been able to successfully explain many of the observed features in the solar system.

This chapter investigates the origin of the planetary systems, including our solar system. It presents scenarios for the origin of different types of planets and techniques to search for planets outside our solar system. The chapter also discusses the habitable zones, the region within a planetary system that could harbor life.

THE ORIGIN OF THE SOLAR SYSTEM

A successful theory for the origin of the solar system must accommodate the following observations (Bennett et al. 2007):

1. All the planets rotating around the sun and moving in the same direction
2. The planets and the sun are almost on the same plane
3. Two different types of planets located at different distances from the sun
4. The presence of comets and asteroids in the solar system and their location

There is ample evidence that stars form in large gas clouds contracting under the force of their gravity (chapter 12). The clouds were very cold with low density, formed from recycled material (in the interstellar medium) over billions of years and hence were rich in heavy elements. As they were extended over large radii, their surface gravity became too weak to have a significant role in their collapse. The initial collapse of these clouds was likely triggered by a shock wave caused by the explosion of a supernova and subsequently, was controlled by the force of gravity. Since the strength of the gravitational force depends on the inverse of the square of radius, as the system collapses (the radius becomes smaller), its inward velocity increases (the system collapses faster) under the increasing force of gravity.

As the gas cloud shrank, its gravitational energy was converted to kinetic energy, increasing the speed of individual particles in the system (and hence its temperature), following the law of conservation of energy. Once the density and temperature at the center sufficiently increased, fusion started and the sun was born at the center of the gas cloud. While the cloud collapsed, it also rotated. As its radius became smaller, the speed of rotation increased to withhold the law of conservation of angular momentum. The result of the rotation was that the material forming the gas cloud spread out, forming the plane of the solar system. During the collapsing process, the clumps in the gas cloud collided and formed larger clouds. As a result, the fragmented clumps obtained the average speed of the rotating system, converting the original clouds into inhomogeneous, flattened and extended rotating disk.

The rotating disk scenario explains all the characteristics of our solar system as listed above. The reason that all the planets orbit the sun in nearly the same direction is that they all were formed on the flattened disk. The direction of the rotation of the disk determines the direction the planets orbit the sun. Furthermore, the orbits of the planets are nearly circular because of the repeated collision between the materials on the disk during the time the disk collapsed (Bennett et al. 2007).

THE DIFFERENT TYPES OF PLANETS

Any theory for the origin of the planets (Box 15.1) should explain the presence of two very different types of planets observed in our solar system: small, rocky planets (these are similar to Earth and are called *terrestrial planets*) and large, gas rich planets (these are like Jupiter and are collectively called *jovian planets*). Given the uniform elemental abundance of the initial gas cloud, a correct theory should also explain how the solar system ended up with two types of planets with different compositions.

BOX 15.1: DEFINITION OF A PLANET

At the International Astronomical Union meeting in 2006, astronomers adopted a set of criteria for classifying a celestial object as a planet. Based on this definition, a planet should satisfy the following characteristics: (1) orbit the sun, (2) be large enough to sustain under its gravity, and (3) attract all the smaller bodies in its proximity. This definition of a planet has nothing to do with science and hence could change.

The terrestrial planets were formed on the inner part of the gas cloud where the temperature was very high while the jovian planets were formed in the cooler outer part. In the regions of the gas cloud where temperature is lower, the gas condenses (atoms and molecules bond without being broken due to the energy caused by high temperature). This process is called *condensation*. Given that different elements condense at different temperatures, the composition of the planets depends on their distance from the sun (figure 15.1). For example, materials consisting of hydrogen compounds (water vapor H_2O and methane CH_4) condense at low temperatures (around 150 degrees Kelvin), while rocky materials, which are in the gas form at very high temperatures, condense and turn to solids at higher temperatures (around 500 to 1,300 degrees Kelvin), and metals (consisting of iron and nickel) condense at even higher temperatures (around 1,000–1,600 degrees Kelvin). As a result, at the very inner part of the solar system, where the temperature was above 1,300 degrees Kelvin, none of the material could solidify and condense. As the temperature reached around 1,300 degrees Kelvin (roughly at the orbit of mercury), metals and some kind of rocks were condensed and turned into solid bodies, while other material were still in the form of gas (figure 15.1). Around the region where Venus, Earth, and Mars are, the temperature reached below 1,300 degrees Kelvin, allowing condensation of rocks and metals. Hydrogen compounds could only condense in the outer region of the solar system and beyond the *frost line*, where the temperature reaches 150 degrees Kelvin. The frost line is between the orbits of Mars and Jupiter and determines the division between the warm inner part and cold outer part of the solar system where the terrestrial (rocky) and jovian (gaseous) planets are separated (figure 15.1). Hydrogen and helium that contain 98 percent of the gas cloud never

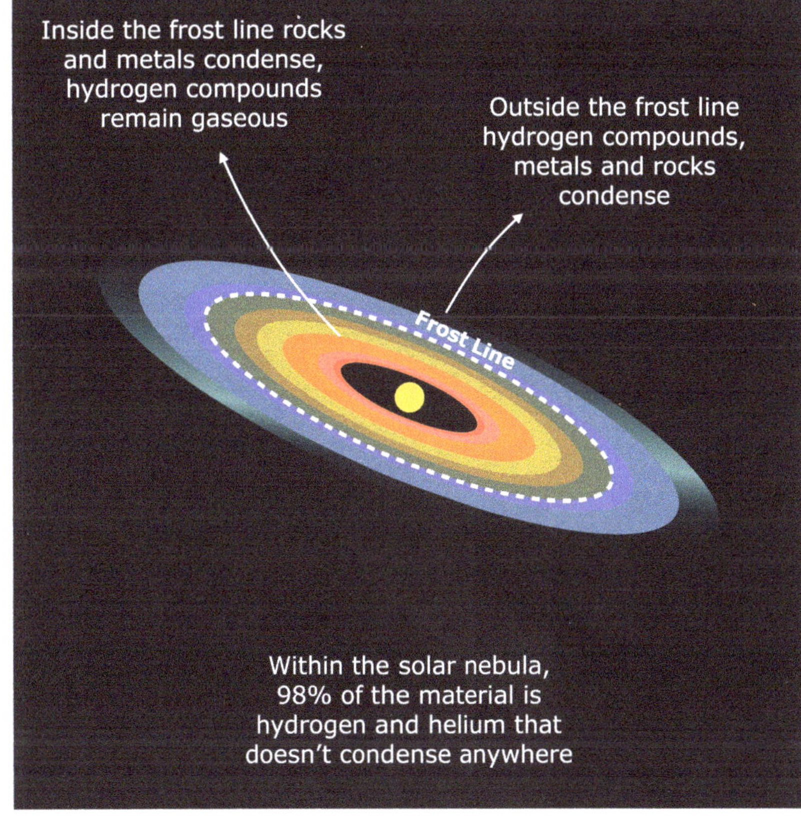

Figure 15.1. Terrestrial (rocky) and jovian (gaseous) planets at different locations on the planetary disk. Rocks and metals condense in the inner part of the disk (where temperature is higher) since metals and rocks can condense at high temperatures. Hydrogen and helium condense in the outer part (lower temperature). The line dividing the high- and low-temperature areas represents the *frost line*.

condense and hence are only found in the form of gas throughout the solar system. Therefore, the condensation process leads to the formation of seeds with different elemental abundances, depending on their distance from the sun and hence their temperature. The seeds will grow by attracting more matter under their gravity, eventually forming the planets we see today (Bennett et al. 2007) (Box 15.1).

THE ORIGIN OF THE PLANETS

The previous section introduced the two types of planets—terrestrial and jovian. Terrestrial planets are small, were formed in the inner parts of the solar system, and are made up of metal and iron. Jovian planets are large, are found in the outer parts of the solar system, and are dominated by gaseous hydrogen compounds. The fundamental distinction between the two types of planets and their different compositions imply that they have had different formation histories (Schneider and Arny 2015).

Terrestrial planets are relatively small because their main constituents (metal and iron) are very rare in the solar system. The seeds formed by condensation of these materials gradually attract other small particles as they orbit the sun and grow in mass and size. Due to their small size, their gravity is weak and does not play a major role in attracting other bodies until the system further grows. At this point the seeds grow and gravity eventually takes over and larger and smaller objects attract one another, growing in size and forming *planetesimals*, meaning "parts of planets" (figure 15.2). Collision between the planetesimals is violent, since they move fast. This results in the destruction of smaller systems and a change in their orbits. As a result, only larger planetesimals survive and orbit the sun (figure 15.2; Chaisson and McMillan 2011). Small planetesimals that were not attracted by larger ones are currently floating in the solar system and are called *meteorites*. Every now and then, some of these meteorites pass through Earth's atmosphere and reach us. They reveal traces of heavy elements.

Jovian planets contain significant amounts of ice as well as metals and rocks. They also attract hydrogen and helium gas (which are the most abundant elements in the solar system) and this is the reason for their large sizes.

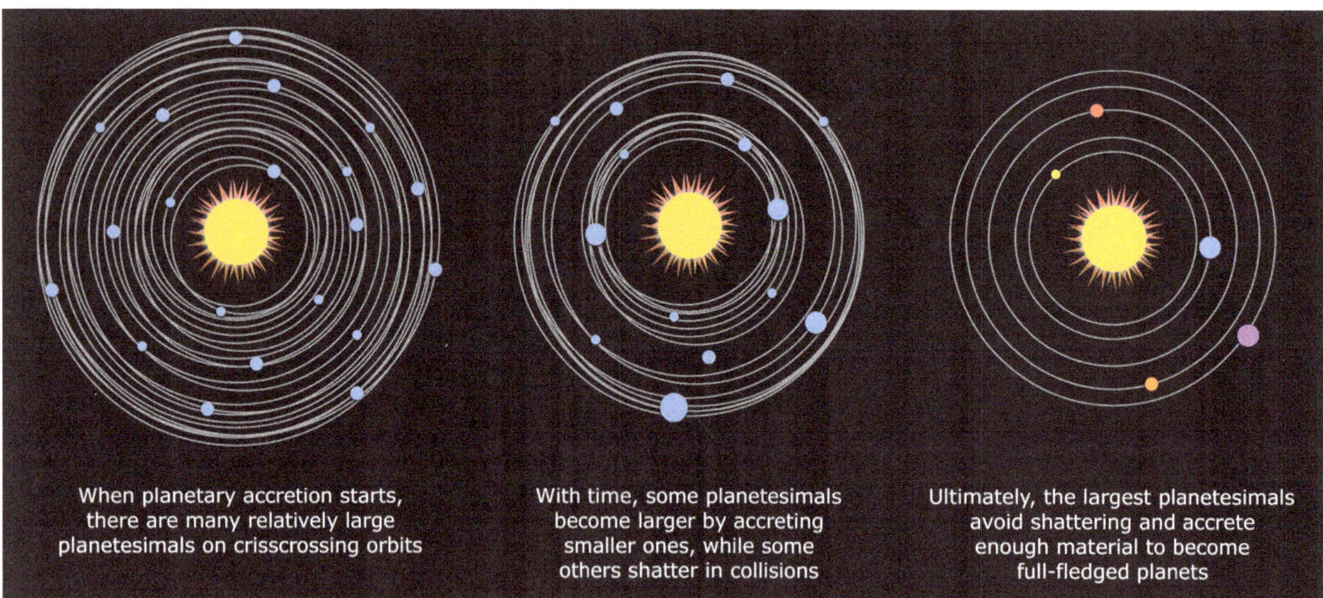

Figure 15.2. From left to right, larger planetesimals accreting the smaller ones and growing bigger. Given the larger mass of the outer planetesimals, they have stronger gravity and attract more of the smaller planetesimals.

The first step toward formation of these planets is the buildup of icy cores. This increases the mass of planetesimals to many times that of the Earth. Because of their large mass, they then attract hydrogen and helium gas and grow larger, attracting more gas. This scenario successfully explains the composition and size of the jovian planets and why they are located in the outer parts of the solar system. The planetesimals that failed to join others to form planets turn to *comets* that are moving within the solar system (Box 15.1).

Today little gas is present in the solar system because the solar winds blow them off. The objects existing in the solar system (apart from two kinds of planets discussed above) are asteroids and comets. Asteroids are rocky objects with similar composition as the terrestrial planets and are found in a region between planets Mars and Jupiter, forming the *asteroid belt*, while comets are found in the outer reaches of the solar system (around Neptune) contain a lot of ice and gas (the same composition as jovian planets) and are in a region called the *Kuiper belt* (Bennett et al. 2007).

THE SEARCH FOR EXTRASOLAR PLANETARY SYSTEMS

Study of the nature and properties of exosolar planets allows better understanding of the evolution of our own planet. By searching for earth-like planets, we could study the conditions under which a planet like Earth formed and hence see the Earth as it was at the beginning of its formation. However, there are a number of challenges one needs to overcome in order to find extra solar planets. First, the extent and size of the planetary systems in the interstellar medium are extremely small compared to cosmic distances, and hence finding them requires wide and deep surveys specifically designed for this purpose. Second, the stars at the center of the planetary systems are significantly more luminous than the planets themselves. As planets don't shine by themselves and only reflect the light from their star, they are relatively faint and one needs to develop special techniques to find them. Third, it is a real challenge to resolve the orbit of a planet that is moving around its star. This requires observations with extremely high resolution from space. In the following we discuss techniques for finding exosolar planets (Bennett et al. 2007; Chaisson and McMillan 2011):

GRAVITATIONAL TECHNIQUE

In this technique the gravitational effect of a planet on its star is measured by monitoring the orbit of the star around which the planet revolves (figure 15.3). Although this becomes extremely complicated in multi planetary systems, it could be reliably applied on single planetary systems. The problem with this technique is the time baseline it requires to monitor the orbit of the star and its movement and the need for extremely

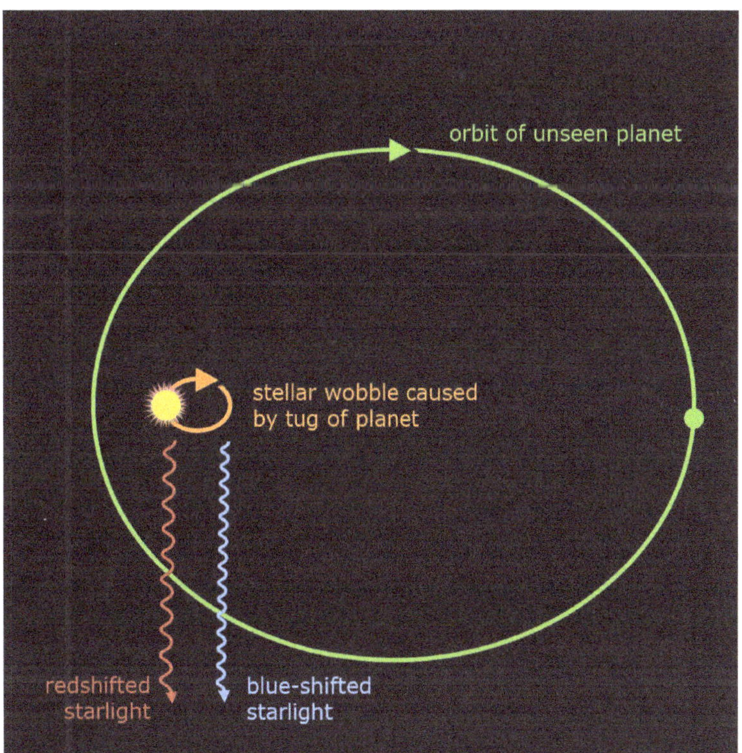

Figure 15.3. Shows perturbations in the orbit of the central star due to an "unseen" planet orbiting it. Due to the force of gravity from the planet, the star's orbit shifts toward and away from the observer due to blueshift and redshift.

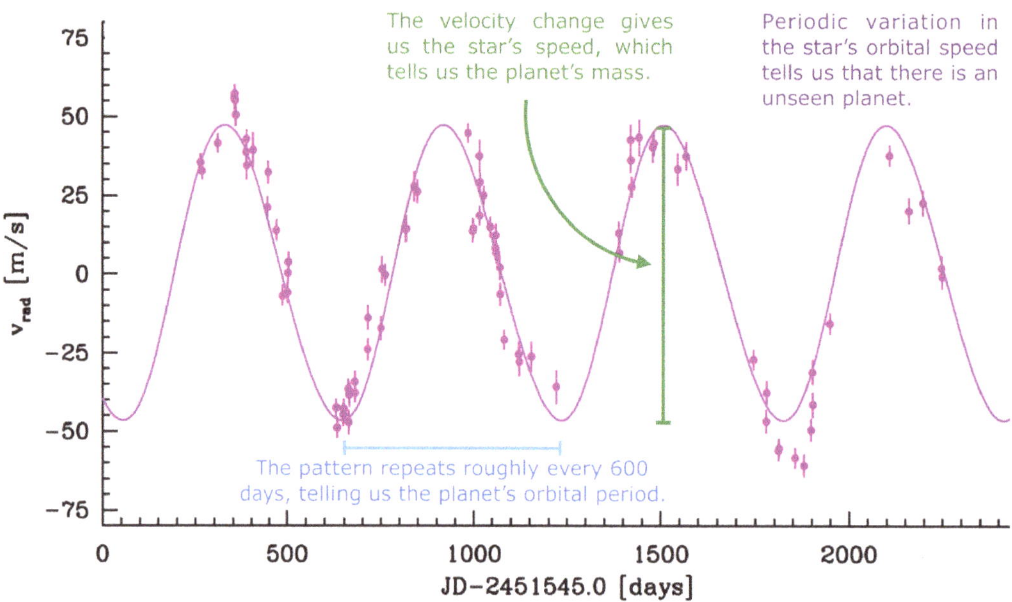

Figure 15.4. Change in the star's orbit due to gravitational pull of the planet orbiting it. The gravity from the planet changes the velocity of the star. This is measured with Doppler effect, which could then be used to estimate the mass of the planet, its distance from the star, and its orbit.

accurate measurement of its location at any given time. Furthermore, depending on the size of the planet's orbit and its distance from the central star, the perturbation in the star's orbit may be too small to be detectable.

DOPPLER TECHNIQUE

This is based on the shift in the spectral lines caused by the central star's motion (Doppler effect). As a planet moves around a star, it causes the star to move alternatively toward or away from us, with its spectral line shifting to the blue and red wavelengths respectively. This technique allows astronomers to measure the speed of the planet, its

BOX 15.2: THE KEPLER SATELLITE AND SEARCH FOR EXOPLANETS

The Kepler satellite (named after German astronomer Johannes Kepler) was launched in 2009 with the aim of finding planets outside our solar system and within the Milky Way galaxy. The photometric detector onboard the Kepler continuously monitored 140,000 main sequence stars, looking for changes in their brightness caused by passing of a planet in front of the star and along our line of sight. Kepler has found 1,042 exoplanets, including systems with multiple planets. The main aim was to identify planets within the habitable zone of stars. So far, four planets are found within the habitable zone of stars with a size close to that of the Earth.

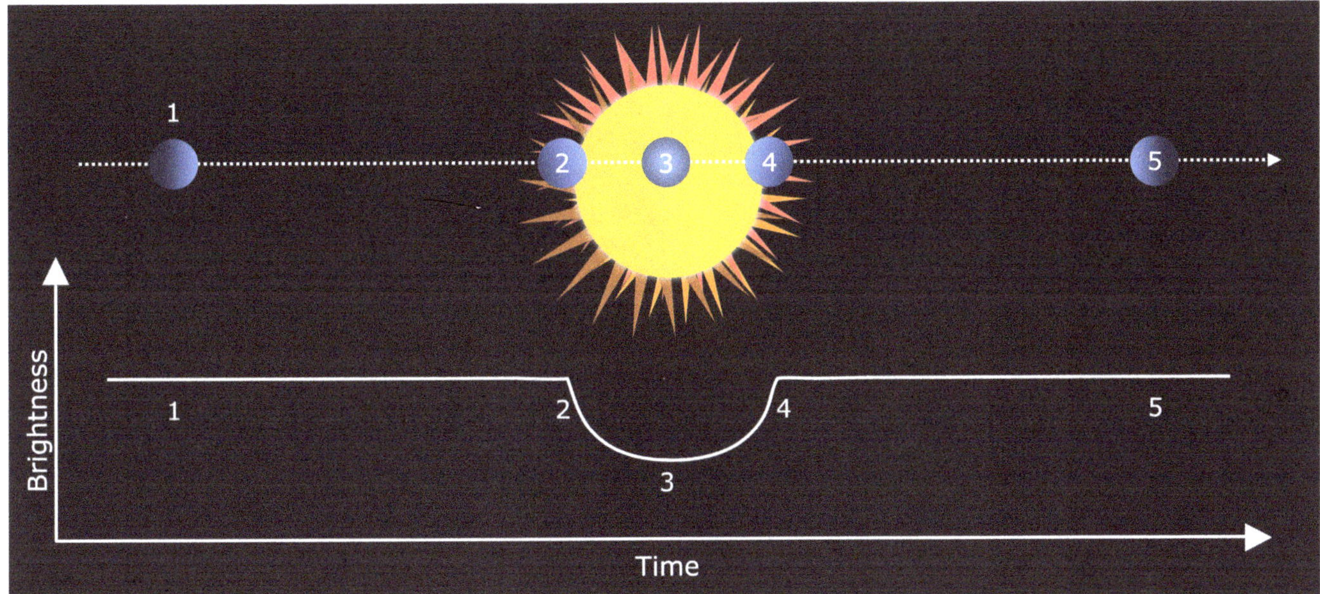

Figure 15.5. Dimming of the starlight when a planet passes between the observer and the planetary system. The size of the drop in the starlight is proportional to the physical size of the planet, whereas its duration measures the speed of the planet. The numbers show the positions of the planet at any given time when it passes in front of the star.

mass, the shape of its orbit and the distance from its star (figure 15.4). The amount of the Doppler shift is used to measure the speed of the planet while the degree of symmetry of its light curve reveals the shape of its orbit. The size of the Doppler shift gives the mass of the planet (more massive planets cause larger shifts).

TRANSIT TECHNIQUE

This technique is widely used and relies on dimming of the starlight when a planet is aligned between the star and our line of sight. This results in slight changes in the luminosity of the star, revealing the presence of a planet moving in front of it (figure 15.5). Periodic change in the brightness of the star indicates the motion of the planet passing in front of the star. The duration of the time when the brightness has dropped (figure 15.5; points 2-3-4) is proportional to the speed of the planet while the amount of the dimming (compared to the luminosity of the star) is proportional to the size of the planet (figure 15.5; point 3).

CONDITIONS FOR PLANETS TO HARBOR LIFE

To harbor life, a planet must satisfy the following conditions:

- It must be associated with a star with a long enough age to allow life to develop. This excludes hot, high-mass stars that have short lifetimes, restricting the search to a class of low-mass stars (a few times the mass of the sun) only.
- It must reside within a region at a distance from its star to allow water in liquid form (that is not too hot or too cold). There is a region called the *habitable zone* where water is found in liquid form (figure 15.6). For low-mass stars the habitable zone is closer to the star, while for higher-mass stars, it is further away. Therefore, the size and location of the habitable zone depends on the luminosity (mass) of the central star. As the luminosity of a star changes with time, the size and location of its habitable zone changes too.

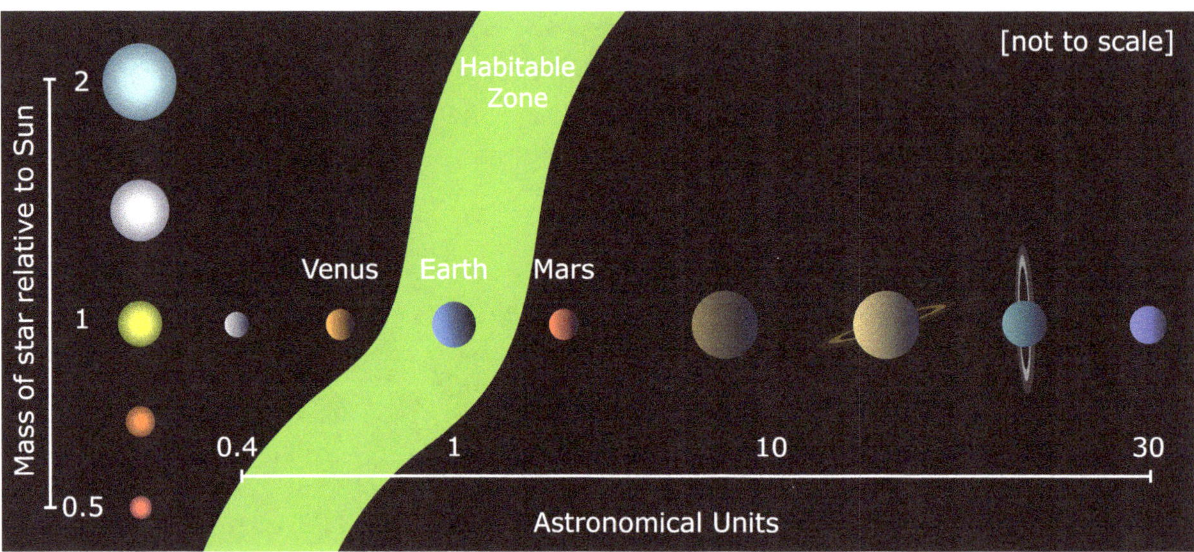

Figure 15.6. The Earth's habitable zone, defined as the region around the Sun where water could be found in liquid form. Therefore, if there is a planet suitable for supporting life (as we know based on our definition of life), it must reside within the habitable zone of its star. The Earth is in the habitable zone of the Sun.

- It must have a mass sufficient to retain an atmosphere. The pressure in the atmosphere should also be sufficient to keep water in liquid form.
- It must contain the chemical composition needed to support life.
- It must be a terrestrial planet and hence closer to its central star.

The Kepler satellite has discovered many planets that resemble the Earth (Box 15.2). For example, Kepler-186f is found to orbit a faint red dwarf star about 490 light-years from Earth. The planet is bigger than the Earth, but the distance from its star and the size of the planet suggests that it could contain liquid water.

SUMMARY AND OUTSTANDING QUESTIONS

The prevailing scenario for the origin of the solar system is the nebular theory that proposes formation through collapse of rotating gas clouds in the interstellar medium. The gas may be pristine (only containing hydrogen and helium and not enriched by heavier elements) or constitute processed material from previous generation of stars rich in heavy elements. The planets were formed within these clouds from the condensed gas collapsed under the force of gravity. These are called the planetesimals that attract other systems, growing in size to form the planets we see today. This scenario explains current observations as to why the planets are all located on the same plane, moving in the same direction and are separated depending on their mass, size, and composition.

There are two types of planets—terrestrial and jovian. The terrestrial planets are formed closer to the sun where, because of the extreme heat (around 1,300 degrees Kelvin), only iron could condense to form the core of the planet. Most of the elements at this temperature are melted and found in gas form. The jovian planets, on the other hand, are located further away from the sun, are much larger than the terrestrial planets, are more massive, and mostly contain gas.

Detailed knowledge is only available for planets in our solar system. For planetary systems beyond our own, the limited spatial resolution to finely separate planets from their central star, dominance of the light by the star (making the planets invisible) and faintness of the planets will make it very difficult to probe their nature. The exosolar planets, as they are called, are found through gravitational interaction with their star or motion of the planet in front of the star, producing a dip in the star's light curve. These techniques have proved very successful in finding exosolar planets.

The Kepler mission launched by NASA in 2009 (Box 15.2) used the planet finding techniques and surveyed 140,000 stars, looking for a change in their luminosity. It found over 2,300 planets, of which about 700 are Earthlike (having the size and the mass of the Earth with a similar distance from their star as Earth is from the sun).

The study of the nature and atmosphere of planets is a hot topic today. Indeed, the first step toward looking for life beyond Earth is to search for planets with conditions that could support life. Of course, we could only find planets within the vicinity of our solar system, given the limited power of our telescopes and detection techniques. The challenge is to find external planets in the habitable zone of their star, where water could be found in liquid form. The size and location of the habitable zone depends on the characteristics of the central star. It is not clear how many Earthlike planets exist or how many of them have atmosphere. There are many fundamental questions to be addressed, among them are: What fraction of the discovered planets so far are in the habitable zone of their central star? What fraction of the planets have an atmosphere and what is the composition of their atmosphere? What is the nature of the planets located in multiplanetary systems? Do planets in other multi-plane systems follow the same pattern as those in the solar system? How biased are the current techniques used to identify planets? These questions and many more will be addressed using observations made by the Transiting Exoplanet Survey Satellite (TESS), which will specifically be looking for bright planets in the habitable zone of their star. Being located in the habitable zone increases the likelihood of the planets having conditions suitable for life, while brighter luminosity makes it possible to study the atmosphere (and composition) of the planet in more detail.

REVIEW QUESTIONS

1. What are the observed characteristics of the solar system that need to be satisfied by a successful theory for the origin of the planetary systems?
2. How did the plane of the solar system form?
3. Explain the characteristics of the two types of planets.
4. Briefly describe the origin of the different types of planets.
5. What are planetesimals?
6. What is the frost line? How does it depend on the characteristics of its associated star?
7. Explain different techniques used for finding exoplanets.
8. How are the physical properties of the exoplanets (mass, orbit, and size) measured?
9. What are the conditions a planet needs to have in order to support life?
10. Explain the habitable zone of planetary systems.

CHAPTER 15 REFERENCES

Bennett, J., M. Donahue, N. Schneider, and M. Voit. 2007. *The Cosmic Perspective*. 4th ed. Boston: Pearson/ Addison-Wesley.

Chaisson, E., and S. McMillan. 2011. *Astronomy Today*. New York: Pearson.

Schneider, S.E., and T.T. Arny. 2015. *Pathways to Astronomy*. 4th ed. New York: McGraw-Hill.

FIGURE CREDIT

- Fig. 15.4: Adapted from Sabine Reffert, et al., "Precise Radial Velocities of Giant Stars II. Pollux and its Planetary Companion." 2006.
- Fig. B15.2: Source: https://commons.wikimedia.org/wiki/File:Kepler_spacecraft_artist_render_(crop).jpg.

CHAPTER LEARNING OBJECTIVES

This chapter will cover:

- Origin of the Earth
- Formation of the internal structure of the Earth
- Origin of the magnetic field of the Earth
- Heavy bombardment

Earth has had a chaotic and often violent history. The young Earth was much less hospitable and a target for meteorites and other celestial bodies. However, the luxury when studying the Earth is that, unlike the universe, galaxies, and stars, resources on the Earth are directly accessible for experimentation in laboratories without the need to invent indirect methods to study them.

Earth is very different from other solar system planets, with a moderate temperature, abundance of water, and an atmosphere and ecosystem that have allowed life to develop and evolve for billions of years. None of the other planets have these combined characteristics. There are many factors influencing Earth to be what it is today. Its distance from the sun and the fact that it is within the habitable zone of the solar system, its mass that is large enough to retain its atmosphere, the chemical composition that contains the chemistry needed for life to develop, have all made Earth to be the hospitable place it is—the only such planet in the solar system, perhaps within a very large radius from it or the entire Milky Way Galaxy! Furthermore, Earth's location with respect to other bodies in the solar system and their combined gravity, has provided it with a stable orbit around the sun. By addressing questions about the early evolution of the Earth, we will indeed be able to find more about the past history and present state of our home planet. This is extremely important, as Earth is a unique planet and the only one that is known to have supported life for billions of years and for sure, hosts intelligent inhabitants. The study of the nature and the physical conditions under which Earth was formed gives us clues toward understanding other planets that may one day host life, as our Earth did.

"Sunlight fell upon the wall; the wall received a borrowed splendor. Why set your heart on a piece of earth, O simple one? Seek out the source that shines forever"

- RUMI

"Personally, I do not know whether humankind is alone in this vast universe. But I do know that we should cherish our existence on this precious speck of matter … the greatest gift that could be bestowed upon us. For all practical purposes, there is only one planet Earth"

- BAN KI-MOON

This chapter studies the formation of the planet Earth, its origin, and its early history. It explores the internal structure of Earth and how it came about, as well as the source of many of its characteristics. The age of the Earth will be estimated and the origin of the Earth's magnetic field will be studied. The chapter will discuss the heavy bombardment and its impact on the Earth.

FORMATION OF EARTH

Earth is a terrestrial planet formed by accretion of microscopic particles in the initial gas cloud. These particles result from condensation of gas in the solar nebula (when the temperature drops as a result of increasing distance from the sun). These then followed an orderly motion around the newly formed sun (figure 15.2). At this point the mass of these particles was too small to affect one another through gravitational attraction and the main force bringing them together was the electrostatic force (the same as "static electricity" generated by rubbing a plastic ruler on our hair attracts pieces of paper). As the size and mass of the individual particles grew, they started to interact through the force of gravity and attracted more such systems, forming *protoplanets* or *planetesimals* (chapter 15). Because of the increase in their surface area, the probability of collision of planetesimals increased and hence, they rapidly grew (within 1 million to 10 million years) in both mass and size, forming planet-like systems.

For the four planets closest to the sun, which includes Earth, the temperature was so high that it led to most of the *volatile material* (materials that are easily converted to gas) evaporating. This, combined with solar winds and radiation pressure from the sun, blew out most of the light elements (hydrogen and helium). Therefore, these planets mostly consist of heavy elements that include rock-forming silicates as well as metals like iron and nickel. Measuring the age of the meteorites, geologists estimated this to have taken place around 4.56 billion years ago.

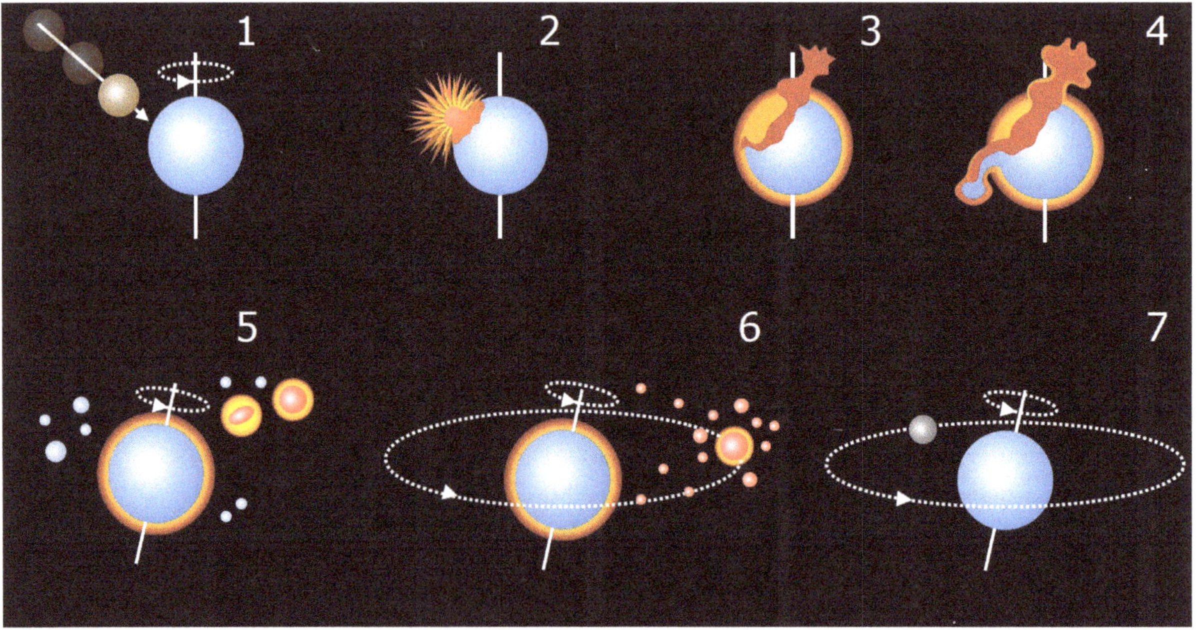

1 & 2.
A planetary-mass object collided with Earth 4.3 billion years ago.

3 & 4.
The impact propelled a giant shower of debris from the object and Earth into space.

5.
The impact tilted the Earth's axis by 23 degrees and sped up its rotation.

6 & 7.
The debris coalesced into the Moon and the Earth re-formed.

Figure 16.1. The giant impact by a Mars-sized planet soon after Earth was formed. This impact put Earth in its present orbit, tilted the orbit of Earth by 23.5 degrees with respect to the vertical axis, and ejected a part of Earth that is now the moon.

As the planetesimals collided and merged with the planet, their kinetic energy was converted to heat, melting the material. This, along with the heat generated through decay of radioactive material (uranium, thorium, cobalt, and potassium) is the origin of the extreme heat now present at the center of the Earth.

Around 4.51 billion years ago, the molten Earth experienced a catastrophic impact by a Mars-sized object. The consequence of this was three major events that dictated the future evolution of Earth and, later, life (figure 16.1). The impact increased the speed of Earth's rotation around the sun and tilted its axis of rotation from vertical (with respect to Earth's orbital plane) to 23.5 degrees inclination (this is why we have different seasons); it also put Earth on its current orbit and ejected a piece of Earth, forming the moon in an orbit around the Earth (this stabilized the Earth's orbit)- (figure 16.1). This is consistent with the age of the moon (from the rocks brought back by the Apollo astronauts), which is estimated as 4.47 billion years (Bennett and Shostak 2006). As catastrophic as this early impact appears to be, it stabilized Earth's orbit and provided the conditions for Earth to be the planet it is. The timeline for the formation of the Earth is presented in Table 16.1.

Table 16.1. Timeline for the early formation of Earth, first column indicates the years after

Time (years)	Main Events
0	A rotating cloud of gas and dust appeared.
10,000	The sun was formed and separated from the cloud.
100,000	Protoplanets were formed through accretion.
1000,000	The inner four terrestrial planets were formed.
1000,000,000	The last protoplanets were cleared; the last major impacts took place.

Source: Wood 1979.

THE AGE OF EARTH

Today, we have different techniques at our disposal to measure the age of the Earth. A straightforward way is to study the mineral grains of zirconium silicate or *zircons* (Box 16.1). Although these reside in much younger sedimentary rocks, radioactive dating based on uranium isotopes in them indicates that some zircons solidified around 4.4 billion years ago. Further studies show that zircons date back to the time when continents started to form, indicating that Earth's crust started to separate from its internal structure about 4.5 billion years ago (Box 16.1).

The rocks brought back to Earth from the moon show an age of more than 4.4 billion years. This is significantly older than the age of the rocks on Earth. The oldest rocks on Earth are found to be 4 billion years old. However, some of the earlier rocks on Earth may have been melted or changed in such a way that they cannot be accurately dated. If the moon was separated from Earth through a collision with a body the size of Mars, as discussed in the last section, it must be slightly younger than Earth, and the difference in the age of their rocks only reveals the changing geological landscape on Earth compared to the moon. This puts the age of Earth in excess of 4.5 billion years.

An upper constraint to the age of Earth is through the age of the solar system as found from the age of the meteorites. These are all found to have the same age, confirming the scenario that they were all formed at almost the same time. As they are the left over material from the very beginning of the solar system, they provide an upper limit to the age of Earth. The age of the meteorites is found to be around 4.57 billion years.

BOX 16.1: THE OLDEST TERRESTRIAL MATERIAL

Geochemists have recently found *zircon* grains with an estimated age of 4.4 billion years in Western Australia. These are among the oldest known terrestrial material. Chemical analysis of zircon shows that they were formed in a cool environment and in presence of water. This finding confirms that Earth had cooled enough to form a crust around 100 million years after the planet was formed following the catastrophic impact. This also confirms presence of water (and oceans) soon after the Earth was formed.

Comparison of the isotopes in Earth, the moon, and meteorites show that Earth and the moon were likely formed about 500 to 700 million years after the first meteorites. This implies an age of about 4.5 billion years for Earth.

THE EARTH-MOON SYSTEM

According to the impact scenario, soon after its separation from Earth, the moon was much closer to Earth. This continued to be the case after the formation of the crusts, water, and oceans on Earth. At these small distances, the moon produced tidal force on Earth's oceans, causing significant tides every twelve hours. Such tidal friction slowed down the speed of Earth, while the gravity of Earth slowed down the rotational speed of the moon. As a result of slowing down Earth's speed, there were more days (defined by the time it takes for one complete rotation of Earth on its axis) in a year (defined as one rotation of Earth around the sun). For example, about 400 million years ago (during the Devonian period), there were four hundred days in a year. The Earth-moon system has been slowing down and moving apart by a few centimeters per year since the planets' formation (Bennett and Shostak 2006).

FORMATION OF DIFFERENT LAYERS OF EARTH

Over a century ago, geologists started to look into Earth's interior by probing earthquake waves—also called seismic waves (from the Greek word *seismos* for "earthquake"). At that point it was realized that Earth's interior was divided into concentric layers of different compositions (figure 16.2). Also, through the work of British physicist Henry Cavendish (1731–1810) in 1798, the average density of Earth was found to be around 5.5 g/cm^3, significantly higher than the density of iron rich rocks (around 3.5 g/cm^3). By going further inside Earth, due to the pressure from upper parts, the rock is squeezed to smaller volumes, further increasing the inner density (Marshak 2012).

The huge energy produced through the accretion and mergers caused the initial material to melt and move freely inside Earth. Since the melting temperature of iron and nickel is lower than the silica (silicon and oxygen; SiO_2), they are found in liquid form. These heavy elements then sank under the force of gravity and, due to high pressure at the center, formed the *solid core* of Earth surrounded by molten (liquid) iron (figure 16.2). A layer of silicate-rich rock surrounds the liquid iron shell—called the *mantle* (meaning "coat" in German). The elements lighter than iron and nickel moved to the surface of Earth and cooled, forming the solid *crust*. This process is called *differentiation* and is responsible for the formation of Earth's three layers: the core, the mantle, and the crust. The size of Earth's layers and main composition of each layer are listed in Table 16.2.

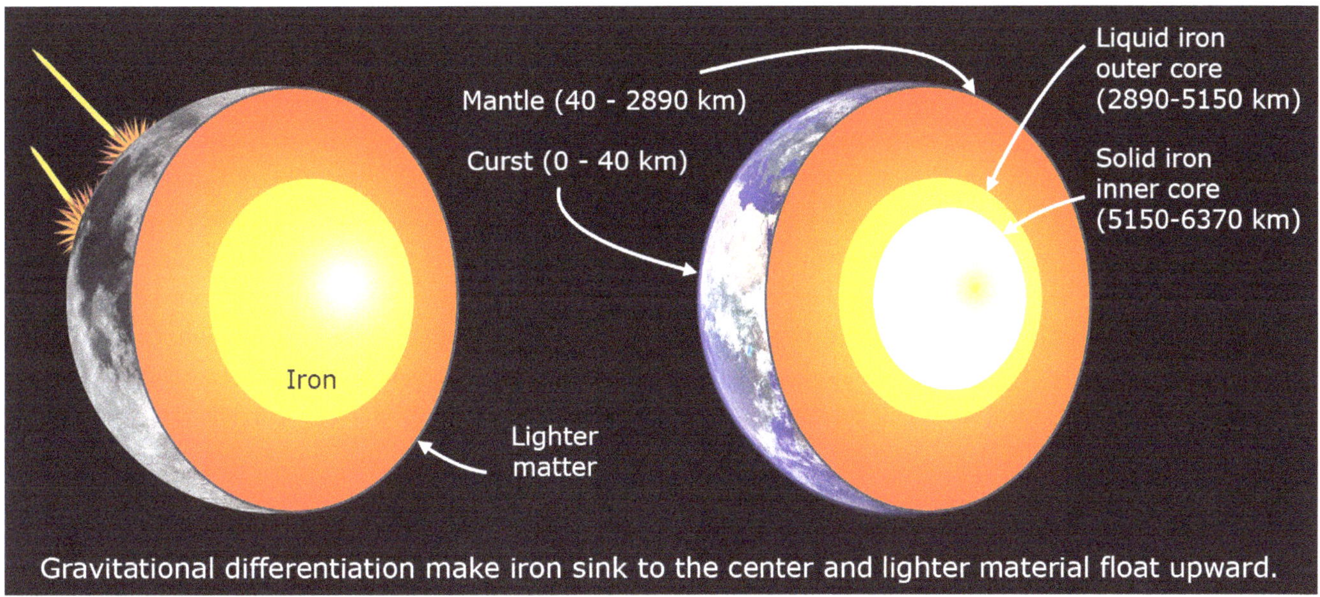

Figure 16.2. The three layers of Earth: the inner iron core and outer molten iron core, the mantle, and the crust.

The solid inner core of Earth mainly contains iron and nickel, within a radius of 1,220 km and a temperature of about 5,000 degrees Kelvin (Earth's temperature increases inward with radius; figure 16.2). This is above the melting temperature of iron and nickel. How could a solid core form at such a high temperature? This is because of the chemical properties of the iron-nickel alloy that is solidified under high pressure rather than low temperature. As one probes inside Earth's mantle, the density increases with depth, not because of the change in the composition of the elements but because of the reduced size of chemical compositions due to increasing pressure. The thickness of the oceanic crust (Earth's surface in the floor of the oceans) is about 7 km, compared to continental crust (the crust forming continents), which is about 40 km. Furthermore, oceanic rocks contain iron and hence are denser than continental rocks, which contain silicates with a low melting temperature. Because the continental crusts are less dense and thicker than oceanic crusts, they ride higher, floating on the denser mantle (Marshak 2012).

To summarize, over 99 percent of Earth's mass is made of only 8 elements, with 90 percent of Earth consisting of only four elements: iron, oxygen, silicon, and magnesium. These are distributed with higher concentration of iron at the core, with oxygen, silicon, and magnesium mostly in the mantle and crust (Table 16.2).

Table 16.2. Properties of Earth's layers

Layer	Depth (km)	Composition
Crust	0–40	Oxygen (46%), calcium (2.3%), magnesium (4%), silicon (28%), aluminum (8%), iron (6%), other (6%)
Mantle	40–3,000	Oxygen (44%), calcium (2.5%), magnesium (22.8%), silicon (21%), aluminum (2.4%), iron (6.3%)
Outer core	3,000–5,000	Oxygen (5%), sulfur (5%), iron (85%), nickel (5%)
Inner core	5,000–6,000	Iron (94%), nickel (6%)

Source: Jordan and Grotzinger 2012.

THE ORIGIN OF THE MAGNETIC FIELD OF EARTH

It is well known from laboratory experiments that electric currents produce magnetic fields in their vicinity. Being an excellent conductor of electric current, molten iron moving in the outer core of Earth is responsible for its magnetic field. How has the magnetic field been sustained over the age of Earth? And how does the extreme heat at Earth's core affect the magnetic field? These questions can be resolved if the magnetic fields are constantly generated. The molten iron in the outer core of Earth is in constant motion as a result of the convection process (movement of hot material to the top and cooler material to the bottom), generating a magnetic field. It is well known that if a conductor moves inside a magnetic field, an electric field is generated. Earth's core is the conductor moving within Earth's magnetic field, constantly generating the electric field, which in turn generates the magnetic field (Marshak 2012).

Like bar magnets, this has two poles, north and south, with invisible magnetic field lines connecting the two poles. Therefore, Earth's field can be represented by a magnetic dipole. This intersects Earth's surface in two points, known as magnetic poles. The magnetic poles are not the same place as the geographic poles but are around 1,500 km from them. For this reason, a compass needle does not point toward the geographical pole but the magnetic pole that is 11 degrees from Earth's spin axis.

Earth's magnetic field protects the Earth against charged particles emanating from the sun in the form of solar wind. These particles are harmful to animals and plants and disturb communication satellites. These charged particles are deflected by the magnetic field of Earth and hence do not hit the surface of the planet. The magnetic field of Earth tends to magnetize rocks (and specially those containing iron) that in turn, are used by geologists to study the behavior of the fields through time (Jordan and Grotzinger 2012).

HEAVY BOMBARDMENT

After the planets were formed, a large number of planetesimals were left in the solar system in the form of asteroids and comets. Many of these "leftovers" were eventually crashed onto the planets during the first hundred million years from the birth of the solar system. This period is called *heavy bombardment*. Earth experienced many such events during its lifetime that caused extinction of 95 percent of the species at any given time (Box 16.2). Today, signatures of heavy bombardments can be seen on the surface of the moon. Because of its larger size and mass and larger gravity compared to the moon, Earth was a more obvious target for such impacts. However, the

---

BOX 16.2: EFFECT OF HEAVY BOMBARDMENT ON LIFE ON EARTH

Upon their impact, celestial bodies generate huge amount of energy. For example, an asteroid 350 km to 400 km large could generate enough heat to vaporize all the oceans on Earth, increasing its surface temperature to 2,000 degrees Celsius. This would have a sterilizing effect, extinguishing all forms of life on Earth. Smaller impacts would vaporize the top few hundred meters of the ocean. In this case life in safer environments (deep in oceans or inside Earth) would likely survive. Therefore, some form of life could have existed around 4.5 billion years ago in the early Earth and may have been extinguished during the heavy bombardment period. It is possible that life arose multiple times, was extinguished, and started all over again.

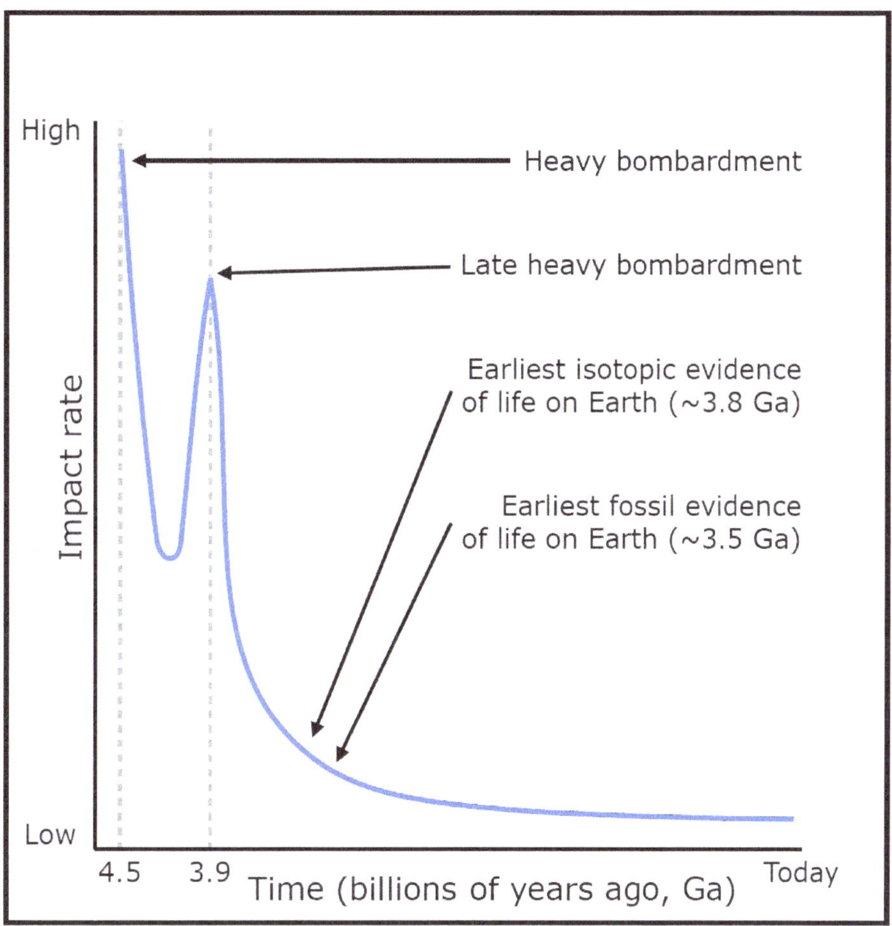

Figure 16.3. Changes in the impact rate with time. There were significant bombardments between 4.5 billion and 3.9 billion years ago. The peak at 3.9 billion years ago is likely due to the planets in the solar system falling in their present orbits. The rate is reduced after that because the combination of gravitational fields of the planets deflected the meteorites away from Earth.

imprints of these collisions were wiped out by erosion and volcanic activity on Earth. Study of the ages of craters on the moon as well as the zircon grains, the oldest element known on Earth, allows measurement of the time when the impact rate dropped. These calculations show that around 3.9 billion years ago, there was a significant increase in the frequency of the impacts (figure 16.3). This is known as late heavy bombardment. The reason for this is likely due to movement of the orbits of the planets caused by gravitational interactions until they settled into more stable orbits. It is likely that this process disturbed the orbits of some planetesimals, causing them to hit the planets (Bennett and Shostak 2006).

SUMMARY AND OUTSTANDING QUESTIONS

Earth was formed by accretion of solid material condensed in the orbiting gas cloud. These came together and grew, initially by the electrostatic force and, once they gained substantial size and mass, through the force of gravity, forming the planetesimals. Through their orderly motion around the newly formed sun, they attracted one another and formed Earth. The heat at the core of Earth is the result of the collision of planetesimals, converting their kinetic energy to thermal energy and heat, as well as decay of radioactive material.

The age of Earth is accurately measured by different independent methods—dating of zircon through uranium isotopes and examination of the rocks from the moon provide lower limits to the age, while the age of the solar system estimated from meteorites provides upper limit. These have now converged to 4.5 billion years. The moon was much closer to Earth soon after it was ejected from the Earth when Earth was hit by a Mars-size object. Through billions of years, it moved away with a rate of a few centimeters per year and will continue to do so.

Earth has three different levels formed through the process of differentiation. This is the process that sinks heavier elements to the center of Earth while lighter elements move to the surface. Different layers include the *inner core*, which consists of solid iron and nickel alloy with a temperature of around 5,000 degrees Celsius. The solidification here is due to the extreme pressure at the center of Earth. The *outer core* contains melted iron with a temperature of about 3,000 degrees Celsius and is in motion through convection. The *mantle* forms the bulk of Earth's structure and consists of whatever material was left after heavier elements sank to the core and lighter elements moved to form the surface crust. It mostly contains silicates and magnesium. The convection process in the mantle moves the heat from the inner part to the surface of Earth. Finally, the crust is formed from the molten material moving to the surface and cooling down. This is the solid part and was formed early in the history of Earth from silicates with lower melting temperature. The motion of the molten iron in the outer core is responsible for the magnetic field of the Earth. This is because iron is a good conductor of electrical current and the known fact that a material with electrical current produces magnetic field in its vicinity.

After its formation, Earth went through a period of heavy bombardment by the leftovers within the solar system. This was very intense from 4.5 billion to 3.9 billion years ago, and the rate reduced after planets in the solar system found their stable orbits and, because of their high mass and strong gravity, deflected the material coming toward Earth.

There are a number of exciting questions yet to be addressed about our Earth, including: How from that initial chaotic state did Earth evolve to become the well-ordered place it is today? Was the Mars-sized impact that happened soon after the formation of Earth truly responsible for the origin of many events leading to subsequent evolution of the planet? What was the impact of solar winds on the Earth and its atmosphere before formation of its iron core? And what is the origin of the element zircon on Earth? Finally, could we use our detailed knowledge of the Earth to study other exosolar planets?

REVIEW QUESTIONS

1. What was responsible for increased probability of planetesimals colliding the early Earth?
2. What is the definition of a volatile material?
3. Why did the lighter materials escape from the early Earth?
4. What is the origin of the heat at Earth's core?
5. What is the age of Earth, and how are the lower and upper age limits estimated?
6. Explain the main effects of the Mars-sized impact on Earth that has continued to the present time.
7. What are the main compositions of Earth?
8. Describe the process of differentiation.
9. Describe different layers of Earth and the characteristics of each layer. Why are the layers located where they are?
10. What is the origin of Earth's magnetic field?

CHAPTER 16 REFERENCES

Bennett, J., and S. Shostak. 2006. *Life in the Universe*. 2nd ed. Boston: Pearson/Addison-Wesley.

Marshak, S. 2012. *Earth: Portrait of a Planet*. 4th ed. New York: Norton.

Jordan T.H., and J. Grotzinger. 2012. *The Essential Earth*. 2nd ed. New York: Freeman.

Wood, J.A. 1979. *The Solar System*. Englewood Cliffs, NJ: Prentice Hall.

TABLE CREDITS

- Tbl. 16.1: John Armstead Wood, "Timeline for the Early Formation of the Earth," The Solar System. Copyright © 1979 by Pearson Education, Inc.
- Tbl. 16.2: Thomas H. Jordan and John Grotzinger, "Properties of the Earth's Layers," The Essential Earth, 2nd Edition. Copyright © 2011 by W. H. Freeman & Company.

17

THE ORIGIN OF CONTINENTS, OCEANS, AND MOUNTAINS

CHAPTER LEARNING OBJECTIVES

This chapter will cover:

- Formation and growth of continents
- Plate tectonics
- The origin of volcanoes
- Formation of mountains and oceans
- The origin and types of rocks
- Evolution of the Earth's landscape

The process of planet formation is very fast. It took about 10 million years for Earth to form. Soon after Earth was formed, it acquired its hard crust. Since then, different processes unraveled with a much slower pace and throughout many billions of years, contributed to shape its present landscape—continents, oceans, and mountains. However, for many years there were no acceptable theories as to how the continents were formed and distributed across the globe, why oceans and mountains are where they are, and in what sequence these were formed.

Around 1925 geologists proposed that the ocean basins were the result of the cracks appearing between broken continents, based on the assumption that the surface of the Earth expanded to at least twice its size. The early hypothesis also suggested that the continents grew where they are located now, formed in isolation by accretion of the land. Mountains then formed at the boundaries between the continental and oceanic crusts by the concentration of thick sediments producing crumpling of the crusts, leading to the rise of the mountains.

The modern theory for the formation of continents was proposed by German meteorologist Alfred Wegener (1880–1930), who hypothesized that the continents were part of the same land and were torn apart and drifted away to form the present distribution of the land. This led to the development of the concept of the continental drift around 1910. Through Wegener's work, the concept of large-scale movement of continents was developed. This started by noting the remarkable

> "The least movement is of importance to all nature. The entire ocean is affected by a pebble"
>
> - BLAISE PASCAL
>
> "In the presence of eternity, the mountains are as transient as the clouds"
>
> - ROBERT GREEN INGERSOLL

similarities of geologic features of the coastlines on both sides of the Atlantic ocean following publication of the first map in 1859, fitting the coast of South America against Africa. Wegener postulated a supercontinent he called *Pangaea* (meaning "all land" in Greek) that was broken up into today's continents, and then drifted away from one another.

This chapter first explains the concepts behind the theory of continental drift. It then describes the discoveries that led to modern scenarios for the origin of the continents, oceans, and mountains. The chapter introduces plate tectonics theory and different types of plate tectonics, the origin of rocks and evolution of Earth's landscape.

THE FIRST CONTINENTS AND CONTINENTAL DRIFT

From the observations of the mountain ranges, rock types, and fossils, geologists map the past distribution of land and sea. The rocks also reveal old episodes of rifting and subduction. From these observations geologists reconstructed the earliest supercontinent called *Rodinia*, formed about 1 billion years ago. Rodinia started to break up around 750 million years ago. The fragments then started to reassemble around 500 million years ago, forming the supercontinent *Pangaea*. The continued movement of the continents changed the land and sea distribution on the face of Earth as well as affecting the climate system. The driving force behind this was the *tectonic plate movement* (see next section) that is resulted by convection below the lithosphere (the outermost layer of Earth), leading to the concept of continental drift, developed by Wegener.

The breaking up of Pangaea started around 200 million years ago, when northern continents (called *Laurasia*) started to separate from the southern *continents* (called *Gondwana*) (Figure 17.1). At about the same time, North America started to drift away from Europe, creating the North Atlantic Ocean, while the Gondwana broke up along the eastern coast of Africa, separating into South America, Africa, India, and Antarctica and creating the South Atlantic Ocean. The present land distribution was formed around 66 million years ago when Australia was separated from Antarctica and India joined the Eurasia (figure 17.1; Jordan and Grotzinger 2012). These all took place through the process called *continental drift*. However, there were arguments against the continental drift concept. For example, this requires the lithosphere of Earth to be able to float and not be rigid, as it was assumed at the time. Also, there were no candidates for the driving force behind moving the continents, as the tidal forces of the sun and the moon were not strong enough to cause the continents to drift apart (Marshak 2012).

In 1921 South African geologist A.L. du Toit produced indisputable evidence in favor of the continental drift concept based on geologic structures, similarities in rock ages and fossils, and climate data between the two sides of the Atlantic. Similar 300-million-year-old fossils were found of the reptile *Mesosaurus* in Africa and South America but no other place, indicating that the two continents were connected when these creatures were alive (figure 17.2). *Mesosaurus* was a freshwater reptile and could swim. If they could swim between South America and Africa, they would likely be able to swim to other places as well. The fact that their fossils are not found in other places indicates that they were separated when the two continents drifted away. Furthermore, evidence based on similar large plant seeds being distributed across the globe (where they could not possibly have been transmitted across the oceans by wind), the finding that present worms in Madagascar more closely resemble those in India than the worms of nearby Africa, similarity of the fossils found in Africa compared to those in India, and the fact that the rocks deposited by glaciers 300 million years ago are now distributed in South America, Africa, India, and Australia imply that these were all part of the same land within the same

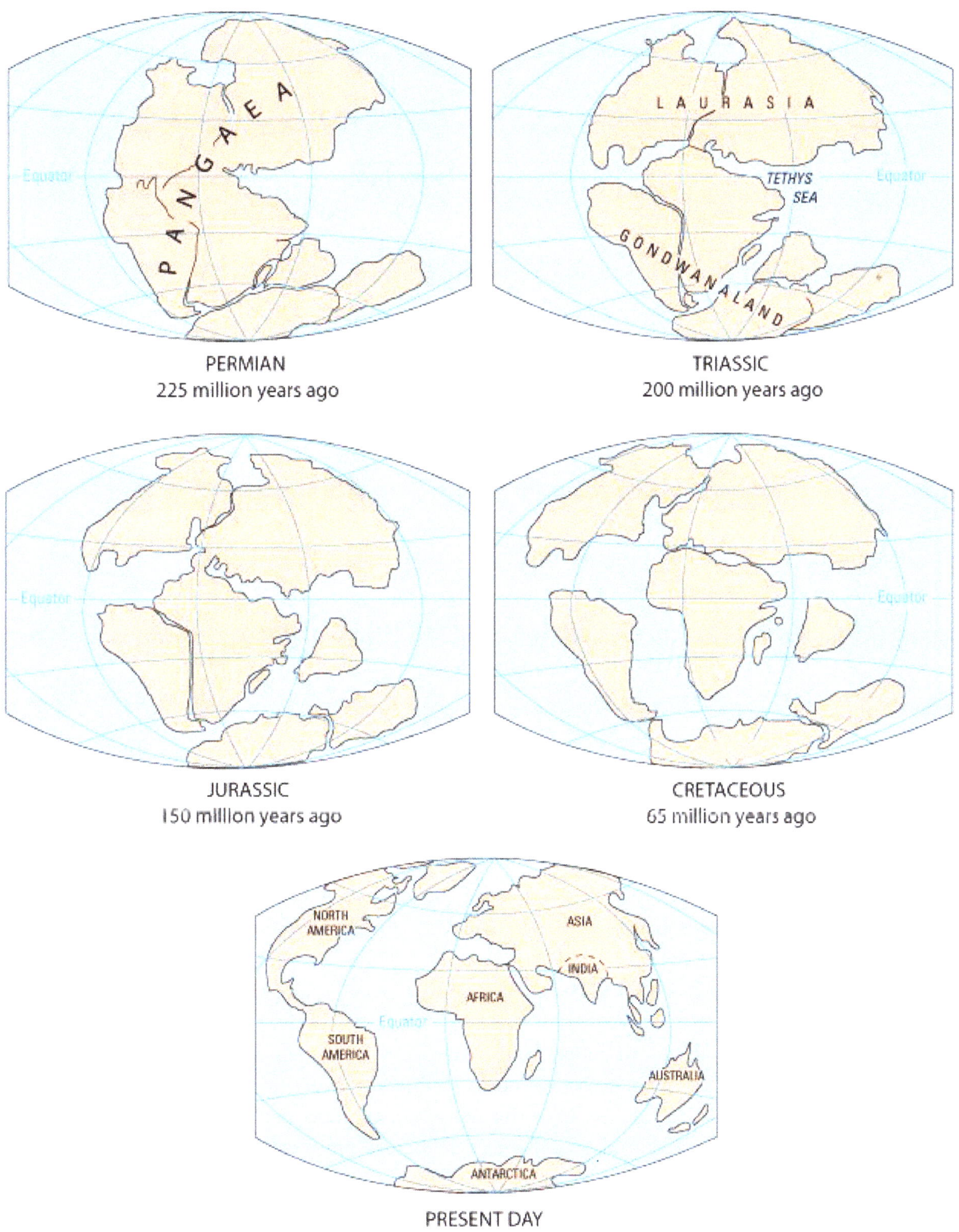

Figure 17.1. Changes with time in the distribution of land on Earth. Continents were part of the same land, Pangaea, and then were broken away around 200 million years ago. The present land distribution came into place over 65 million years ago.

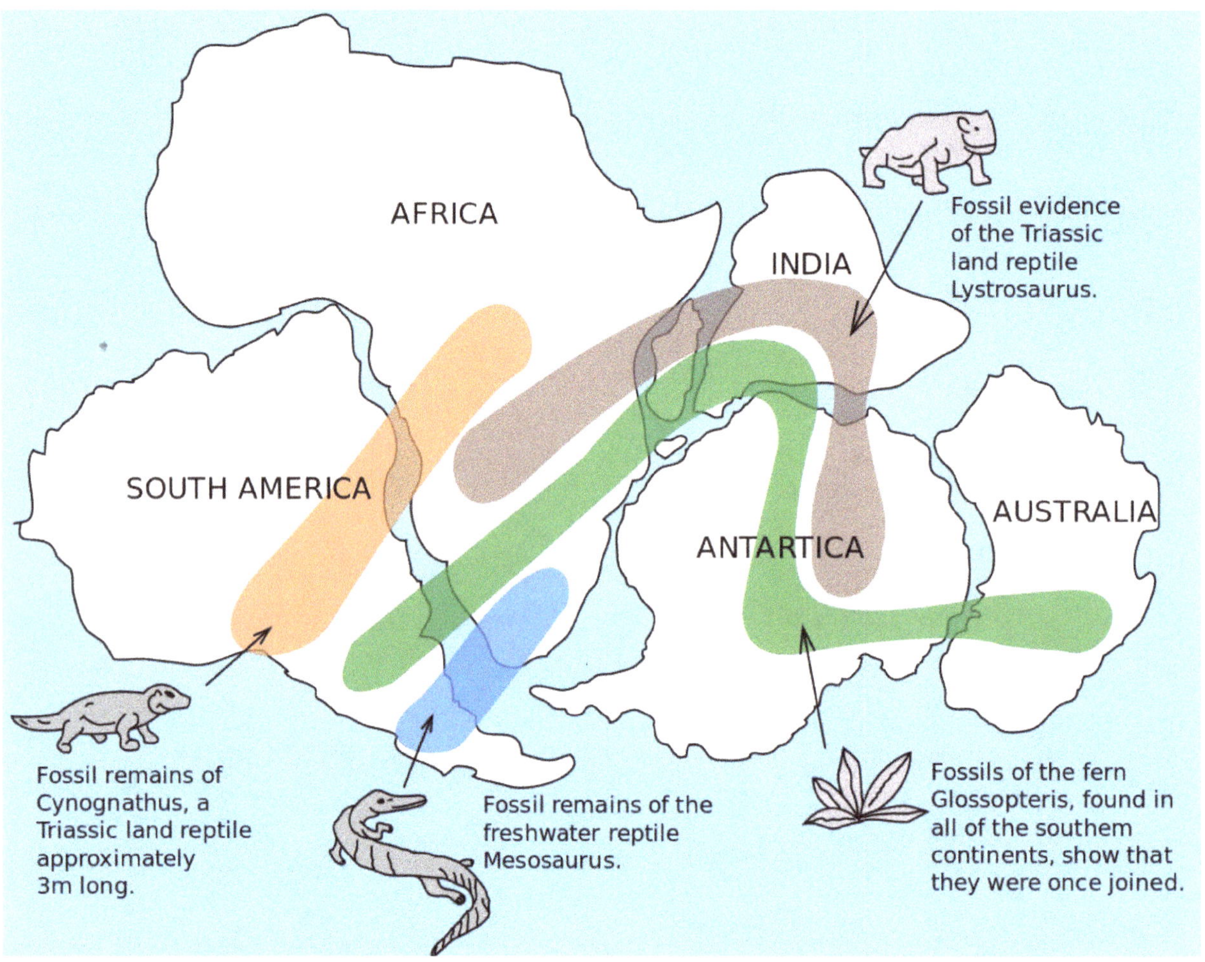

Figure 17.2. Plant and animal fossil evidence and the discovery of certain plant seeds in different continents confirm that the continents were all part of the same land that started to break up around 200 million years ago.

continent (figure 17.2; Marshak 2012). The fundamental question now is: How did Pangaea break apart and different pieces of land started to move?

THEORY OF PLATE TECTONICS

The plate tectonics theory, developed during 1960s, states that outer layers of Earth, the *lithosphere*, are formed from thirteen rigid plates that move over Earth's surface with respect to one another—slide parallel, move toward or away against each other. These are called *tectonic plates* (*tectonic* comes from the Greek word *tekton*, meaning "builder"). The plates move over the weaker *asthenosphere* (the less solid layer below the lithosphere) with a speed of 1 cm to 15 cm per year (figure 17.3). The largest plate is the *Pacific Plate* that covers most of the Pacific Ocean. The North American Plate extends from the Pacific coast of North America to the middle of the Atlantic Ocean, where it meets the Eurasian and African Plates (Jordan and Grotzinger 2012).

The fractures on the lithosphere occurred as a result of stresses generated by mantle convection, with the resulting plates moving over the mantle (figure 17.3). Geological activities take place at the border of plate boundaries. This is where earthquakes happen, volcanoes are located, mountains are formed, and rifts are observed.

SEAFLOOR SPREADING

The force that could move continents apart is caused by convection in Earth's mantle (figure 17.3). This forms cracks on the lithosphere of Earth that would move tectonic plates apart, forming new oceanic crusts through the process called *seafloor spreading*. In this process, the *magma* (hot, molten rock in the mantle) comes out of Earth, cools down, and solidifies (figure 17.4). Evidence for this comes from the discovery of the *Mid-Atlantic Ridge*, an opening in the seafloor of the Atlantic Ocean surrounded by young *basalts* (dark, fine-grained rock) and not old granite. Moreover, mapping the ocean floor around the Mid-Atlantic Ridge showed the presence of a deep valley (or a rift). The age of the crust increases away from the ridge, showing that the crust has been spreading on the floor of the Atlantic Ocean as the magma is coming out of Earth's asthenosphere layer (figure 17.4). Geologists further found that all earthquakes in the Atlantic take place around this point, at the faults between tectonic plates. This confirmed that the rift is an active boundary between two plates. The seafloor spreading theory therefore confirms that continents can move apart through the creation of new lithospheres at mid-ocean ridges. About two-thirds of Earth's surface and its entire oceanic crust were all produced by seafloor spreading over the last 200 million years (Jordan and Grotzinger 2012).

The question now is if the material coming out of Earth could recycle and go back in Earth. If this were not the case, the surface of Earth would increase as more lithosphere is formed. This problem was resolved by considering tectonics to be rigid plates moving over Earth's surface. Furthermore, it was found that the process of rocks forming and evolving—folded, compressed or sheared—all happened near these boundaries. These findings confirmed the plate tectonics theory (Jordan and Grotzinger 2012).

The magma contains silicon and oxygen and varying proportions of aluminum, calcium, sodium,

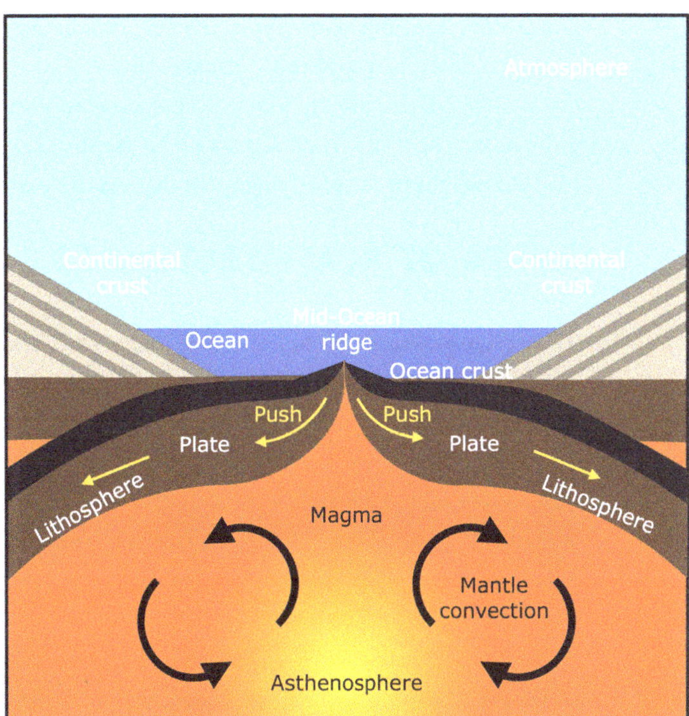

Figure 17.3. Convection in the mantle (movement of the material with different temperatures). This generates the force that breaks up the lithosphere of Earth, resulting in plate tectonics.

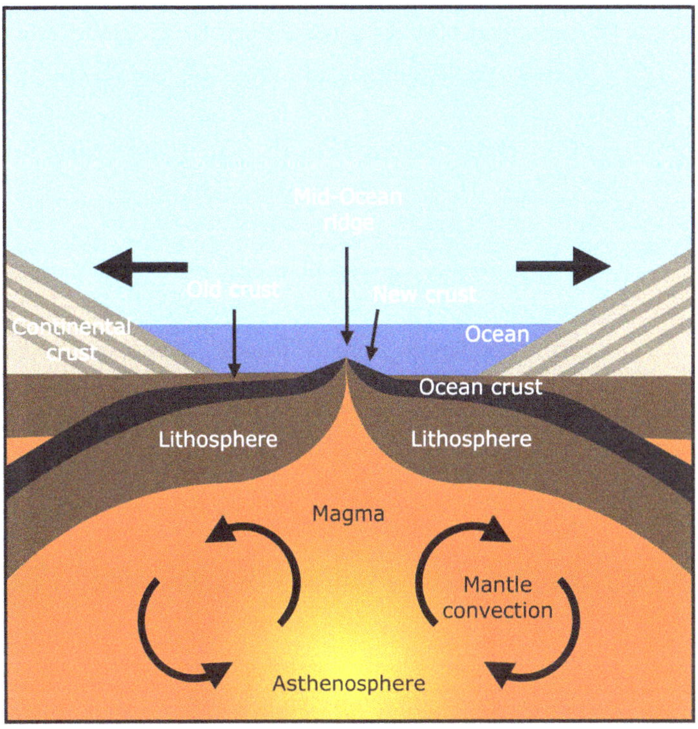

Figure 17.4. Magma from the asthenosphere of Earth creates the mid-ocean ridge. This produces the ocean crust that spreads on the floor of the ocean. Divergent boundaries are responsible for seafloor ridge formation (between two oceanic plates) or ridge valleys (between two continental plates).

potassium, iron, and magnesium. The magma coming out of Earth contains volatiles such as water, carbon dioxide, nitrogen, hydrogen, and sulfur dioxide (SO_2). These come out of Earth through the volcanic process and in the form of gas. Therefore, they contain the water molecules (that eventually reside in the atmosphere) as well as the minerals in the rocks. The magmas rise to the surface of Earth because it is less dense than the rocks surrounding it and the weight of the overlying rocks that creates pressure at the depth, squeezing magma upward.

TYPES OF PLATE TECTONICS

Plate tectonics move along, over and under each other and are responsible for earthquakes, volcanoes, creation of mountain ranges and ridge valleys. There are three types of plates, defined by their direction of movement with respect to each other. As the plates move away from one another at some location (diverge), causing seafloor spreading, they move toward each other at another place (converge), creating mountain ranges. While seafloor spreading (from the material moving out of the Earth) increases the surface of Earth, the convergence process reduces it. These two processes, on average, have equal effects, keeping the surface of Earth constant. The plate tectonics movements are discussed below.

Divergent boundaries: This is where plates move apart and new lithosphere is created (figure 17.4). When happening at seafloors, the magma coming out of the interior of Earth spreads out, forming the oceanic crust and the floor of Earth's oceans. This also forms mountain ranges in the bottom of oceans. The island of Iceland is a part of the Mid-Atlantic Ridge formed through divergence of the plates (Jordan and Grotzinger 2012).

Divergent boundaries could also happen in continents, producing *continental rifts* characterized by valleys, volcanisms, and earthquakes. The Red Sea and Gulf of California are examples of continental rifts where the continents have separated enough for the seafloors to form. In other occasions, the continental rifts started but have not quite finished, like the Great Rift Valley of East Africa.

Convergent boundaries: These are formed when lithospheric plates come together. This leads to a number of geological features as summarized below.

Ocean-ocean convergence: This is when the lithospheres of two oceanic plates converge (figure 17.5a). This results in one plate moving underneath the other in a process called *subduction*. The subducting plate sinks to the asthenosphere and into the mantle, producing a deep-sea trench (deep and narrow ditch). Only the oceanic plates subduct as they are denser than continental plates. As the cold lithosphere further merges into the mantle, the pressure will build up and melting above the downing plate takes place, generating magma. The magma forms a chain of volcanoes and islands behind the trench. The example here is the Hawaiian Islands and deep-sea Mariana Trench of the Western Pacific, where the ocean reaches a depth of 11 km.

Ocean-continent convergence: This is when a continental plate and an oceanic plate converge (figure 17.5b). Since continental crusts are lighter, the oceanic plates are *subducted*. The result is deformation of the continental crust as it is compressed by force of convergence. This leads to uplifting of the rocks from the continental crust and formation of mountain chains parallel to the deep-sea trench. The water carried by the oceanic subducting plate will cause the mantle to melt, creating volcanoes. This leads to strong earthquakes deep inside Earth. Example of this is the Andes chain of mountains in South America.

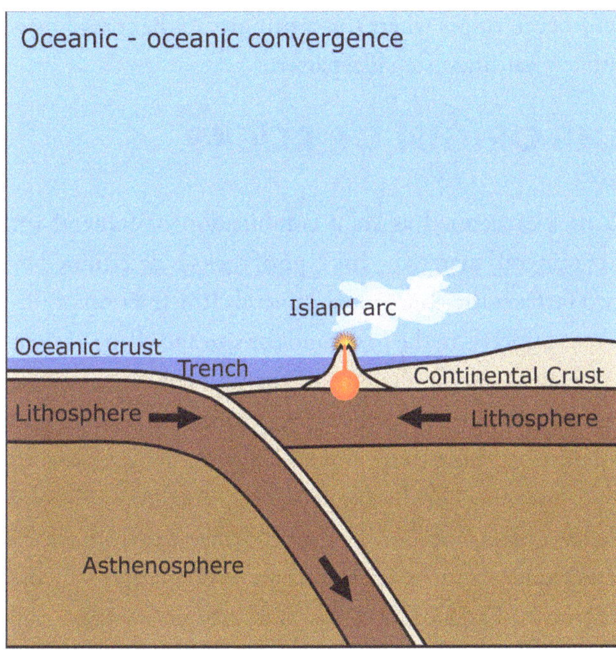

Figure 17.5a. Shows two oceanic plates converging below the ocean. In the place they merge, a trench is formed under the ocean, with volcanic activity creating new islands as the lava builds up, and volcanic mountains are created, extending above the ocean surface. Examples for this are the Hawaiian Islands and the volcanoes in Hawaii that are built by the convergence of two oceanic plates.

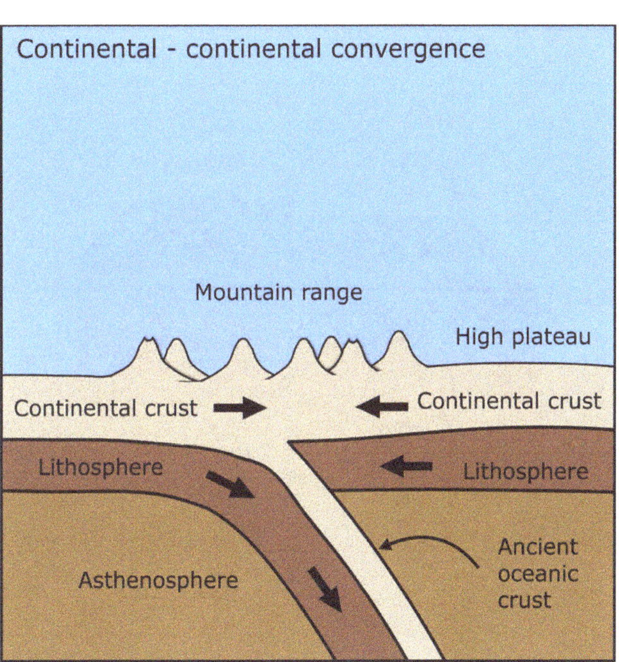

Figure 17.5c. Two continental plates converging. This results in an upheaval of land, resulting in a plateau (for the plate that is on the top) on one side and a mountain range on the other. Plates that are colliding in the middle of continents do not often result in volcanic mountains. Examples of continental plate convergence are the Himalayan mountain range, the Ural Mountains (both in Asia), and the Alps (in Europe).

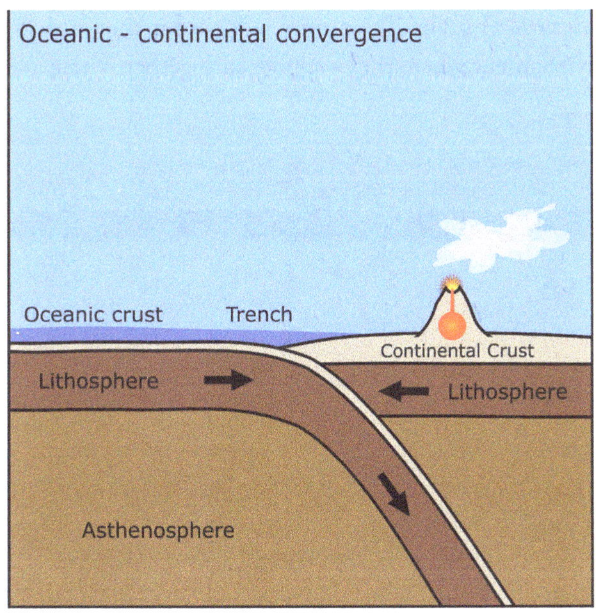

Figure 17.5b. Shows an oceanic crust moving under a continental crust. This creates coastal mountain ranges that are mainly volcanoes. Examples of this are found along the Pacific Ocean in South America (Peru) and in North America (Alaska).

Continent-continent convergence: This results from convergence of two continental plates (figure 17.5c). The subduction process does not happen in this case. Both plates end up floating above the mantle while one plate will override the other. The collision of the Eurasian and Indian Plates (both continental plates) created double thickness of the crusts, forming the highest mountain range in the world, the Himalayas, and the Tibetan Plateau. Because of the crumpling of the crust strong earthquakes happen as a result of continent-continent convergence.

Transform boundaries: These are where plates slide past each other, with lithospheres not created or destroyed (figure 17.6). In this case the plates move alongside each other, causing earthquakes. The San Andreas Fault in California is the result of the Pacific Plate sliding against the North American Plate. Transform boundaries are found along

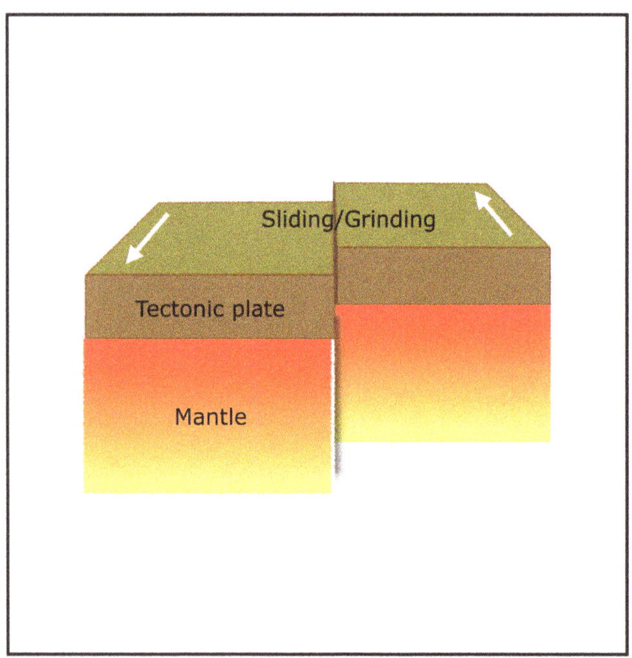

Figure 17.6. Two tectonic plates moving side-by-side (transform boundaries), causing earthquakes.

mid-ocean ridges where the continuity in the spreading is broken, causing a step like pattern.

THE ORIGIN OF ROCKS

A rock is defined as solid combination of mineral (or nonmineral) materials. The type of rocks is determined by two factors: the constituent minerals (the fraction of different minerals in the rock) and the size and shape of their mineral crystals and grains (texture). The magma coming out from the interior of Earth hits colder temperature and solidifies, turning to *igneous* rocks (from the Greek word *ignis*, meaning "fire"). The type of igneous rocks depend on the composition of the material (magma) from which it is formed and its environment. When magma in the interior of Earth cools slowly, microscopic crystals form. In this case some of the crystals have enough time to grow to several millimeters in diameter before the whole mass is solidified as igneous rock. When magma is released through volcano eruption, it cools down rapidly at the surface of Earth. In this case individual crystals have no time to form, with smaller crystals forming simultaneously. This explains the origin of the shape and type of different igneous rocks (Box 17.1).

When preexisting rocks break and the resulting grains are transported by wind or water and deposited in new environment, the grains are cemented together, forming *sedimentary* rocks. The properties of sedimentary rocks again depend on the composition of the grains and the environment where they are pressed together. These are

---

BOX 17.1: TYPES OF ROCKS

In geology, rocks are divided into three basic types:

Igneous rock: This is made from molten rock that cools and solidifies.

Metamorphic rock: This is produced under high pressure and heat, which are high enough to change its structure and chemical composition but not high enough to melt it.

Sedimentary rock: This is produced under high pressure by compression of sediments mainly at the bottom of the seas.

Rocks could change from one type to other. For example, an igneous rock can change under heat and high pressure to metamorphic rock while both these could change to sedimentary rocks by erosion. As a result, the type of rock will not reveal much about its composition. Instead, each rock contains a mixture of different crystals with each individual crystal representing a mineral with a particular chemical composition. The above types indicate how a rock is built while the minerals composition reveals what it is made of. Therefore, the rocks are further divided into subclasses. Examples of subclasses of igneous rocks include ***basalt***, which is dark and dense igneous rock produced by undersea volcanoes, rich in iron and magnesium-based silicate; and ***granite***, which is lighter in color and less dense than basalt, common in mountain ranges, and largely composed of quartz.

found on the land or under the sea. Sometimes high pressure and temperature change the properties of preexisting rocks. This could also happen through the squashing or stretching of the rocks. The result is the *metamorphic* rocks (from the Greek words *meta*, meaning "change," and *morphe*, meaning "form"). The temperature for metamorphic transition is below the melting point of the rocks (about 700 degrees Celsius) but high enough for the rock to still change by chemical reactions (about 250 degrees Celsius). Starting from igneous rocks, all other types of rocks could be generated, with each rock type formed under specific tectonic activity conditions (Box 17.1).

THE CHANGING FACE OF CONTINENTS, OCEANS, AND ROCKS

Within the framework of plate tectonics, processes that lead to the formation of continents, mountain ranges, and oceans are all correlated. The opening and closing of ocean basins, the appearance and disappearance of mountain ranges and evolution in the size and mass of the continents continuously take place in a cycle that approximately lasts 250 million years. This is called the **Wilson cycle** after the Canadian geologist Tuzo Wilson (1908–1993) and has the following stages, as also shown in figure 17.7 (Jordan and Grotzinger 2012).

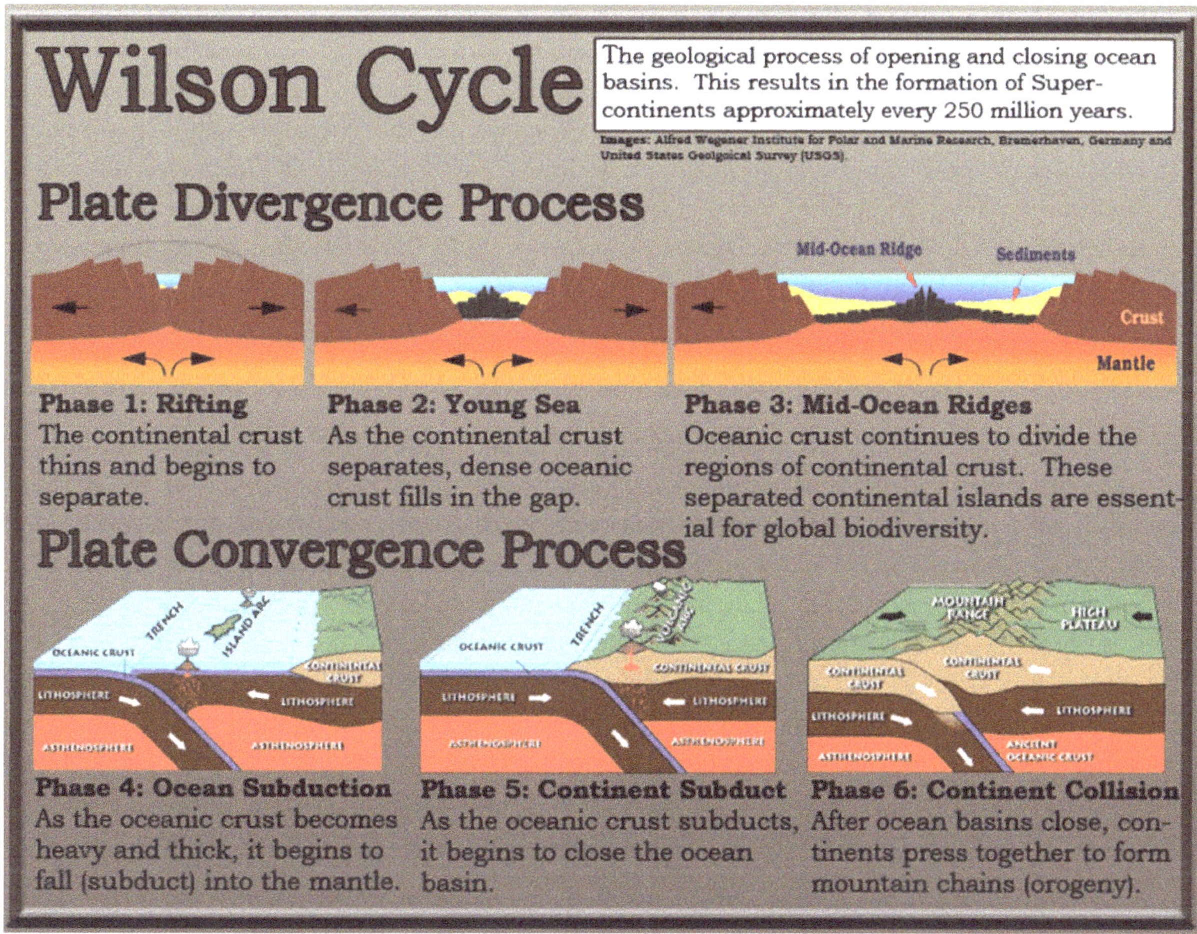

Image courtesy of the Alfred Wegener Institute for Polar and Marine Research, Bremerhaven, German and United States Geological Survey.

Figure 17.7. The breaking of the continents, opening of the ocean basins, and formation of mountains, all explained by the Wilson cycle that is repeated every 250 million years.

The rifting within a continent leads to breaking of the crust and opening of a new ocean basin and formation of an oceanic crust. This process is initiated because of thinning of the continental crust by erosion and weathering. As the continental crusts separate, the ocean basin becomes larger, with an ocean being formed between the two, now separate, continents. After the ocean was created, convergence begins with oceanic crust subducted under a continental crust, leading to formation of volcanic mountains at that position. As the subduction progresses, the ocean becomes smaller while making the continental crust thicker. As continents collide, the crust in the interaction region becomes thicker, leading to the formation of mountain ranges and new supercontinents. The supercontinent then erodes with its crust becoming thinner due to erosion, and the cycle starts again (figure 17.7).

During the Wilson cycle different types of rocks are also produced or one type is converted to the other, called the *rock cycle*. When the cycle starts by rifting of the continent, sediments erode from continent and are deposited in the rift basin, forming sedimentary rocks. As seafloor spreading starts in the newly formed ocean basin, magma rises in the mid-ocean ridge, forming igneous rocks. The closing of the ocean basin leads to formation of mountain ranges and collision of the continents. This generates high temperature and pressure needed to form metamorphic rocks. The mountains created through this process make the moisture in the air to rise and cool. This results in the weathering followed by erosion that would strip away high mountains. At the same time, streams transport the material (and sediments) from collision zones to the oceans where they are deposited as sedimentary rocks.

Finally, there are two processes that lead to continental growth. The low-density silicate-rich rocks move from the mantle to the crust of Earth. The continental crusts are often formed in subduction zones from the magma produced by melting of the subducting lithosphere. These magma move to the surface and form the crust of Earth. The other process responsible for continental growth is accretion, caused by the material previously separated from the mantle coming together during the plate movement, increasing continental masses. Once formed, continental crusts are seriously altered due to the mountain building process and folding. When two continents collide, they experience horizontal compression, increasing the thickness of the crust. This results the rocks in the lower crust to melt, generating magma that rises and forms the upper crust.

SUMMARY AND OUTSTANDING QUESTIONS

Modern scenarios for formation of the continents, oceans, and mountains started to take shape in the 1960s, when the concept of continental drift was developed. Using the evidence based on the continental coastlines as well as the fossil records across the world and geological dating, geologists concluded that all the continents were initially part of a large land, named *Pangaea*. This was then started to break up around 200 million years ago and has been evolving since then. The present-day distribution of the continents took shape around 65 million years ago.

To explain the process of continental drift, the tectonics theory was developed. According to this theory, the lithosphere of Earth is divided into thirteen solid plates that move with respect to each other. The cause of the breaking of the outer crust of Earth was the force produced due to convection of the mantle (moving of the hot and cold material because of their temperature difference, with hot material lying at the top). This is responsible for the formation of all the structures on the surface of Earth. The plate tectonics scenario was confirmed by the discovery of Mid-Atlantic Ridge where hot material coming out of Earth's mantle cooled down and spread out on the ocean floor forming the solid oceanic crust. The ridge at mid-Atlantic appeared at the intersection between two plates. The material closest to the ridge consists of young basalt, with the age of the crust increasing away from the ridge.

When two tectonics plates move apart from each other, a rift is formed (divergent boundaries). If this happens with oceanic plates, the hot magma comes up and forms the crust of the ocean, which would grow and form mountains in the bottom of the oceans. In case of continental plates, ridge valleys will be formed. When two plates move toward each other (convergent boundaries), depending on whether they are two oceanic plates, an oceanic and a continental plate or two continental plates, they end up in volcanoes and island arcs, trenches and volcanoes or mountain ranges respectively. When the oceanic and continental plates converge, given that the oceanic plates are denser, they move under the continental plates and in the lithosphere where the material is melted forming volcanoes.

The breaking of the continents, creation of ocean basins, and formation of mountain ranges are all explained by plate movements through the Wilson cycle. This also explains the origin of different types of rocks and conversion of them from one type to the other. Once the continents are broken into separate lands, they drift around. The concept of continental drifting has gained support from various observations today and is responsible for the land and sea distribution in the world. Today's ecosystem is a result of continents breaking up and moving as separate lands. Location of different lands in different latitudes allows the ecosystem suited to different climates to develop.

Many of the things that took place and led to the present world are interrelated, as can be seen from the Wilson cycle. These affect the development of the ecosystem needed to support life, like a stable temperature, presence of nutrition, and the material needed for life. The outstanding questions therefore include: How does plate tectonics affect the circulation of nutrients and necessary material for the development of life on Earth? How have these conditions been sustained for so long on our planet? And what would the present drifting of the continent lead to 50 million years in the future? What are the observable effects of plate tectonic movement that one should look for when studying life in other planets? Movement of the land because of tectonics activities over the last one billion years (since the time of super-continent Rodinia) has had serious effects on the development and evolution of life on Earth. This has caused different climates and the required conditions for plants and animals to grow. The oceans so generated, harbor the first life on Earth. These events are all interrelated.

REVIEW QUESTIONS

1. Explain the process of continental drift and the observations supporting that.
2. What were the arguments against continental drift?
3. All the continents started from a single land. How long ago was this? What was this land called?
4. How long ago did the present distribution of land take place?
5. Explain the theory of plate tectonics.
6. How many tectonic plates exist, and how did they form?
7. What is seafloor spreading?
8. What is the significance of the Mid-Atlantic Ridge?
9. Explain different types of plate tectonics.
10. How were the Hawaiian Islands formed?
11. What is the subduction process?
12. Explain which of the oceanic or continental plates are thicker and why.
13. How did the Himalayan mountain range form?
14. Explain the different types of rocks and their origin.
15. What is the Wilson cycle?

CHAPTER 17 REFERENCES

Marshak, S. 2012. *Earth: Portrait of a Planet*. 4th ed. New York: Norton.
Jordan, T.H., and J. Grotzinger. 2012. *The Essential Earth*. 2nd ed. New York: Freeman.

FIGURE CREDITS

- Fig. 17.1: Source: http://creationwiki.org/File:Continental_drift.jpg.
- Fig. 17.2: Source: https://en.wikipedia.org/wiki/File:Snider-Pellegrini_Wegener_fossil_map.svg.
- Fig. 17.7: Copyright © Alfred Wegener Institute for Polar and Marine Research/United States Geological Survey (CC by 4.0.)

THE EVOLVING EARTH:
A DYNAMIC HISTORY

CHAPTER LEARNING OBJECTIVES

This chapter will cover:
- Different stages in the evolution of Earth
- Timeline for the formation of oceans and the atmosphere
- Evidence for the first living cells
- Development of life through geological time
- Mass extinction events

Earth's crust developed around 4.4 billion years ago. Since then the surface of Earth has gone through dramatic changes due to the tectonic motions, continental formation and drift, seafloor spreading, mountain building, and many other processes, as I discussed in chapter 17. This is before secondary effects like erosions, earthquakes, and life started to reshape the planet. Understanding the evolution of Earth and the processes responsible for it are essential for studying the conditions needed for the emergence of plant and animal life. The problem, however, is that we only know of one planet like Earth, and therefore there is nothing to compare our findings with (unlike the stars and galaxies, for which we could find close analogs). As a result, it is difficult to view the early Earth. The only way is to extract information from the fossil and rock records preserved from the past. The problem with this approach is that the available records from the early Earth are not complete, as the material containing those records may have been eroded or converted to other forms.

At its very early stages, the surface of Earth was molten with no solid crusts formed. At the time, Earth was subjected to massive bombardment by meteorites, with its atmosphere containing no oxygen but mostly nitrogen, methane, ammonia, carbon dioxide, and water. The moon had about half its present distance from Earth and exerted significant tidal effect on the planet's surface. No geological records could survive the harsh conditions of the very early Earth. Therefore, our knowledge of the first 500 million years of Earth's history is very

"Oh threats of Hell and Hopes of Paradise!
One thing at least is certain—
This Life flies;
One thing is certain and the rest is Lies—
The Flower that once has blown forever dies"

- RUBAIYAT OF OMAR KHAYYAM

"It suddenly struck me that that tiny pea, pretty and blue, was the Earth. I put up my thumb and shut one eye, and my thumb blotted out the planet Earth. I didn't feel like a giant. I felt very, very small"

- NEIL ARMSTRONG

limited. About 80 percent of Earth's history, from 4.5 billion years to 700 million years ago, was in *Precambrian or Cryptozoic* (the Latin word for "hidden life") time. Information from this time is found from the igneous and metamorphic rocks. However, many of these early rocks are severely deformed and lack index fossils. Later history of Earth, including formation of the early atmosphere and seawater is revealed through the study of sedimentary rocks. Development of isotropic dating significantly changed our understanding of the chronology of the events on Earth (Marshak 2012).

This chapter presents the history of Earth at different geological times (Box 18.1) and investigates the conditions under which each event was led to the next. It studies the sequence of the events that led to Earth being the planet it is today. In this chapter units of time are defined in terms of giga-years (10^9) ago (denoted by GYA) and million years (10^6) ago (denoted by MYA).

STUDYING THE EARLY HISTORY OF EARTH

Early history of the Earth could only be studied indirectly using whatever evidence is left behind from that time. To perform such studies geologists look for the following signatures (Marshak 2012):

- Identifying early *orogens* (mountain belts) that have now been eroded and looking for rock records left behind by deformation (folds and faults), metamorphism, and igneous activity.
- Studying the age and growth of the continents. Using age dating techniques, geologists measure the age of the crusts at different points, the age and characteristics of the rocks, and when they were affected by orogenic activities. This reveals the tectonic environments under which the crusts formed.
- Studying the sedimentary rocks deposited in a location reveals the type of the sediments accumulated at that location. For example, the change in sea level can be measured by identifying the environments where rocks containing marine fossils were deposited.
- Looking for fossils and rocks formed at different latitudes, geologists look for plants that could grow in certain climate conditions. For example, finding tropical plants near the poles and dating them implies that at a specific time, the poles had a warmer climate and atmosphere.

BOX 18.1: GEOLOGICAL TIMES

Geologists have divided Earth's history into a number of time intervals. These intervals are not equal in length since they are based on geological events and biotic history of Earth. Examples of the events used to divide the geological times include appearance of animals with hard parts or extinction of plants and animals at a given time. The geologic times consist of the following.

Eons: These are the largest intervals of geologic times. They are hundreds of millions of years in duration.

Eras: Eons are divided into smaller eras. The boundaries of eras are determined by significant events in Earth's history. They are several hundred million years long.

Periods: Eras are divided into periods. The boundaries of periods are not as well defined as those for the eras. These are tens of millions of years long.

Epochs: These are smaller sub-divisions of periods and are only used for the more recent events. This is because geologic fossils at earlier times are eroded with many of their characteristics disappeared because of long-term processes on Earth. These are millions of years long.

TIMELINE OF THE EVOLUTION OF EARTH

Based on the latest estimates, Earth is 4.57 billion years old. Broadly speaking, the timeline for Earth's evolution is divided into two categories based on its biotic history: the period before advanced life, called the *Precambrian* eon (*Hadean*, *Archean*, and *Proterozoic* eons), and the *Phanerozoic* eon, consisting of *Paleozoic* (ancient life), *Mesozoic* (middle life), and *Cenozoic* (recent life) eras, indicating the period of Earth's history when life started to evolve and advanced life appeared (figure 18.1; Box 18.1). In the following sections I discuss different eons in the history of Earth (Marshak 2010).

PRECAMBRIAN EON (4.57 GYA–542 MYA)

This eon covers 80 percent of the age of Earth and is subdivided into three different eons: *Hadean*, *Archean*, and *Proterozoic* (figure 18.1). These span the history of our planet before complex form of life was developed.

THE HADEAN EON (4.57–3.85 GYA)

Geologists have no geological record between the time Earth was formed (about 4.57 GYA as measured from the age of planetesimals and meteorites), and the appearance of the oldest rocks (about 4.05 GYA) or formation of continental crust (about 3.85 GYA). The time interval between the formation of Earth and formation of the continental crust (4.57–3.85 GYA) is called the Hadean (meaning "the underworld" in Greek) eon. During this time Earth faced continuous collisions with planetesimals and grew larger. The kinetic energy due to these collisions was converted to heat. This extreme heat, combined with the thermal

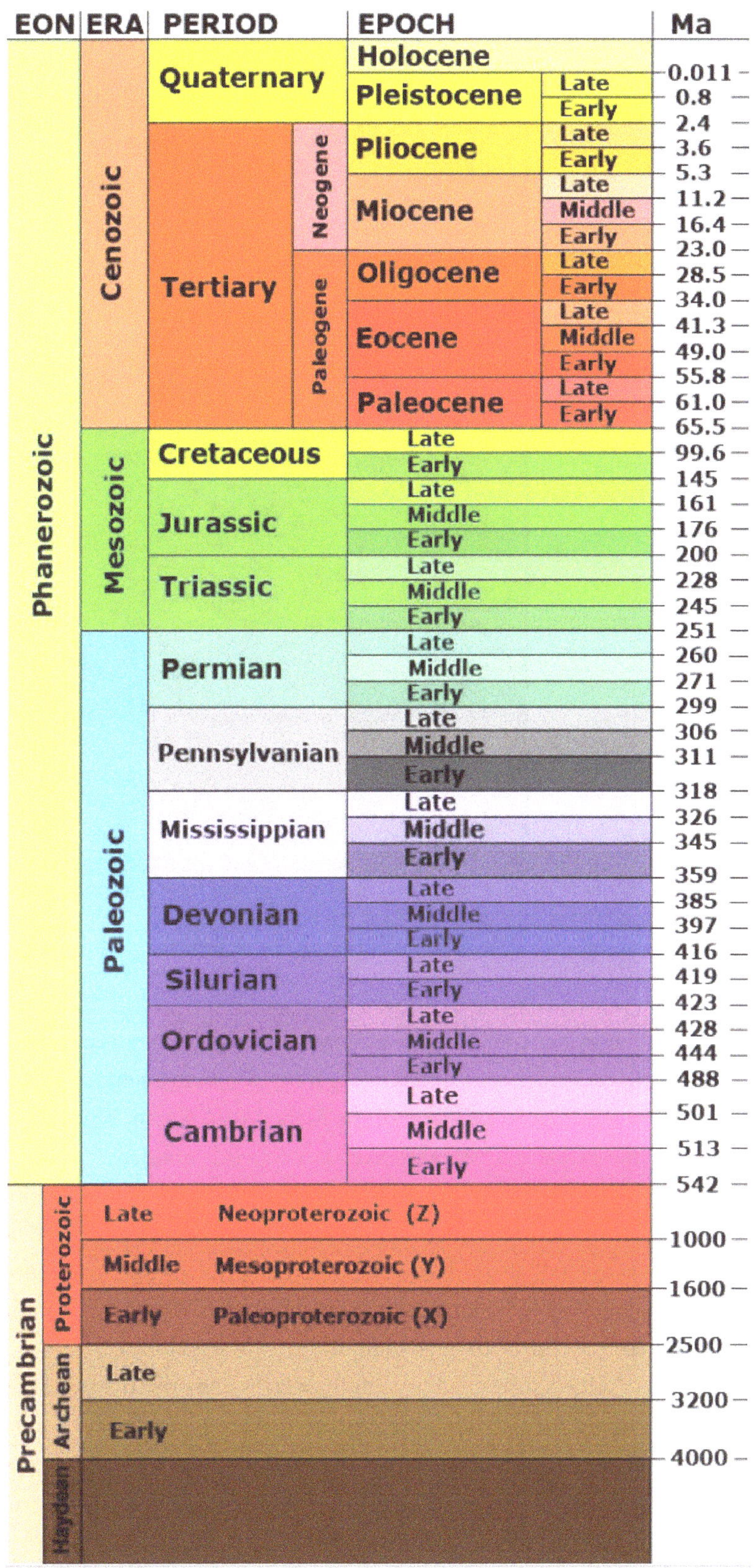

EON	ERA	PERIOD	EPOCH		Ma
Phanerozoic	Cenozoic	Quaternary	Holocene		0.011
			Pleistocene	Late	0.8
				Early	2.4
		Tertiary (Neogene)	Pliocene	Late	3.6
				Early	5.3
			Miocene	Late	11.2
				Middle	16.4
				Early	23.0
		Tertiary (Paleogene)	Oligocene	Late	28.5
				Early	34.0
			Eocene	Late	41.3
				Middle	49.0
				Early	55.8
			Paleocene	Late	61.0
				Early	65.5
	Mesozoic	Cretaceous	Late		99.6
			Early		145
		Jurassic	Late		161
			Middle		176
			Early		200
		Triassic	Late		228
			Middle		245
			Early		251
	Paleozoic	Permian	Late		260
			Middle		271
			Early		299
		Pennsylvanian	Late		306
			Middle		311
			Early		318
		Mississippian	Late		326
			Middle		345
			Early		359
		Devonian	Late		385
			Middle		397
			Early		416
		Silurian	Late		419
			Early		423
		Ordovician	Late		428
			Middle		444
			Early		488
		Cambrian	Late		501
			Middle		513
			Early		542
Precambrian	Proterozoic	Late	Neoproterozoic (Z)		1000
		Middle	Mesoproterozoic (Y)		1600
		Early	Paleoproterozoic (X)		2500
	Archean	Late			3200
		Early			4000
	Hadean				

Figure 18.1. Different eons, eras, periods, and epochs in the 4.5-billion-year history of Earth.

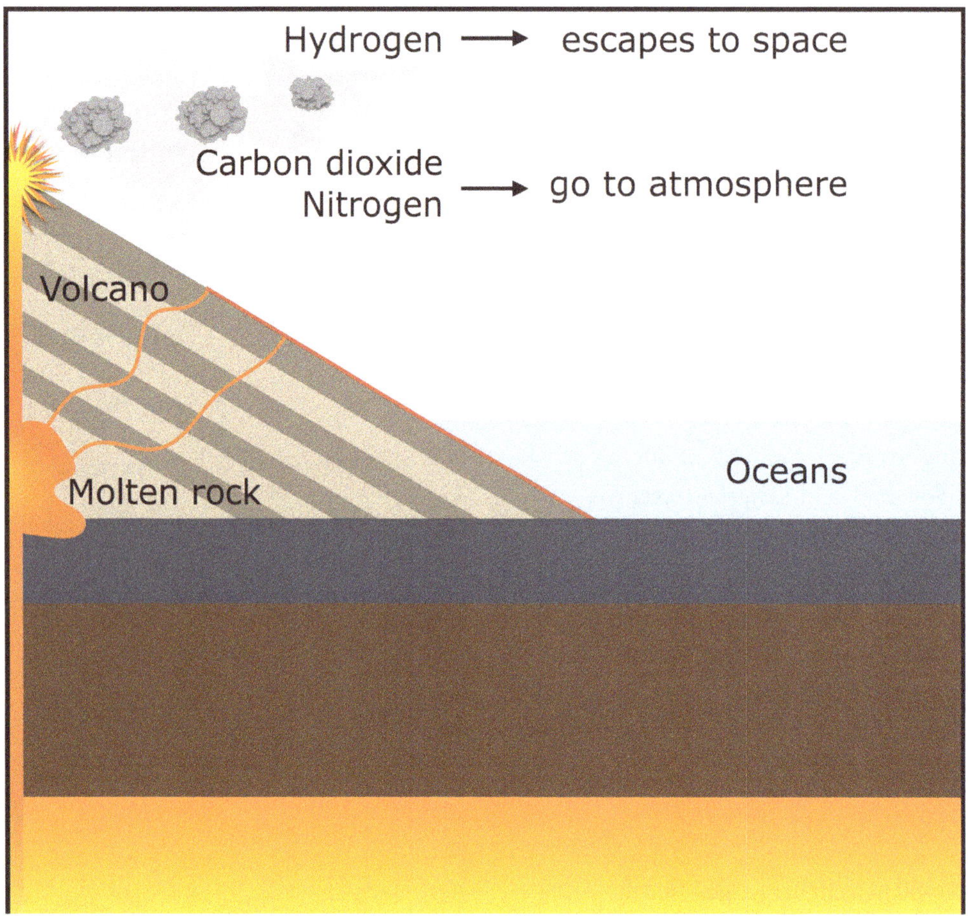

Figure 18.2. Shows the outgassing process by the volcanoes that results in the release of the material and chemical elements from Earth's center. This is responsible for the atmosphere and build up of water vapor which would subsequently condense and form rain filling the oceans.

energy generated due to radioactive decay of chemical elements, kept Earth in a molten state. Around 4.5 GYA Earth underwent differentiation by iron being sunk to the core of Earth due to its gravity. At about the same time, an object the size of the Mars hit the Earth. This increased the speed of Earth in its orbit around the sun and released debris that were subsequently condensed and formed the Moon (chapter 16; Figure 16.1) around 20,000 km from Earth (for comparison, Earth-Moon distance today is 384,000 km). Earth was glowing until 4.4 GYA, when the heat generated by radioactivity reduced because the radioactive elements with short half-life were all decayed. At this point the crust of Earth was formed and solidified. Evidence for this comes from the discovery of minerals called *zircon* found in Western Australia, initially formed in igneous rocks and estimated to have an age of about 4.4 billion years (Marshak 2012).

During the Hadean eon, materials from the mantle of Earth were released through the *outgassing* process (figure 18.2). This is the process of releasing material from the center of the Earth to the outside environment through volcanic activity. The outgassing is responsible for build up of gases in the Earth's atmosphere and for generating water in oceans. These volatiles (elements that could turn to other elements often in the form of gas) contained water (H_2O), methane (CH_4), ammonia (NH_3), hydrogen (H_2), nitrogen (N_2), carbon dioxide (CO_2), and sulfur dioxide (SO_2). It is likely that the water vapor released from the mantle was converted to liquid water

BOX 18.2: MAJOR EVENTS OF THE HADEAN EON

This is the time during which Earth was formed from the collision of planetesimals followed by accretion of dust and gas. The extreme heat produced by collision of extraterrestrial bodies produced molten rocks. When cooled, heavier elements like iron sank to the core and lighter elements like silicon moved to the surface of Earth. Discovery of zircon grains in Australia, dated 4.4 billion years ago, confirms that the crust of Earth was solidified during this time. Furthermore, traces of water were found in the zircon grains. These support the hypothesis that stable land was first formed during the Hadean time, as well as the first oceans and the atmosphere (from outgassing of gases).

(in the cooler environment outside the interior of Earth) and formed rain and hence the water in the oceans (figure 18.2). Therefore, it is possible that the first oceans were formed around 4.4 GYA. There are traces of water in the zircon grains found today implying that water indeed existed about 4.4 GYA. The first evidence for water (and possibly oceans) on Earth comes from the Hadean eon, around 3.85 GYA, from marine sedimentary rocks found in today's Greenland. The formation of oceans needs a cold and solid crust for Earth. During this time, Earth was also the target of heavy bombardment that would evaporate any existing water or destroy its young atmosphere formed from the outgassing process. This continued until 3.9 GYA when the intensity of the heavy bombardment diminished, so did the heat generated by the radioactive material (Marshak 2012; Box 18.2).

THE ARCHEAN EON (3.85–2.5 GYA)

The Archean (in Greek meaning "beginning") eon started when the solid crust of Earth was formed at the end of the Hadean eon about 3.85 GYA. During the Archean eon, the continents were formed, reshaping the surface of Earth. At the beginning of this eon, Earth was too hot to support solid crusts, but later on the molten surface was solidified into crusts and the land that turned into continents. The continents and volcanic arcs were formed around 3.2 to 2.7 GYA by the smaller lands formed in subduction zones. By the end of the Archean eon, about 80 percent of the continents had been formed (Marshak 2012). Also, the first rocks were formed in this eon (Box 18.3).

BOX 18.3: ROCK TYPES IN THE ARCHEAN EON

The crust of Earth formed during the Archean eon as the temperature decreased. This contained five rock types. Therefore, any such rocks found today, provide information about that eon. These include (Marshak 2012):
- **Gneiss:** Found in collisional zones from relics of Archean metamorphism.
- **Greenstone:** Formed from ocean crust between colliding continental crusts. Basalts formed in early continents are of this rock type.
- **Granite:** Formed from magmas generated by melting of crusts in continental volcanic arcs.
- **Greywacke:** A mixture of sand and clay eroded from volcanic areas and deposited in oceans.
- **Chert:** Formed from precipitation of silica in deep sea.

Undoubtedly, the most important event during the Archean eon was the development of the first form of life. The evidence comes from the rocks that preserve fossils of bacteria or *archaea* cells and by biomarkers, only produced by living organisms. A convenient tracer of living organisms is ^{12}C, associated with living things. The $^{12}C/^{13}C$ ratio in carbon-rich sediments is therefore used to look for early forms of life. Geologists have found rocks as old as 3.8 billion years that contain signature of living organisms. Archean rocks dating back to 3.2 GYA have been found that contain *stromatolites*, sediments produced by *cyanobacteria* (believed to be the first living thing on Earth). These are found in shallow waters in tropical environments. Therefore, the evidence indicates that the earliest life started deep in warm waters of the oceans during the *Archean* eon around 3.2 GYA while there is evidence that photosynthetic organisms first appeared around 2.7 GYA (chapter 21).

By the end of the Archean eon, Earth's crust was completely formed, the first continents were in place and primitive life had started in warm and deep waters. Plate tectonics were active at this time and continental collisions started to form mountain ranges. CO_2 in the atmosphere was resolved in water in the oceans, reducing the fraction of this gas in the atmosphere. This eon lasted until 2.5 GYA (Box 18.4).

There is evidence that primitive atmosphere and oceans emerged early in this time. Furthermore, the earliest sign of life in the form of bacteria and green algae are found in 3.5-billion-year-old rocks. The atmosphere was anoxic during this time. Volcanic activities produced much of the H_2O and CO_2 in the atmosphere but not much free oxygen. The bulk of the free oxygen in this eon was produced through photosynthesis of carbon dioxide and water by anaerobic *cyanobacteria* (blue-green algae), releasing oxygen as a by-product. Oceans were formed by condensation of water vapor from the outgassing material released by the volcanoes.

THE PROTEROZOIC EON (2.5 GYA–542 MYA)

The Proterozoic (in Greek meaning "earlier life") eon lasted for about 2 billion years and continued until the Cambrian period (where a big explosion of the species happened). The main events during this eon were the continued assembly of the continents from fast moving tectonic plates to more stable and larger constructions and enrichment of the atmosphere by oxygen. During this eon around 90 percent of the crust of the continents was formed. The continents collided and merged, eventually leading to the formation of a supercontinent, called *Rodinia*, around 1 GYA. This eon witnessed the rapidly changing map of Earth. It is hypothesized that around 750 MYA, the lands constitute today's India, Australia and Antarctica were separated from the supercontinent and floated around.

Another important event during the Proterozoic eon was the continued evolution of life from single-celled organism without a nucleus, called *prokaryotic* cells (archaea and bacteria), to more complex cells with nucleus, called *eukaryotic* cells. Evidence from biomarkers and fossils found in rocks confirms that the first *eukaryotic* cells existed as early as 2.1 GYA in rocks. Therefore, the eukaryotic cells which are the basis for the complex multicellular organisms today, first appeared during the Proterozoic eon. There is significant fossil evidence that these cells emerged in abundance around 1.2 GYA. Toward the end of the Proterozoic eon (around 565 MYA), more complex forms of multicellular organisms appeared on the scene. Fossils of these organisms have been seen well into the Cambrian before they were extinct. Also, for the first time simple organisms with the ability to move were developed. These include organisms such as jellyfish and worms.

Once life in its primitive form evolved, it affected the development and composition of the atmosphere. Before life arose, there was little oxygen in the atmosphere. Sometime during the Archean eon, oxygen was first emanated into the atmosphere by *cyanobacteria* that released oxygen as a waste product (oxygen was toxic to these bacteria). Later on, with the emergence of photosynthetic organisms that take CO_2 from the atmosphere and release O_2, oxygen entered the atmosphere in large fractions. The oxygen was first absorbed in the minerals

Figure 18.3. Traces of iron oxide (red regions) produced through the banded iron formation (BIF).

and after those could no longer absorb the oxygen, it was released to the atmosphere, leading to the *great oxygenation event* around 2.4 GYA (Marshak 2012). This event had huge impact on life and its evolution by allowing more complex multicellular organisms to develop. The atmospheric oxygen allowed formation of *ozone* (O_3) in the atmosphere that could block the harmful ultraviolet radiation from the sun. This made life on the land possible.

There were other effects the oxygenation had on the world. Before the great oxygenation event took place, whatever iron existed, was in the form of chemicals soluble in seawater. Once oxygen was initially released, it entered into reaction with iron, changing it to chemical forms like iron oxide that could not be dissolved in water and would sink as sediments to the floor of the oceans. Eventually, this became rock-type sediment called *banded iron formation (BIF)* that is a result of layers of iron oxide minerals. Traces of BIF are found among minerals and are the source of iron today (figure 18.3). The BIF is used to estimate the time the oxygenation of the atmosphere took place. These studies indicate that the oxygenation of the atmosphere completed by 1.8 GYA. Apart from generating ozons, the increase in the oxygen in the atmosphere made it easier for multicell organisms to generate metabolic energy more efficiently.

At the end of the Proterozoic eon, the climate on Earth became cold—both on land and sea. The ice layers on the oceans cut the supply of oxygen to oceans resulting in the extinction of sea life. The glaciers age came to an end when CO_2 gas was built up in the atmosphere due to volcanic activities and could not be absorbed by oceans because of surface ice. The increased CO_2 in the atmosphere caused *greenhouse effect*, warming Earth, resulting in melting of the ice on land and in oceans (Box 18.5).

BOX 18.5: MAJOR EVENTS OF THE PROTEROZOIC EON

One of the most important events of this eon was the accumulation of oxygen in the atmosphere. Oxygen generation started as early as the *Archean* time but the amount released to the atmosphere was insignificant until all the sulfur and iron that absorb oxygen were oxidized (by absorbing oxygen). After this point, oxygen was released to the atmosphere around 2.3 GYA. This eon also witnessed formation of continents and significant tectonic activities. While *stromatolites* prosper during this time, the first complex cells, eukaryotes, and multicellular forms of life emerged at this eon.

The Phanerozoic eon (in Greek meaning "visible life") is the most recent eon and, as a result, has left many signatures that can be used to find details about this time (figure 18.1). During this eon the continents changed, mountain ranges raised and diversity of organisms appeared with all imprinted in fossils today. This eon is divided into three eras: *Paleozoic* (in Greek meaning "ancient life"), the *Mesozoic* (in Greek meaning "middle life"), and *Cenozoic* (in Greek meaning "recent life") (Box 18.6). The following describes each of these eras.

Early Paleozoic era (542 MYA–444 MYA): This era covers the Cambrian and Ordovician periods (figure 18.1). At early Phanerozoic era (Cambrian period) the continent Rodinia broke into smaller continents (among these are *Laurentia* that is composed of North America and Greenland and *Gondwana* composing of South America, Africa, Antarctica, India, and Australia, Figure 17.1). This followed by an increase in sea levels, generating shallow waters within which marine life could evolve. The sea level later reduced during this era due to the accumulation of sediments.

The fossil records from this era show a diversification of organisms. This diversification that took place after the Cambrian period is called the *Cambrian explosion* that lasted for 20 million years. The reason this happened is likely because of the breaking up of the continents with each piece of land developing its own ecological environment. The animals during the Cambrian period were shell-like and by the end of this period *trilobites* (Chapter 23) were living on the seafloor.

The first vertebrate animals (in the form of jawless fish) lived in the early Paleozoic era (during the Ordovician period). There were no land organisms for most of this era, with our earliest evidence for a land plant or animal being in the late Ordovician period and in the form of algae. Life started to move to land from the sea soon after the ozone formed in the atmosphere and protected Earth from ultraviolet radiation from the sun. There was a mass extinction at the end of the Ordovician period because of a decrease in temperature that led to glaciation.

There are distinct differences in the marine life between the Cambrian and Ordovician periods. The most obvious is the diversity in both plant and animal life. During the Cambrian period the *faunas* were simple in structure with *trilobites* and sponges covering the bottom of the oceans. This was dramatically changed during the Ordovician period where there were a number of organisms reaching over half a meter above the seafloor. There were only 150 families of animals during the Cambrian period, increasing to 400 families in the Ordovician period. The Cambrian faunas with low diversity were simple in structure (Marshak 2012).

Middle Paleozoic era (444–359 MYA): This era covers the *Silurian* and *Devonian* periods. The glaciation ended during the Silurian period because of the green house effect. The CO_2 gas, generated by volcanoes, could not be absorbed by water because the oceans were covered by ice. As a result the CO_2 was trapped in the atmosphere, causing climate to warm up that, in turn, led to the melting of the ice and raising the sea levels. Orogeny (mountain building) continued during this time.

Because of the melting of the ice on the sea, the marine life started again and new species replaced those that were disappeared during the mass extinction of late Ordovician period. For the first time plants constituting woods, with seeds and veins lived on the land. The plants grew bigger as they evolved and, at the end of Devonian period, there were thick forests on the land. Around this time, insects, spiders, and scorpions appeared on land and sharks and bony fish mastered the oceans. Late in this era, *amphibians* moved to the land. The first animal to walk on the land is likely to be *Tiktaalik*—a kind of fish that lived in oxygen poor environment of shallow waters in late Devonian period, about 375 MYA. The *Tiktaalik* represents the first evidence for transition from fish to amphibians who developed lungs and started to inhale air (Figure 23.13). They are likely the ancestors of today's reptiles, birds, and mammals (Box 18.6).

Late Paleozoic era (359–251 MYA): This era cover the *Carboniferous* and *Permian* periods. The climate cooled down and the water levels on the seas reduced. During the Carboniferous period the seas gave way to continents, creating coastal regions and river deltas where sand and organic material would reside. Part of these lands that moved close to the equator grew large wood forests (because of the tropical weather) that eventually turned to coal after being buried for many millions of years. An important event during this period was the continued collision of the continents, resulting in a super-continent called *Pangea*. During this time some of today's structures on Earth were formed (what is Africa today collided with southern Europe, while China and Siberia connected) (Box 18.6).

Biological evolution continued during late Paleozoic era. At the end of the Carboniferous period insects such as cockroaches came to the scene while amphibians and reptiles appeared on the land during the Permian period. An important event here was a new way of reproduction, with the reptiles laying eggs and not having to go to water to reproduce. This allowed the animals to inhabit areas that were previously nonlivable. There was a mass extinction event during late Paleozoic era. This is likely caused by extensive volcanic activities that blocked the sunlight and changed the chemistry of water in oceans, extinguishing up to 95 percent of the species on Earth (Marshak 2012).

BOX 18.6: MAJOR EVENTS OF PALEOZOIC ERA

This started 542 MYA and ended around 251 MYA. There was a diversification of life during this era that is known as Cambrian explosion. This was caused by breaking up the continents and hence, development of different ecosystems depending on land distribution. Given the abundance of oxygen in the atmosphere, multicellular organisms were developed. Furthermore, formation of the ozone layer allowed the living organisms to move from sea to the land. Fossil evidence indicates that the first animal walking on land was a fish that lived in shallow waters around 375 MYA. This was also the first known transition from fish to amphibians. They developed lungs and are ancestors of reptiles, birds, and mammals. During this era the continents collided and merged, and biological diversity continued. A new way of reproduction was developed with reptiles laying eggs without the need to reproduce in water. For the first time new types of plants populated the land. The Paleozoic era witnessed many mass extinctions caused by glaciation and accumulation of CO_2 in the atmosphere produced by volcanic activities.

THE MESOZOIC ERA (251–64 MYA)

Early and middle Mesozoic era (251–145 MYA): This era consists of *Triassic* (248–206 MYA) and *Jurassic* (206–144 MYA) periods. A significant effect during this era was the breaking up of the super continent, Pangea, due to rifting. As a result, North America was separated from Europe and Africa. Subsequent to that, the Atlantic Ocean started to grow and mid-Atlantic ridge was formed. During this time the climate was relatively warm, becoming cooler toward late Jurassic period. Another significant event in the Mesozoic era was the creation of volcanic island arcs by subduction. There were these islands that were merged with existing continents generated from breaking up of the Pangea, resulting in an increase in the size of the continents.

Following the mass extinction during the Permian period, early Mesozoic era witnessed appearance of swimming reptiles in the sea, first turtles on the land, and flying reptiles in the air. These animals diversified during this time. Dinosaurs were first appeared at the end of the Triassic period. By the end of the Jurassic period, huge dinosaurs were ruling Earth. Dinosaurs weighted in excess of 100 tons, were warm-blooded, with their legs located under their bodies unlike the reptiles. At this time the first feathered birds, *Archaeopteryx*, appeared on the skies (Figure 23.20). At the end of the Triassic period, the earliest ancestors of mammals—a creature looking like a rat—came into existence (Box 18.7).

Late Mesozoic era (145–64 MYA): This mainly covers the *Cretaceous* period. The Pangaea further broke, with Africa and South America separated from the Antarctica and started to move away from one another, increasing the size of the Atlantic ocean. Also, what is today's Australia was separated from the Antarctica, with India moving toward the "mainland" Asia. Sea spreading happened fast, with increased frequency for volcanism. The CO_2 gas released by volcanic activities increased the temperature of the atmosphere through green house effect, raising water levels of the sea, causing severe flooding of the continents. During the Cretaceous period a new type of fish with short jaws ruled the seas along with giant turtles. On the land, dinosaurs were occupying all the habitats. Birds were diversified and mammals acquired larger brains, still remaining small in physical size.

At the end of Cretaceous period, about 65 MYA, Earth experienced a huge impact by a 13-km-wide meteorite at where is now the Yucatan Peninsula in Mexico. The evidence from this comes from a rapid change in the composition of strata between the late Mesozoic era (Cretaceous period) and early Cenozoic era (Tertiary period). This is termed "KT" from the first letters of Cretaceous and Tertiary (figure 18.4). Fossil dating techniques confirm that this difference happened abruptly, supporting a sudden and catastrophic event. Geologist Walter Alvarez found that a thin layer of clay was deposited between the deep-sea limestone layers of the Cretaceous and Tertiary periods (figure 18.4). Cretaceous plankton (microorganisms consisting of bacteria drifting on the surface of lakes) shells were found in the limestone below the clay layer, while Tertiary limestone (with no plankton) was found above the clay layer. This indicated that at some point suddenly all the plankton disappeared between these two periods with the only material settling outside the sea being the clay (Marshak 2012). Further analysis of the clay revealed that it contained other unusual ingredients like wood ash and shocked quartz (grains of quartz subjected to intense pressure). They also found that the clay contained an extremely heavy element called *iridium* that is only found in extraterrestrial objects. Geologists analyzing the clay between these two periods soon found that all the clay in the world from that period constituted this element. The only way to explain these combined findings was by a major collision of an extraterrestrial body with Earth. The ash resulted from burning of the forests, with the force of the impact causing the shocked quartz grains. This catastrophic collision happened 65 MYA and led to the extinction of dinosaurs and a mass extinction of plants and animals at the end of the Cretaceous period. The force of this impact led to tsunamis causing continents to go under the seawater, produced so much heat that blazed the forests, evaporated the

Figure 18.4. The dark band indicative of the approximate boundary between the Cretaceous (below) and Tertiary (above) age rocks. This constitutes a layer of clay, also containing coal.

BOX 18.7 MAJOR EVENTS OF THE MESOZOIC ERA

This era witnessed division of the supercontinent Pangaea into smaller continents. Nonavian dinosaurs ruled Earth for over 135 million years and reached their peak during the Jurassic period. Birds also first appeared in the Jurassic period, having evolved from the dinosaurs. First mammals emerged during this time but their sizes were small. This era ended with a mass extinction 65 MYA that is believed to have killed the dinosaurs and extinguished animal and plant life.

water in oceans and ejected debris to the atmosphere. The debris blocked the sunlight for months, decreasing the atmospheric temperature. This, combined with the acid rain produced by the interaction of the chemicals with water, slowed down the photosynthesis, breaking the food chain and causing mass extinction (Marshak 2012) (Box 18.7).

THE CENOZOIC ERA (65 MYA–PRESENT)

This is the closest era to the present time (from the Greek *ceno* meaning "new" and *zoi* meaning "life"—"new life"). As such, more detailed information is available from this era, and therefore, it can be further divided into smaller time segments (periods and epochs) (Box 18.8).

During this era, the Pangaea was further divided and evolved, resulting in the formation of the orogeny and the distribution of the continents we observe today. In this era, Australia was separated from Antarctica, Greenland was separated from North America, and the North Sea was formed between Britain and continental Europe. Due to seafloor spreading, the Atlantic Ocean became bigger, causing North America to move westward from Europe. India and a number of volcanic islands collided with Asia to form the Himalayas and the Tibetan Plateau. Similarly, Africa and some volcanic islands collided to form the Alps while convergent boundaries in South America formed the Andes (Marshak 2012).

BOX 18.8: CENOZOIC ERA—THE MOST RECENT GEOLOGIC TIME

The Cenozoic era consists of two periods: Tertiary (65 to 1.8 MYA) and Quaternary (1.8 to the present).

The **Tertiary** period is further divided into two epochs: **Paleogene** (a Greek word for "old origin"), covering 65 MYA to 24 MYA; and **Neogene** (a Greek word for "new origin"), covering 24 MYA to 1.8 MYA (this divide is specific to the Cenozoic era).

The Paleogene is subdivided into three epochs: **Paleocene** (a Greek word for the "old recent"), covering 66 MYA to 56 MYA; **Eocene** (from the Greek word for "dawn of new"), covering 56 MYA–34 MYA; and **Oligocene** (meaning "few recent" in Greek), covering 34 MYA to 23 MYA.

The Neogene period is subdivided into two epochs: **Miocene** (23 MYA to 5 MYA—meaning "less recent" in Greek), and **Pliocene** (5 MYA to 2.5 MYA—meaning "more new" in Greek).

The **Quaternary** period is divided into two epochs: **Pleistocene** (2.5 MYA to 11,000 years ago—meaning "most new" in Greek), and **Holocene** (11,000 years ago to present—meaning "wholly new" in Greek).

BOX 18.9: MAJOR EVENTS OF THE CENOZOIC ERA

During this era the continents evolved to the configurations we see today. Cooling of the climate resulted in a decrease in water level in the oceans and creation of land bridges between the continents, leading to migration of animals on the land. Many of the landmarks of today's world, including mountain ranges and volcanic islands, were generated during this time. The Cenozoic era is also known as the age of mammals. The apelike primates first appeared in this era, including the very first appearance of our ancestors over 2 MYA.

Climate became cooler during the Cenozoic era, and by the early *Oligocene* epoch (33 MYA), Antarctic glaciers were formed. This cold trend continued until the Late Miocene epoch (about 11 MYA). This led to freezing of the oceans that resulted into recession of the water, with the land being exposed to air. This resulted into "land bridges" between different continents, connecting for example, Alaska to East Asia, providing migration routes for humans and animals. Similar routes were formed between Australia and southeastern Asia, leading to migration of people and animals to Australia. The climate started to warm up around eleven thousand years ago in the *Holocene* epoch that extended to the present climate.

At the time the world settled after the mass extinction at the end of the Cretaceous period, the plant life started and forests grew. The grasses spread on the land by the middle of this era and the tropical temperatures at the Cenozoic era led to the spread of forests. The fossil evidence confirms that most of the mammals around us today, first appeared in the beginning of the Cenozoic era. In particular large mammals appeared during this time but were extinct over the past ten thousand years. It was in the Cenozoic era that apelike primates, our distant ancestors, first appeared and diversified in the Miocene epoch about 20 MYA. This was followed by the appearance of the first human-like primates about 4 MYA and the members of the human family, *Homo*, around 2.4 MYA, all in Africa. Our own species, *Homo sapiens,* obtained its identity about five hundred thousand years ago. The first appearance of modern humans was around two hundred thousand years ago. All these happened during the Pleistocene and Holocene epochs of Earth's history (Box 18.8), when the climate went through a rapid change (Marshak 2012) (Box 18.9).

MAJOR MASS EXTINCTIONS IN EARTH'S HISTORY

During its life, Earth witnessed a number of devastating mass extinctions that wiped out many species of plants and animals. After each of these extinctions, life started again, creating new species, depending on the environment and conditions at the time. The fact that life rebuilt itself on the planet after each of these extinctions means that the conditions for life are reproducible. In other words, the cause of life wasn't a single event in the distant past or was not spontaneous. The known extinction events in chronological order include:

- Early Cambrian (512 MYA) was the earliest known mass extinction, wiping out 50 percent of all marine species.
- End-Ordovician (439 MYA) extinction removed 85 percent of marine species, including many trilobites
- Late-Devonian (365 MYA) extinction led to the disappearance of 70 to 80 percent of plant and animal species, including corals and brachiopods.
- Permian-Triassic (251 MYA) was the biggest Earth extinction, with 96 percent of marine and terrestrial species disappearing.

- End-Triassic (199 MYA) event led to the extinction of 76 percent of mostly marine species, including sponges, gastropods, cephalopods, insects, and vertebrates.
- End-Cretaceous (65MYA) extinction is known as the event that led to the demise of dinosaurs. Over 80 percent of the species on land and in the sea vanished. Because this is the most recent extinction, more details are available about it.

The cause of the extinctions varies. Many of them did not happen at an instant but were result of a gradual process taking millions of years. For example, the end-Ordovician extinction was caused by changes in sea level affecting only the marine life (since this was the only place life could exist at that time). The late-Devonian extinction happened over a period of 20 million to 25 million years, caused by global cooling (as it only affected the species in warm regions). The Permian-Triassic extinction was the most devastating event and was caused by two extinction events separated by 10 million years. It took Earth about 20 million years to recover from the effect of this extinction. This event changed the history of life on Earth. The cause of this extinction is not clear but is likely a number of different events including: sudden release of CO_2 and methane to the atmosphere, volcano activity, and asteroid collision. The last major extinction (End-Cretaceous) was caused by collision of a meteorite or comet of at least 10km in diameter. The compelling evidence for this comes from the excess of iridium (an element which is known to come from extraterrestrial objects) in sediments from the Cretaceous-Tertiary period. The location of this impact was found to be a crater in Yucatan Peninsula in Mexico with 180 km in diameter and 20 km in depth (Prothero and Dott 2010).

SUMMARY AND OUTSTANDING QUESTIONS

Earth has had a dynamic history, with its map constantly changing. Development of accurate isotopic dating techniques for determination of the ages of rocks led to detailed study of the evolutionary history of Earth. A sequence of sedimentary rocks lying horizontally on top of each other, with each layer younger than the layers beneath it and older than the layers above it, determines the order of the events in the history of Earth. The fossils found in each of these layers tell us about the organisms present at the time. The inconsistencies and gaps between the sequences (when no rocks were deposited or existing rocks were eroded) indicate deviations from continuous evolution such as mass extinctions. This allowed geologists to divide geologic time into eons, eras, periods, and epochs. The geologic time intervals are not equal but are divided based on major events taking place at these times.

Because of the extreme heat soon after the formation of Earth (4.57 GYA), no rock records exist from the first 600 million years of Earth's history. The continental crust was formed around 3.85 GYA, when the primitive atmosphere emerged with very little oxygen. The earliest fossil records show that the first form of life started during this time in the form of bacteria and archaea. The solid and permanent crusts formed during the Archean eon collided and merged, resulting in orogenic belts as well as the oceans. Around 2.5 GYA, photosynthesis by simple organisms (such as cyanobacteria) increased the oxygen level in the atmosphere. Around 1 GYA, continental crusts merged, producing the supercontinent Rodina. This was broken into smaller continents that, in turn, collided and coalesced again, leading to the formation of mountain ranges and a new supercontinent, Pangaea, around 500 MYA. During this time, many plants and insects appeared on the land, invertebrates with shells and jawless fish in the sea. Around 200 MYA, Pangaea broke into smaller lands, leading to the formation of the Atlantic Ocean, among other things. For about 135 million years dinosaurs ruled Earth until their extinction 65 MYA. Around this time, continents collided and eventually settled to the land distribution we see today. The first primates appeared around 4 MYA with the first of the human family appearing 2.4 MYA. Homo sapiens started to appear on the scene around five hundred thousand years ago.

There are outstanding questions as to whether this history is unique to the Earth or may be repeated in other planets. Could we use the information obtained from Earth observations to identify planets at the same stages of evolution as the Earth billions of years ago? What are the signatures that provide information about conditions on other planets and the existence or nonexistence of life? Could we increase temporal resolution of our observations, finding more details about earlier times in Earth's history? It is not yet clear when the first oceans appeared. Data based on zircon grains indicate the first evidence for water on Earth, or the first oceans, 4.4 GYA. However, there are also alternative interpretations for those data.

REVIEW QUESTIONS

1. Geologists divide Earth's history into two broad categories. What are these, and what is the main criterion used for this division?
2. What is the definition of *volatiles*? Give a few examples of volatiles.
3. When was the first evidence of water found, and how was this discovered?
4. How did the Earth's crust form and solidify?
5. When were the continents formed?
6. How was the earliest form of life discovered? In what form was the earliest life?
7. How oxygen first emerged in the atmosphere?
8. Explain the main events during the Proterozoic eon.
9. Explain the great oxygenation event.
10. What is the likely reason behind *Cambrian explosion*?
11. What are the differences in marine life between the Cambrian and Ordovician periods?
12. What caused the end of glaciation during the Silurian period?
13. How and at what period was the Pangea supercontinent formed?
14. What was the cause of the mass extinction during the Paleozoic era?
15. What were the major geological events during the early and middle Mesozoic era?
16. Explain how land bridges between different parts of the world were formed. What was the result of these land bridges?
17. When did the first mammals appear?
18. What were the causes of mass extinctions throughout the history of the Earth?

CHAPTER 18 REFERENCES

Marshak, S. 2012. *Earth: Portrait of a Planet*. 4th ed. New York: Norton.
Prothero, D.R., and R.H. Dott. 2010. *Evolution of the Earth*. 8th ed. New York: McGraw-Hill.

FIGURE CREDITS

- Fig. 18.1: Source: https://commons.wikimedia.org/wiki/File:Geologic_time_scale.jpg.

- Box 18.3: Stephen Marshak, "Rock Types in the Archean Eon," *Earth: Portrait of a Planet*, Fourth Edition. Copyright © 2012 by W. W. Norton & Company, Inc.

- Fig. 18.3: Copyright © Graeme Churchyard (CC by 2.0) at https://en.wikipedia.org/wiki/File:Banded_iron_formation_Dales_Gorge.jpg.
- Fig. 18.4: Source: https://en.wikipedia.org/wiki/File:KT_boundary_054.jpg.

EMERGENCE OF THE CONDITIONS FOR LIFE

CHAPTER LEARNING OBJECTIVES

This chapter will cover:

- The origin of Earth's atmosphere
- The origin of the oxygen and ozone in the atmosphere
- Why water is so essential for life
- The origin of seawater
- The Ocean-Earth, Atmospher regulatory system
- The greenhouse effect

During its lifetime, Earth faced many catastrophic events, each time leading to the extinction of 70 percent to over 90 percent of its species (chapter 18). Each mass extinction event led to the planet being uninhabitable for millions of years after which life started again. This indicates that after each mass extinction the right conditions for life were achieved. In other words, Earth readjusted itself repeatedly to allow emergence and evolution of life again and again. There are many factors that are essential to support and sustain life. These are correlated, producing a complex chain of events, all needed in order for life to sustain. The creation of the atmosphere, the oxygen and ozone levels in the atmosphere, the presence of certain gases that could retain heat, the regulator systems, and the production of water and biochemical material are all substances needed for life to start and sustain.

However, to study the ingredients required for life, we first need to define what is meant by "life." We only know of one life, and that is the life on the planet Earth. Therefore, any knowledge about life, its ingredients and its origin, depends on our own experience and on the environments and conditions we ourselves have been exposed to. Similarly, when we look for conditions suitable for life in other places in the universe, we look for the biosignatures similar to the ones that are known to be supporting life here. Given these boundary conditions, if there is a different kind of life out there in other planets, formed and sustained with different ingredients and conditions, we will not be able to detect it because we are looking

"A story has no beginning or end: arbitrarily one chooses that moment of experience from which to look back or from which to look ahead."

- GRAHAM GREENE

"If all difficulties were known at the outset of a long journey, most of us would never start out at all."

- DAN RATHER

for "different" biosignatures. Apart from the ingredients essential for life (either plant or animal life), we also need conditions to support and sustain it and the "right" environment to allow life to evolve and prosper.

This chapter discusses the conditions required for the emergence of life on Earth. These include the origin of the atmosphere, the oxygen buildup in the atmosphere, the origin of water on Earth, and the creation of the ozone layer, the Earth-Ocean-Atmospher regulator system and green house effect.

THE ORIGIN OF EARTH'S ATMOSPHERE

Earth's atmosphere was formed well before life could migrate from the sea to the land. Once life emerged, it influenced the atmosphere. It is logical to assume that the primitive atmosphere of the early Earth was similar to the atmosphere of the less-developed solar system planets today, or that it contained the same gases as the meteorites that are likely to resemble the early terrestrial planets. However, the only other solar system planet (apart from the Earth) that has retained its atmosphere is the Jupiter, which has a very different composition in its atmosphere than Earth. Similarly, the meteorites are found to contain different gases compared to Earth's atmosphere. Therefore, there is no direct analog to Earth's atmosphere and one should seek other explanations for its origin, independent of other solar system planets (Prothero and Dott 2010).

The leading hypothesis for the origin of Earth's atmosphere is the process in which the gas from Earth's interior is transferred to its surface through volcanic activities, called the *outgassing*. This transfers steam, carbon dioxide, nitrogen, and carbon monoxide from Earth's interior (figure 18.2). The process could well explain the abundance of atmospheric nitrogen, helium, argon, and water vapor. The steam (water vapor) condenses once it hits a cooler environment and turns to rain.

The production of the atmospheric gases took place through chemical reactions between the primitive gases that appeared in the atmosphere through the outgassing process. The atmosphere then was similar to that of Jupiter consisting of methane (CH_4) and ammonia (NH_3). In this scenario the current composition of Earth's atmosphere was produced through the following steps:

1. The UV radiation from the sun dissociated water vapor into hydrogen and oxygen: $2H_2O + UV\ light\ energy \rightarrow 2H_2 + O_2$. The hydrogen generated this way will escape to space and will not be retained by Earth because of its low atomic mass.
2. The oxygen molecule formed in the above process will interact with the existing methane, forming carbon dioxide and water vapor: $CH_4 + 2O_2 \rightarrow CO_2 + 2H_2O$.
3. Oxygen also enters into reaction with ammonia to form nitrogen and water: $4NH_3 + 3O_2 \rightarrow 2N_2 + 6H_2O$.

The nitrogen (N_2) and carbon dioxide (CO_2) in the atmosphere are produced through these reactions. Once all the methane and ammonia are consumed, accumulation of O_2 will accelerate as more water vapor is dissociated with nothing left to enter into reaction with it. This explains the abundance of oxygen, nitrogen, CO_2, and water in the atmosphere (Bennett and Shostak 2005).

THE ORIGIN OF THE OZONE IN THE ATMOSPHERE

The presence of the ozone (O_3) in Earth's atmosphere is of outmost importance for the development of life. It protects Earth from high-energy ultraviolet radiation emitted by the sun. Ozone is produced by breaking of oxygen molecules (O_2) to free oxygen atoms (O) by the ultraviolet radiation (with energy $h.\nu_{uv}$ where h is Planck's constant and ν_{uv} is the frequency of the ultraviolet radiation) from the sun. The free oxygen then combines with the molecular oxygen to form ozone, as shown below:

$O_2 + h\nu_{uv} \rightarrow 2O$

$O + O_2 \rightarrow O_3$

The ozone is unstable, and as soon as it is hit by ultraviolet light, it breaks into an oxygen molecule (O_2) and an atom (O). Through this process, the energy of the ultraviolet radiation from the sun (which is biologically harmful) is converted into heat energy.

$O_3 + h\nu_{uv} \; (from\ the\ sun) \rightarrow O + O_2$

The oxygen atom so produced, combines again with an existing oxygen molecule, producing the ozone. This oxygen-ozone cycle continues. The ozone started to form when the oxygen level in the atmosphere reached a critical level (about 10 percent of its present abundance). The ozone occupies an altitude of about 15 km to 30 km above Earth (Bennett and Shostak 2005).

THE ORIGIN OF SEAWATER

In the outgassing hypothesis, the water vapor from Earth's interior condenses when it encounters colder temperatures upon its release. This converts the water vapor to liquid water, filling the oceans. Therefore, the rate of accumulation of seawater is directly proportional to the atmospheric production of water vapor. Additional water was also added to the oceans with a slower pace through volcanic activities. It is also possible that some comets have carried water with themselves into Earth (Prothero and Dott 2010).

The source of the chemical composition of the seawater is connected to the rocks. For example, the salt existing in the seawater originated from the rocks due to streaming rivers washing and carrying it to the seas. Over time, the concentration of the salt and other chemicals increased until a state of equilibrium reached when ocean water could no longer dissolve any chemicals. Similarities between the fossils of sea life and the living organisms today indicate that the change in the composition of seawater slowed down around 600 million years ago.

Study of the zircon grains indicates that liquid water existed on earth more than 4 billion years ago, with oceans forming near the same time. This required the existence of an atmosphere at that time. This implies that most of the outgassing of the atmosphere was probably completed by that time. It was also found that zircons contain minerals indicating plate tectonic activity around 4 billion years ago. This absorbed vast amount of atmospheric carbon dioxide, reducing the greenhouse effect and leading to a cooler temperature for earth, formation of solid rocks, and life.

The Mars-size impact more than 4.4 billion years ago (Chapter 16) led to the melting of the rocks. This rock vapor eventually condensed and resulted in a carbon dioxide atmosphere as well as hydrogen and water vapor. Because of the high CO_2 atmospheric pressure, liquid water oceans existed that early in the history of earth despite a surface temperature as high as 230 degree Celsius. As the earth cooled (because of the removal of greenhouse gases from the atmosphere), CO_2 was dissolved in water, further reducing the greenhouse effect and the temperature.

THE ORIGIN OF OXYGEN IN THE ATMOSPHERE

Oxygen constitutes 21 percent of the atmospheric gases today, the second-most abundant element in the atmosphere after nitrogen. However, the percentage of oxygen was not always this high, with the accumulation of oxygen in the atmosphere being a gradual process taking billions of years. Oxygen was first released to the atmosphere over 3 GYA (during the Archean eon) by microorganisms called *cyanobacteria,* which are among the first living creatures on Earth. These microorganisms were *anaerobic,* thrived in the absence of oxygen, and obtained their energy from sulfate. These consist of algae, green plants, and some kinds of bacteria and produce oxygen through the photosynthesis process—taking energy from the sunlight and converting water and carbon dioxide to carbohydrate and oxygen. Oxygen was released to the atmosphere with the carbohydrate (sugar) stored as an energy source by the organism. This process took place around 2.7 GYA. The oxygen so produced entered into reactions with iron that was dissolved in seawater, producing iron oxide minerals that subsequently sank to sea level. These minerals are found in thin layers of sediments called banded iron formation (BIF; chapter 18). This process continued until all the dissolved iron was used, and as a result, oxygen could no longer be absorbed. From this point onward, oxygen was released to the atmosphere, building up the oxygen level (Cranfield 2014).

Once the level of oxygen molecules increased in the atmosphere, ozone could form and generate a protective shield against the high-energy ultraviolet radiation from the sun. This allowed the emergence of plants and animals and the development of life on land. The plants then took away carbon dioxide from the atmosphere and replaced it with the atmospheric oxygen through the photosynthesis process. While ozone provided the protection for all the land life, the organisms themselves generated most of the oxygen from which the ozone was formed through the feedback process explained above (Cranfield 2014).

There is evidence that transformation from an oxygen-poor to oxygen-rich atmosphere took place between 2.4 and 1.8 GYA. This was confirmed by observations of the BIF, a process that could only happen in an oxygen-rich atmosphere. Furthermore, observations of sand stones show well-formed *pyrites* (iron sulfide) in sediments from a time before 1.8 GYA. This could only happen in an oxygen-poor environment, since in oxygen-rich atmospheres pyrites are oxidized and rust and hence do not survive long enough to remain as sediments. By the great oxygenation period, the amount of oxygen had risen to 3 percent of the gases in the atmosphere and stayed at that level until 0.6 GYA. At this point all the oxygen-absorbing material in rocks and ocean waters were saturated, and whatever oxygen that was produced was released into the atmosphere. This rapidly increased the percentage of oxygen in the atmosphere, reaching around 12 percent at the end of the Proterozoic era. The increasing number of photosynthesis organisms contributed to the rapid increase in the oxygen percentage. This has kept the percentage of oxygen in the atmosphere constant over the last 500 million years. The complex interaction between life and its environment (atmosphere and the oceans) is one reason for the stability of the atmosphere.

CHEMISTRY AND PROPERTIES OF WATER

Water is an essential component of life. Indeed, water protected the primitive life against the harsh environment outside before the ozone layer formed and land became habitable. It carries nutrition for plants and animals and plays a major role in controlling the ecosystem by shaping our environment and balancing its temperature. Water carries chemicals between different locations on Earth and nutrition to the oceans. It is the most abundant molecule on our planet and the first thing to search for when looking for life in other planets.

Each water molecule consists of an oxygen atom sharing electrons with two hydrogen atoms by forming covalent bonds. The molecule has a V-shape, with the oxygen atom at the bottom and the hydrogen atoms at the two sides (figure 19.1). The oxygen's positively charged nucleus attracts the negatively charged electrons toward itself. As a result, the oxygen has a slightly more negative charge, and the hydrogen atoms connected to it have a slightly positive charge. Because of this configuration, water molecules are *polar*, and as a result, water molecules orient themselves so that the negative and positive sides face one another (figure 19.1). The relatively positively charged hydrogen atom of one water molecule and relatively negatively charged oxygen atom of the adjacent water molecule are then attracted to each other by hydrogen bonds (a detailed discussion of chemical bonds is given in chapter 20). Here I discuss the main properties of water, what makes it so special and why water molecules have the properties they have.

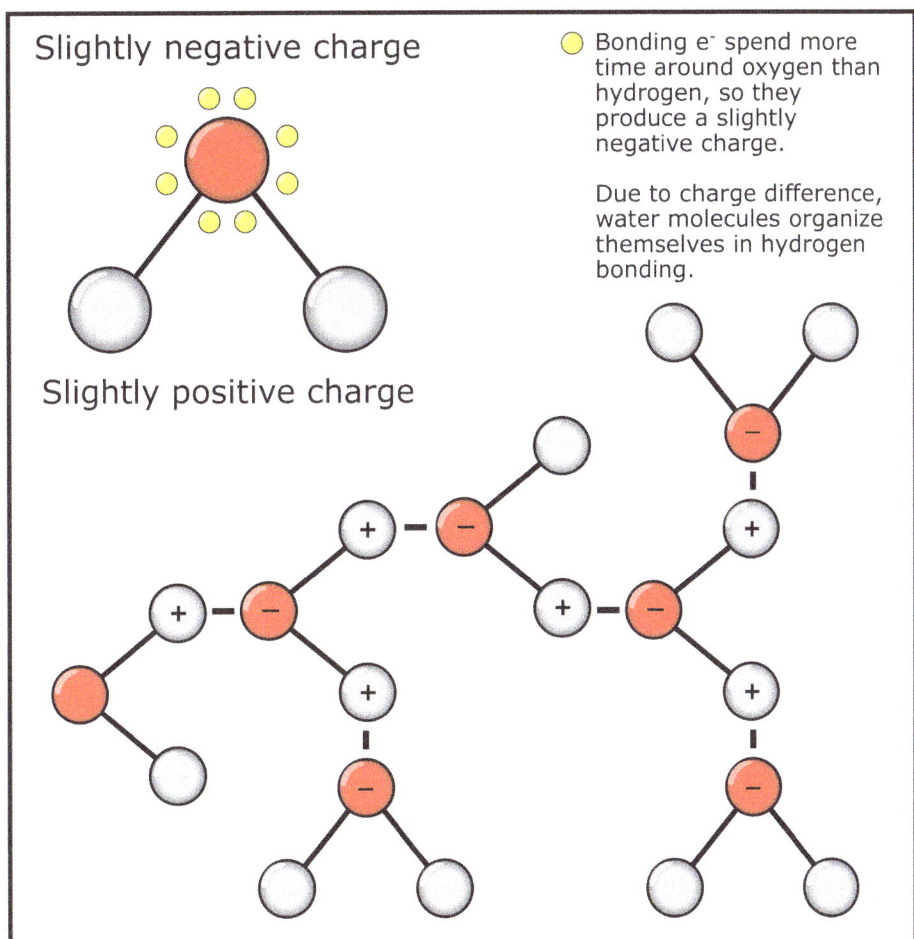

Figure 19.1. An example of a water molecule with two hydrogen atoms (gray circles) connected to one oxygen atom (red circle) through an angle of 104.5 degrees. Because of the higher atomic mass of oxygen, electrons are located closer to oxygen, giving it a relatively more negative charge (electronegativity). The horizontal/vertical solid bars are hydrogen bonds between water molecules.

Cohesion: The hydrogen bonds link water molecules together (figure 19.1) giving water high surface tension. This property is called cohesion. This is responsible for water molecules sticking together and flowing, and for the water to exist as liquid form at the temperature and pressure on Earth. The cohesion property of water is also responsible for the transport of nutrition within animal's vessels or transport of water in plants.

Large heat capacity: The hydrogen bonds that connect the water molecules together give water the property to absorb heat without a great change in its temperature. The temperature is defined as the average speed of the molecules. Therefore, the energy generated by the heat will break the hydrogen bonds (that would immediately be restored) and not spent on increasing the speed of the molecules (the temperature). As a result, the increase in the temperature of water, when subjected to an external heat source, is slow. This is important and responsible for the organisms maintaining their internal energy.

Good solvent: The polarity of water makes it a good solvent, facilitating chemical reactions. Molecules that attract water are called *hydrophilic*, while those that do not attract water are called *hydrophobic*. Because of this property, water could move chemicals from mountains to the oceans.

Low density in solid form: Because of the V-shaped structure of water molecule and hydrogen bonding, water becomes most dense at 4 degrees Celsius. At temperatures below this, the vibration of water molecules dominates, and hydrogen bonds become more open, meaning that water expands (becomes less dense) as it freezes. In other words, the V-shaped water molecule bonds with four adjacent molecules through hydrogen bonds, forming a lattice in which the molecules are held slightly further apart than in the case of the liquid. This means that ice is less dense than liquid water. This is the reason ice floats on water. If this wasn't the case, ice would sink in oceans during winter times and, after many years, layers after layers of ice would fill oceans, destroying marine life and significantly changing the ecosystem on Earth. Instead, since ice stays on top of water, it insulates the water underneath and hence will keep its temperature above the freezing point, saving marine life.

THE OCEAN-EARTH-ATMOSPHERE REGULATORY SYSTEM

The crust of Earth, the atmosphere, seawater, and living organisms undergo complex chemical exchange that regulates the global environment. These processes combined, keep the global temperature within a constant range that makes life possible (see Box 19.1). The distribution of carbon and its compounds plays a vital role in this process. Most of the carbon (in the form of CO_2) is released to the atmosphere by volcanic activities. This does not accumulate in the atmosphere forever but is removed by a process called *chemical weathering* where it is absorbed by rocks at Earth's surface or is washed out by rainwater. A particularly important property of chemical weathering is its dependence on temperature, taking place faster at higher temperatures. This relation is responsible for regulating the atmospheric temperature as I explain below (Prothero and Dott 2010). Suppose for some reason the temperature of the atmosphere becomes high, resulting in an increase in the weathering rate and, as a result, more efficient removal of CO_2 from the atmosphere (by rainwater or absorption by rocks or in oceans). This decreases CO_2 concentration in the atmosphere, reduces the green house effect (see next section) and therefore, decreases the temperature. Now, when the temperature is close to the freezing point, the concentration of CO_2 released to the atmosphere by tectonic activities starts to increase, as there is no liquid water present to remove it (less efficient weathering). This continues until the temperature rises to the level where ice melts, liquid water appears, and weathering starts again. The carbon cycle (Box 19.1) keeps a balance, forbidding all the carbon to enter the atmosphere or end up in rocks and, as a result, regulates and sustains the temperature of Earth (Cranfield 2014).

It takes 100 million to 200 million years for carbon to move between rocks, the atmosphere, and oceans. Apart from being responsible for fixing the temperature of the atmosphere, the carbon cycle has other observable effects. At the point where the atmosphere meets ocean surface, the carbon dioxide from the atmosphere dissolves in water and releases hydrogen, making the water more acidic (figure 19.2, panels 1 and 2). This results in carbonic acid formation (figure 19.2, panel 2). The hydrogen then reacts with carbonate resulted from rock weathering to generate bicarbonate ions (figure 19.2, panel 3). The carbonate is formed when the carbon in the atmosphere dissolves in rainwater and forms a weak acid—called carbonic acid (figure 19.2, panel 2). This then, through chemical reactions (chemical weathering process), dissolves rocks. The result of this reaction is calcium, magnesium, potassium, or sodium ions. These materials are then carried into the oceans by rivers and enter into reaction with bicarbonate ions to form calcium carbonate, mostly made by shell-building organisms. When these organisms die, they form layers of sediments and after millions of years form rocks. This is the origin of the carbons in rocks and limestone (figure 19.2; Cranfield 2014).

The energy from the sun combined with the continual reuse of the existing atoms helps the living systems to sustain life, to grow and to reproduce. In this process inorganic molecules are combined to form organic material needed by living things. The complex organic compounds are then converted to inorganic material that will in turn be used in nature to reproduce organic compounds. For example, there are decomposer bacteria that break down organic material from dead animals and plants to inorganic material, reused by other organisms to remake organic material. This recycling of material takes place in many of the atoms required by living systems. In the following I summarize the most important recycling processes (Tillery, Enger and Ross 2019).

The Carbon Cycle: Most of the carbon is found in rocks, with the rest residing in the atmosphere, oceans, plants and soil. Carbon exchange between these reservoirs is called the *carbon cycle*. Directly or indirectly, plants are the main source of carbon emission to the atmosphere. They combine carbon dioxide from the atmosphere and water (along with energy from the sun light) to form complex organic molecules such as sugar ($C_6H_{12}O_6$) through photosynthesis process. They then combine oxygen with sugar to produce water, carbon dioxide (CO_2) and energy needed for their growth. Animals eat plants and break the complex organic compound into simpler compounds (amino acids and sugar) they need, releasing carbon dioxide in the process. Furthermore, when plants and animals die, they release their carbon into the atmosphere through the carbon cycle. The presence of plants is tightly related to the CO_2 content of the atmosphere. There is a seasonal dependence of the fraction of the CO_2 in the atmosphere. During the winter months, when there are no plants, the level of CO_2 in the atmosphere increases while in the spring and summer when plants grow, and help the carbon cycle, the fraction of the CO_2 reduces. This is responsible for fixing of the temperature of the atmosphere.

The Nitrogen Cycle: Nitrogen is essential for the construction of the amino acids, needed to form proteins, nucleic acids responsible for genetic material and to generate the energy needed in the living systems (Chapter 20). Nitrogen molecule (N_2) forms 80% of the Earth's atmosphere however, what is needed for the organisms to function is the Nitrogen atom (N). Some bacteria can convert Nitrogen molecules to nitrogen atoms. These bacteria live in the soil or are in the form of cyanobacteria functioning as Nitrogen fixers, having the ability to convert nitrogen molecules to nitrogen atoms used by plants and animals to make amino acids and proteins (Chapter 23). All plants and animals take their nitrogen needs from their food. The proteins in their food break down to their amino acid components during digestion. The amino acids could then be reassembled to new proteins that build their bodies. The decomposer bacteria act upon the dead plants and animals and releases nitrogen in the form of ammonia (NH_3) that is taken up by the new plants and animals and converted to nitrogen used by these organisms. In summary, in the nitrogen cycle, nitrogen from the atmosphere goes through the organisms (many of them bacteria) and eventually is released to the atmosphere to be recycled again.

The Phosphorous Cycle: Like nitrogen, phosphorous is an essential element in building and sustaining life. The biological molecules that make the structure of the cells or the genetic material all consist of phosphorous (Chapter 20). The main source of the phosphorous atoms is rock. The phosphorous is released by the erosion of rocks and dissolved in water that is then absorbed by plants to construct the molecules they need. When an organism dies, decomposer bacteria recycle phosphorous compounds back to the soil. These are then dissolved in water and end up as deposits in oceans (providing the phosphorous need of aquatic systems) as sediments or in rocks. These will become available for living things through geological processes. The phosphorous need of animals is supplied by their food, consuming other animals or plants.

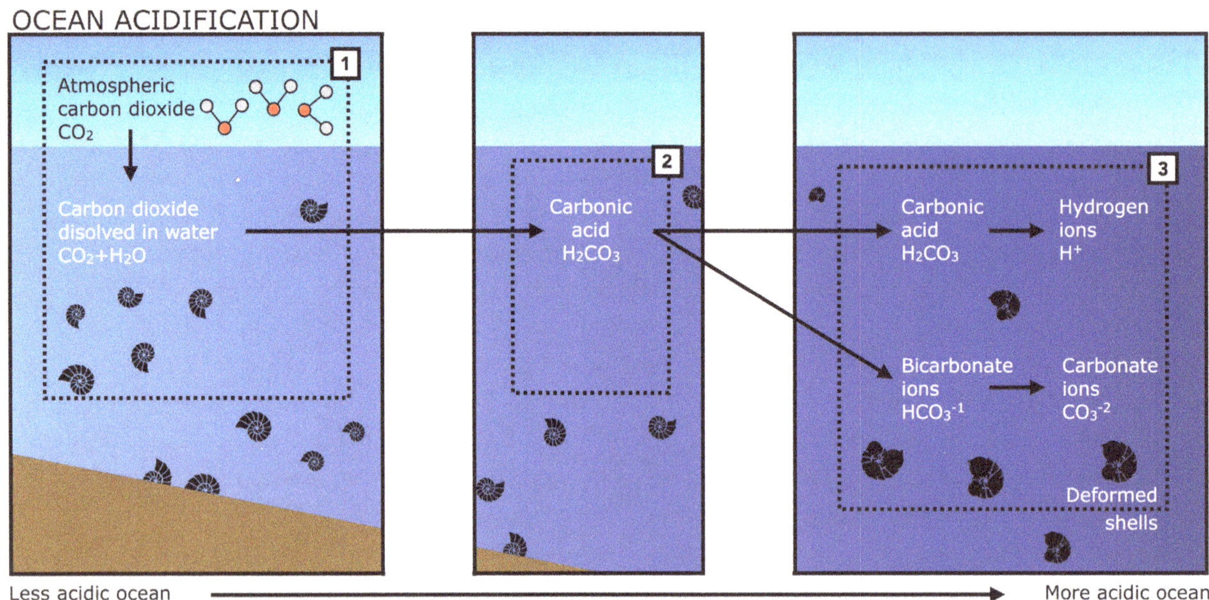

Figure 19.2. The atmospheric carbon dioxide (CO_2) interacts with seawater (process 1) and hydrogen, generating carbonic acid (CO_3H_2) and a more acidic environment (process 2). The hydrogen reacts with carbonate generated by rock weathering producing bicarbonate ions (HCO_3^-: process 3). The bicarbonate ions enter into process with calcium, magnesium, sodium ions (produced from rocks, washed by rain water and moved to oceans by rivers). This leads to formation of carbonate ions and calcium carbonate that is the constituent of the shells.

THE GREENHOUSE EFFECT IN THE ATMOSPHERE

Carbon dioxide, methane, and halocarbons are gases that absorb the infrared energy (heat) from the sun and re-emit them in all directions. A percentage of this re-emitted radiation hits the Earth and warms up its surface, creating the *greenhouse effect*. Without these greenhouse gases, the temperature of Earth would decrease to −18 degrees Celsius. With too much greenhouse gas, the temperature of Earth would reach 400 degrees Celsius. The amount of greenhouse gases in the atmosphere is sustained within the range needed to keep the planet habitable.

Over 50 percent of the greenhouse gas in the atmosphere is in the form of water vapor, compared to 20 percent that is in the form of carbon dioxide. An increase in the temperature of Earth causes ocean waters to vaporize and increase the water vapor in the atmosphere. When Earth cools down, the water vapor condenses and forms the rain. On the other hand, carbon dioxide can remain as gas in a wider range of temperatures in the atmosphere and is the primary gas responsible for heating the atmosphere and keeping the fraction of the water vapor constant. When carbon dioxide percentage drops, Earth cools and water vapor is removed from the atmosphere (due to rain), reducing the contribution to the temperature by water vapor. Similarly, increasing the percentage of carbon dioxide results in warmer temperature and more water vapor in the atmosphere. Therefore, while the percentage of carbon dioxide in Earth's atmosphere is less than water vapor, it controls the heating due to the greenhouse effect (Bennett and Shostak 2005).

SUMMARY AND OUTSTANDING QUESTIONS

The planetesimals that collided and formed the Earth 4.57 billion years ago brought with themselves chemical elements that sank to the core of the planet. After Earth was formed, these chemicals (such as carbon dioxide, nitrogen, and methane) found their way out through the outgassing process as a result of volcanic activity. Except for the lightest of the elements (hydrogen) that escaped to space, the rest were retained by Earth's gravity and formed the atmosphere. The helium and nitrogen abundances in the atmosphere agree well with predictions based on this scenario.

Evidence from zircon grains found in Western Australia suggests that water first appeared on Earth around 4 billion years ago. The water vapor most likely came from the interior of Earth through the outgassing process and turned to liquid form once exposed to a cooler climate. This generated the water in the oceans. Oxygen was generated in the atmosphere by the breaking of water molecules by energetic UV radiation from the sun. The oxygen, being a highly reactive element, then entered into reactions with the methane and ammonia in the atmosphere, producing carbon dioxide and nitrogen as well as water. Once all the methane and ammonia were used, the oxygen would not be consumed in chemical reactions anymore and, as a result, would accumulate in the atmosphere. The increase in the oxygen level in the atmosphere resulted in the formation of ozone (O_3) that was produced by breaking oxygen molecules (by the ultraviolet light from the sun) and then combining an oxygen atom and oxygen molecule. Once the ozone formed, it would shield Earth from the harmful UV radiation by the sun and hence made the land habitable for plants and animals. At this point life in its primitive form migrated from the sea to the land. Exactly when this happened is still unclear. The first evidence for oxygen in the atmosphere came from cyanobacteria that released oxygen as a waste product over 3 billion years ago. Once plants started to grow on land, they increased the oxygen level through the photosynthesis process. The atmosphere reached its present level of oxygen around 500 million years ago.

The presence of water is essential for life. Water molecules stick together through hydrogen bonds and could flow (i.e., are cohesive) and transport chemicals. It is one of the very few substances that could be found in the gas, liquid, or solid forms within the range of temperature of Earth (high heat capacity). It can resolve chemicals and transport them (solvent) and has lower density when is in solid form, playing an important role in protecting the ecosystem.

The constant interaction between Earth, the oceans, and the atmosphere has sustained the present ecosystem and has balanced the temperature of the atmosphere to the level that living organisms can prosper. At warm temperatures when water is found in liquid form, CO_2 gas is removed from the atmosphere by rainwater or is absorbed by rocks. This decreases the atmospheric temperature (because of removal of greenhouse gas) to the level that water freezes out, slowing down the rate with which CO_2 is removed. Meanwhile, the CO_2 is supplied to the atmosphere by volcanic activities, increasing the temperature, melting the ice and hence increasing the efficiency of the processes that work toward removing CO_2. This cycle continues and is responsible for controlling the temperature of the atmosphere.

The outstanding questions here are: How the temperature of Earth has been kept for such a long time within the narrow range needed to sustain life? There have been debates that the initial CO_2 content of the atmosphere was too low to warm up the Earth. The problem becomes more complicated with the fact that the sun was less luminous and hot during the first hundreds of million years of Earth's life. A likely explanation is that methane that existed in abundance during early times played the role of the greenhouse gas or that the albedo of the early Earth was much lower than it is today, increasing the temperature.

REVIEW QUESTIONS

1. What chemical elements were present in the atmosphere of the early Earth?
2. How are the nitrogen (N_2) and carbon dioxide (CO_2) in Earth's atmosphere produced?
3. Explain the process that fixes the oxygen abundance in the atmosphere.
4. Why is the ozone layer essential for life on Earth?

5. How did seawater acquire its chemical composition?

6. What is the evidence for the transition from an oxygen-poor to an oxygen-rich atmosphere?

7. Explain the main properties of water and how these are essential to support life.

8. What is chemical weathering?

9. Briefly explain how the temperature of the atmosphere is controlled and kept within its present range.

10. Explain the greenhouse effect and how it has played a major role in preserving and controlling the biosphere.

CHAPTER 19 REFERENCES

Bennett, J., and S. Shostak. 2005. *Life in the Universe.* 2nd ed. Boston: Pearson/Addison-Wesley.

Cranfield, D.E. 2014. *Oxygen: A Four Billion Year History.* Princeton, NJ: Princeton University Press.

Jordan, T.H., and J. Grotzinger. 2012. *The Essential Earth.* 2nd ed. New York: Freeman.

Prothero, D.R., and R.H. Dott. 2010. *Evolution of the Earth.* 8th ed. McGraw-Hill.

Tillery, B. W.; Enger, E. D. and Ross, F. C. 2019, *Integrated Science.* 7th Edition: McGraw-Hill

THE BASIC INGREDIENTS OF LIFE

20

CHAPTER LEARNING OBJECTIVES

This chapter will cover:

- The definition of life
- Why carbon is important constituent of life?
- The nature of genetic material
- The chemistry of life
- The structure and function of DNA and RNA
- Protein synthesis
- The genetic code
- The energy producing processes

"Life is really simple, but we insist on making it complicated"

- CONFUCIUS

"Start with what is right rather than what is acceptable."

- FRANZ KAFKA

In the previous chapter I discussed the conditions that are essential for life to develop and sustain. Although these were required to support life once it started, the circumstances under which life began, and the material needed to initiate and sustain life, have been among the most fundamental topics to study. A large number of very delicate events must all have come together under the right circumstances to result in what we call "life". By understanding the nature of life, we get to know ourselves better as well as other forms of life, if they exist. The essential functions required to initiate life ultimately depend on the chemical and biological properties of certain organic molecules. Therefore, by reducing the living organisms to their basic components, we can study the fundamental material responsible for life as we know it. However, before doing this, one needs to address the following questions: What is the definition of life? How different ingredients came together to start life? And where those ingredients themselves came from? I explore the first two questions in this chapter and the third question in the next chapter where I study different scenarios for the origin of life.

The main processes responsible for life have their roots in the laws of nature, while the environment and evolution also play significant parts. Looking around ourselves, we see a diversity of living things, with tens of millions of species. Despite this huge diversity, the chemistry of life is only based on a handful of molecules, made from a few chemical elements. These elements interact with one another based on a number of basic rules. The way the existing diversity in nature is resulted from a few chemical elements and a handful of rules is by itself a fascinating subject to study.

The first step in our journey to decipher life is to find the ingredients all the "living" things have in common. For example, what is the difference between the materials forming the tissues in my body and those forming the table I am sitting at? Are they made from completely different ingredients, or are they the same material arranged in a different way, coexisting with some catalysts that helped initiate life in me but not in the table? How did chemistry transition to biology in nature? Scientists have just started to address these questions.

This chapter is about the main ingredients without which life would not exist. By deconstructing living organisms into their constituent ingredients, I study how they all came together to create the living things we see today. I then review the chemistry that binds molecules together and discuss the biochemistry of life. Finally, I study the complex interaction between the basic material responsible for life: how they combine to form the genetic material, to synthesize protein and to generate the energy required for a living system to function.

WHAT IS LIFE?

To define life in terms of a few characteristics we have observed on our planet, will miss other forms of life that may exist but based on entirely different materials and principles. Despite all the amazing diversity discovered in the terrestrial life, it only provides a single example. To develop a general theory of living systems, one needs more than one example (Cleland and Chyba 2002). As a result, defining life with a few words is insufficient and often misleading. Given this, I am cautious in coming up with a few restrictive conditions to call an organism alive. Instead, I discuss the common characteristics of all living organisms. We may use these as the baseline to develop a more general theory for life. The main observations that distinguish the living and nonliving worlds include (Sadava et al. 2014):

- Cells are the basic components of all living organisms.
- Cells produce energy by taking nutrition from the environment and converting it to what is needed for biological functioning.
- All cells (from plants or animals) have similar structures.
- Living organisms consist of common chemical compounds, including carbohydrate, fatty acids, nucleic acids, and amino acids.
- The process of protein synthesis and transfer of genetic code is the same in all living organisms .
- Living organisms can reproduce and pass their genetic information to their offspring.
- By self-regulating their internal environment, living organisms maintain the conditions that allow them to survive.

These observed characteristics confirm that all forms of life on Earth have had a common origin, with the present diversity of living organisms all resulting from one form of life. Given the amazing similarity of the cells, the energy production process, the chemistry of amino acids and nucleic acids, it is hard to imagine multiple origins for life on our planet. In the rest of this chapter I use the existing knowledge from the living organisms on Earth to explore the story of life starting from the basic chemistry that is essential in building the scaffolding of living systems.

CHEMICAL BONDS

Chemical bonds are responsible for combining simple atoms to form molecules and combining molecules to form complex compounds essential for the birth and functioning of the living systems. First lets review the structure of atoms and the various chemical bonds that are so fundamental for the living systems to exist and to function.

I recall that the simple way to visualize the configuration of an atom is through the Bohr model (after the physicist Niels Bohr). In this model electrons move on different shells, representing different energy levels, around the nucleus of atoms. Since the negatively charged electrons are attracted by the positively charged nucleus, the electrons need energy to resist the force of attraction. By gaining more energy, electrons move to higher shells and in return, they give back that energy. The first shell could have a maximum of up to two electrons, with each additional shell allowed to accommodate a maximum of eight electrons. This is called the *octet rule*. The outermost shell is referred to as the *valance shell*. It determines atom's chemical

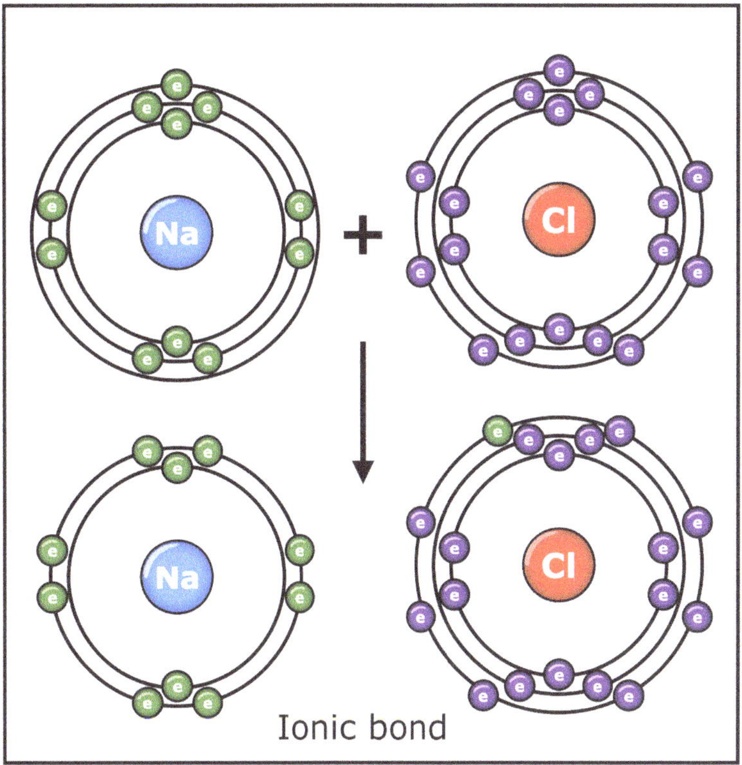

Figure 20.1. Ionic bonding. These bonds result from the transfer of one electron from one to another atom, making the valance shells of both atoms complete. Here, one electron from the outer layer of sodium (Na; green circle) is transferred to the outer layer of chlorine (Cl: blue circles), forming NaCl (table salt). This makes the sodium positively charged (losing an electron: Na^+) and chloride slightly negative (gaining an electron; Cl^-). The NaCl is formed when a number of these molecules and compound material are combined (Na^+ and Cl^- attracting one another)

properties and its appetite to enter into reactions. Two atoms enter into a chemical reaction when, after their reaction, both have a stable outer shell (containing maximum number of the electrons that shell could accommodate). If an atom has more than one shell, the octet rule holds. This states that the outermost shell in an atom is most stable when it has eight electrons. All the inert elements (elements that do not enter into interaction with any other element in nature) have eight electrons in their last shell with the exception of helium, which has only one orbit occupied by two electrons. Two or more atoms combine through chemical bonds to form new molecules. The chemical bonds responsible for different reactions are listed in Box 20.1 and are explained below:

IONIC BONDS

An example of an ionic bond is the formation of sodium chloride (table salt). Sodium has only one electron in its outer (third) valance shell and chlorine a total of seven electrons in its final (second) valance shell (figure 20.1). Therefore, if the valance shell electron in sodium atom is transferred to the last shell of chloride, both atoms become stable. Sodium loses one electron, and hence its last (second) shell is complete with eight electrons and chloride receives an electron, completing its last shell with eight electrons. In this process, sodium attains a positive charge (as it loses one electron and hence has one proton more), and chloride becomes more negative (as it now has one electron more than its number of protons). These charged elements are called *ions*. Ionic bonds are effective by the strong attractive force between the positively and negatively charged ions. When sodium and chlorine interact this way, the ionic compound sodium chloride is produced (figure 20.1).

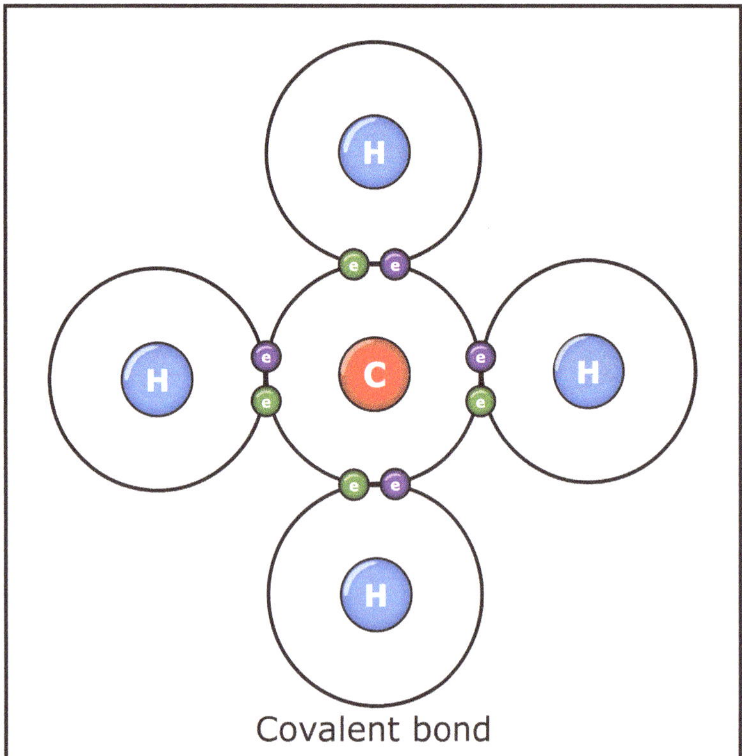

Figure 20.2. Shows covalent bond between a carbon atom (with four electrons in its valance shell—purple circles) and four hydrogen atoms (each with one electron in the valance shell—green circles). Sharing one electron with each of the hydrogen atoms makes the outer shells of both carbon and hydrogen complete, forming methane (CH_4).

This results when two atoms share electrons so that both attain the maximum number of electrons allowed in their valance shell (figure 20.2). For example, a hydrogen atom with only one electron in its outermost shell could share that electron with another hydrogen atom to complete their last shell, each atom ending up with a complete shell, thereby forming a hydrogen molecule, H_2 (H–H meaning that they share a single electron between them). Depending on the configuration in the outermost shell, some atoms may share two or more electrons to complete their valance shell. An example is the oxygen atom, with six electrons in the outer most shell. Two oxygen atoms, each sharing two electrons, increase the number of electrons in their last orbit to eight, resulting in oxygen molecule, O_2 (O=O meaning that they share two electrons between them). Also, carbon with four electrons in its valance shell could share four more electrons with four hydrogen atoms, increasing the number of electrons in its valance

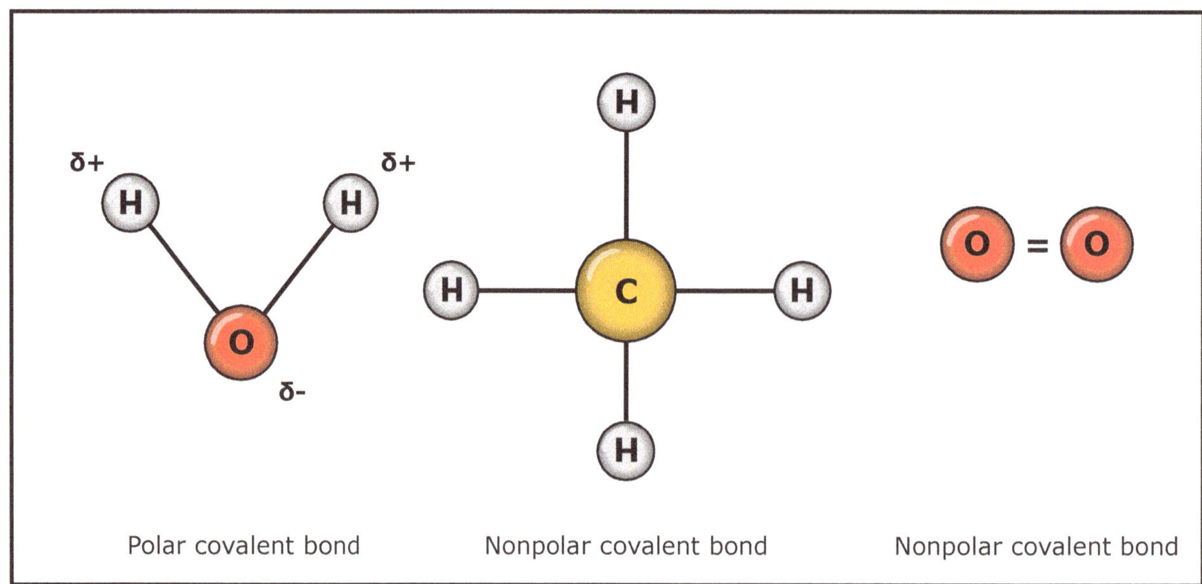

Figure 20.3. Examples of polar and nonpolar covalent bonds. The electronegativity is non-zero when the atoms are not located in opposing directions from each other (the example here is water—leftmost panel). When the bonds are symmetrical, the electronegativity is canceled out (the example here is methane and molecular oxygen—middle and right panels). The lines indicate covalent bonds between two atoms (each line is one bond representing the sharing of an electron).

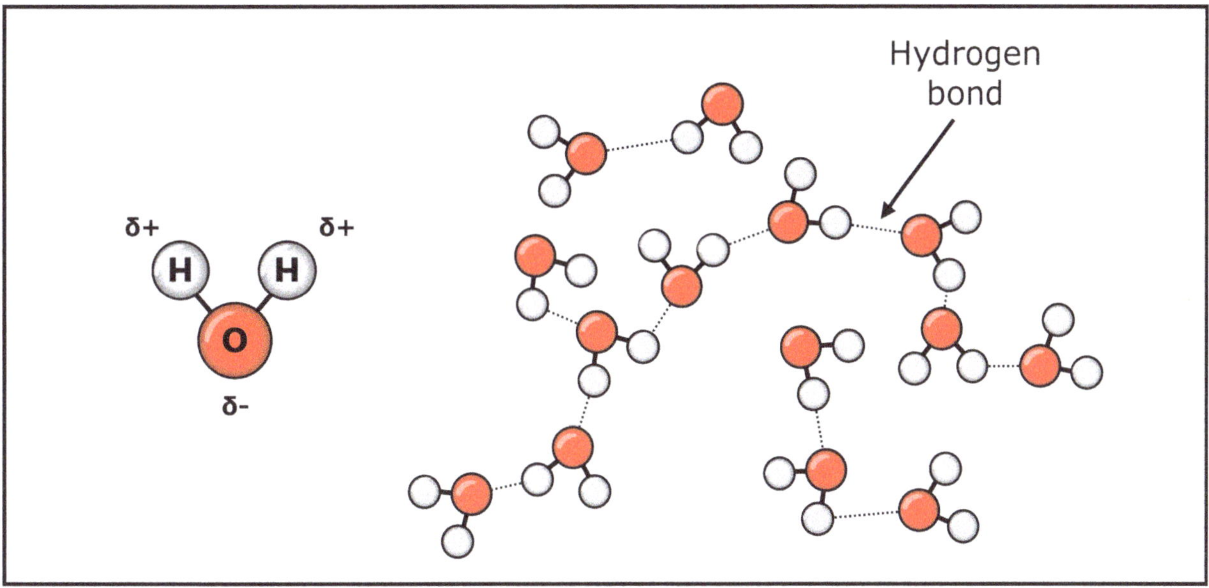

Figure 20.4. The hydrogen bonds (dotted lines) between hydrogen (gray circles) and oxygen (red circles). The presence of electronegativity results in the hydrogen bonds between water molecules, giving water its cohesive property.

shell to eight, while each of the hydrogen atoms ends up with two electrons in their first shell (maximum allowed), forming methane (CH_4; figure 20.2).

When sharing of the electrons between two atoms is equal, a *nonpolar* covalent bond is formed (Figure 20.3). If one atom is able to attract an electron more than the other atom, it becomes more negative—has more *electronegativity* (Figure 20.3). The ability of an atom to be able to attract more electrons depends on the number of protons in the nucleus of that atom. Greater the number of protons, stronger attraction of the electrons and more electronegativity. When electrons are not shared equally, the bond is a *polar* covalent bond. An example of electronegativity is the water. In symmetrical structures, the electronegativity is canceled, making the molecule non-polar, like oxygen molecule O_2 (O=O). However, in case of water where two hydrogen atoms make an angle with the central oxygen (figure 20.3), the molecule is not symmetrical, and hence the polar bonds do not cancel each other, making water a polar molecule. The polarity of molecules determines how they interact with other molecules. Another polar molecule is the amine group (containing NH_2).

BOX 20.1: TYPES OF CHEMICAL BONDS

- A ***chemical bond*** between two atoms is the sharing or transferring electrons between them. The chemical bonds keep atoms together and lead to new substances.
- An ***ionic bond*** is when one electron is transferred between the valance shells of two atoms, making both the valance shells complete.
- A ***covalent bond*** results when two atoms share electrons.
- A ***polar covalent bond*** results when two atoms unequally share electrons. In this case the shared electrons are attracted closer to the heavier atom, making it slightly negative. This property is called *electronegativity*.
- A ***hydrogen bond*** results from the interaction of a hydrogen ion with slightly positive charge and an electronegative ion.

This is caused by the attraction of slightly positive hydrogen to a slightly negative atom in its vicinity. For example, in a water molecule, because of its polarity (and nonsymmetrical covalent bonds between hydrogen and oxygen), the hydrogen has a slightly positive and oxygen a slightly negative charge. As a result, hydrogen ions are attracted by oxygen ions in their vicinity, generating a hydrogen bond (figure 20.4). Hydrogen bonds are weak and could easily be broken. This is an important property of these bonds that plays a major role in the formation of the molecules of life. We will see later in this chapter that this property of the amine group is very important in forming the observed structure of some of the most important molecules.

CARBON: THE ELEMENT OF LIFE

The most abundant elements in the universe are hydrogen and helium. On the Earth's solid crust, the dominant elements are silicon, oxygen, aluminum, and calcium. However, the four elements that are most essential for life collectively make up 94 percent of the mass of a typical cell (excluding water): carbon (47 percent), oxygen (30 percent), hydrogen (9 percent), and nitrogen (8 percent). The main element responsible for the cell's structure and the chemical compounds essential for life is the carbon. The main reason carbon forms the basis of life on Earth is that it can easily enter into chemical reactions (through chemical bonds) with many different elements, making it one of the most flexible elements capable in building complex molecules. Because of the number of electrons in their valance shell, hydrogen can only chemically bond to one element while oxygen can bond to

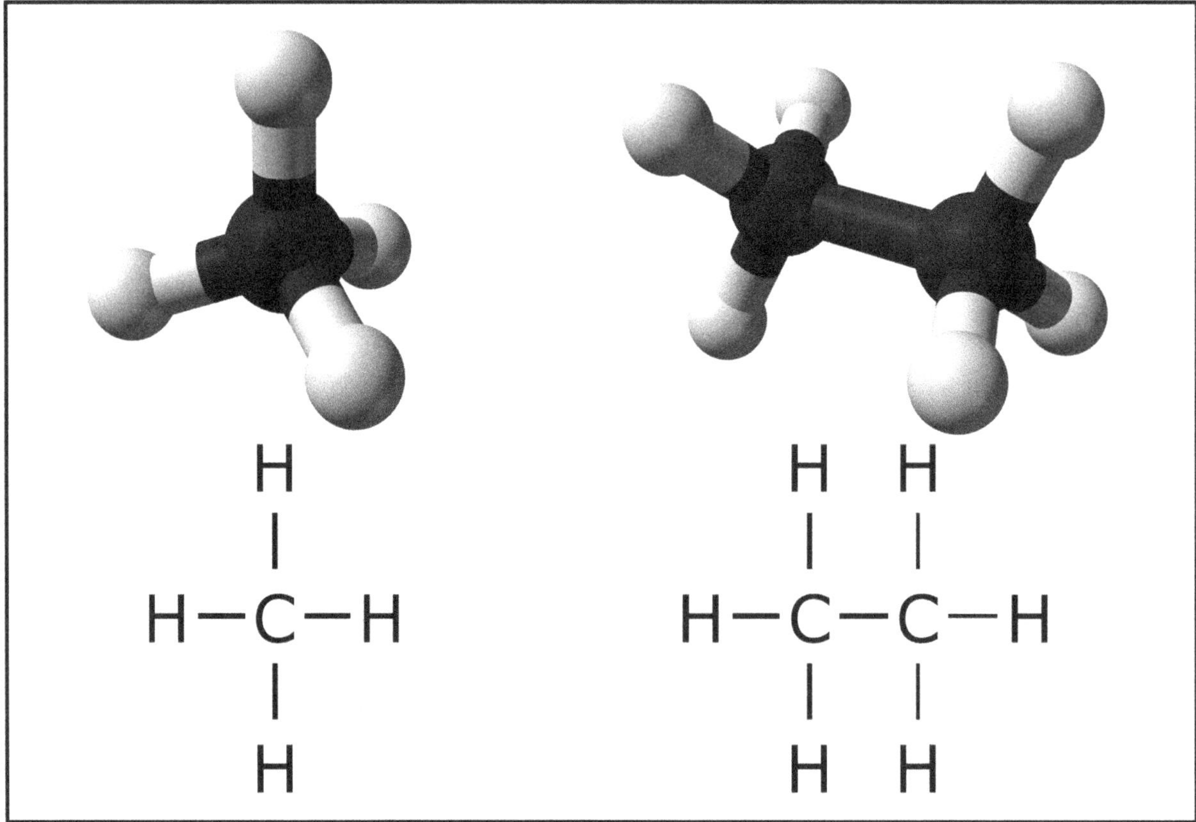

Figure 20.5. Shows the three-dimensional structure of methane (CH_4) and ethane (C_2H_6)–top panels–and the chemical covalent bonds that form them (each represented by a single line connecting the symbols). The ability to rotate and form different structures is clear from the 3-D structures.

BOX 20.2: IS CARBON UNIQUE?

Is there any other element with carbon's versatility and ability to form multiple bonds? If life were discovered in other planets, would it again be carbon based? The other chemical element that has four valance shell electrons (and hence the ability to simultaneously connect to many other elements) and is abundant in nature is *silicon (Si)*. Its position in the periodic table is just below the carbon. Silicon might provide an alternative to carbon as the chemical basis of life. However, there are some serious issues that question this hypothesis as listed below (Bennett and Shostak 2007):

- The bonds formed by silicon are significantly weaker than the bonds formed by carbon, making complex silicon-based molecules more fragile than carbon-based molecules. Also, silicon-based compounds cannot survive in water for long, a requirement for the compounds that are essential for life.
- Silicon is only found in molecules that are covalently bound to oxygen. It is, for example, found as silicate (SiO_2) on Earth's solid crust. The same is true for silicon-containing molecules in other solar system planets (like Mars) and meteorites. This limits the number of silicon-based compounds.
- Silicon cannot make double bonds, unlike carbon it only makes single bonds. This limits the number of interactions silicon can engage in, as well as the structure of the molecules it can produce (there are about one thousand silicate-based minerals on Earth, compared with millions of carbon-based compounds).
- Carbon is more mobile, as it can be found in gaseous form (like carbon dioxide). Silicon is not found as gas (silicon dioxide is only found in solid form like quartz).
- Silicon is one thousand times more abundant on Earth than carbon. Despite that, we only have carbon-based life. If silicon-based life were possible, we would have had it on Earth.

two. Carbon, at the same time, can chemically bond to four elements, making it very versatile and able to make a wide variety of molecules and compounds (see also Box 20.2).

The ability of a carbon atom to simultaneously make four covalent bonds (Figure 20.2), the spatial orientation of these bonds (Figure 20.5), and the fact that the bonds could rotate freely, all contribute to the diversity of carbon-based molecules and their ability to form long and complex chains. The versatility of carbon and its ability to form millions of molecules is the basis of *organic chemistry*. The same atoms could form different molecules with varied structures depending on their spatial arrangements. Molecules with the same chemical formulae but different spatial structures are called *isomers.*

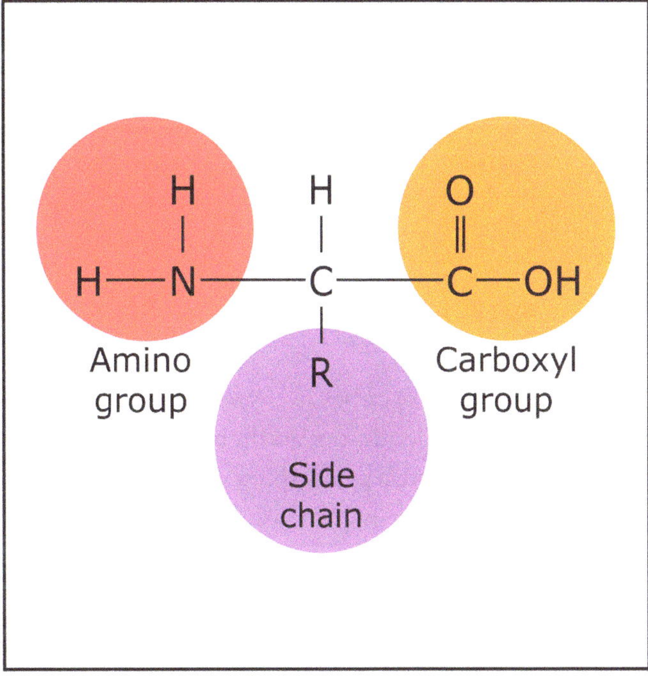

Figure 20.6. General formula for an amino acid consisting of a central carbon atom covalently linked to an amino group (*NH₂*), a carboxyl group (*COOH*), a hydrogen atom (*H*), and a side chain (*R*).

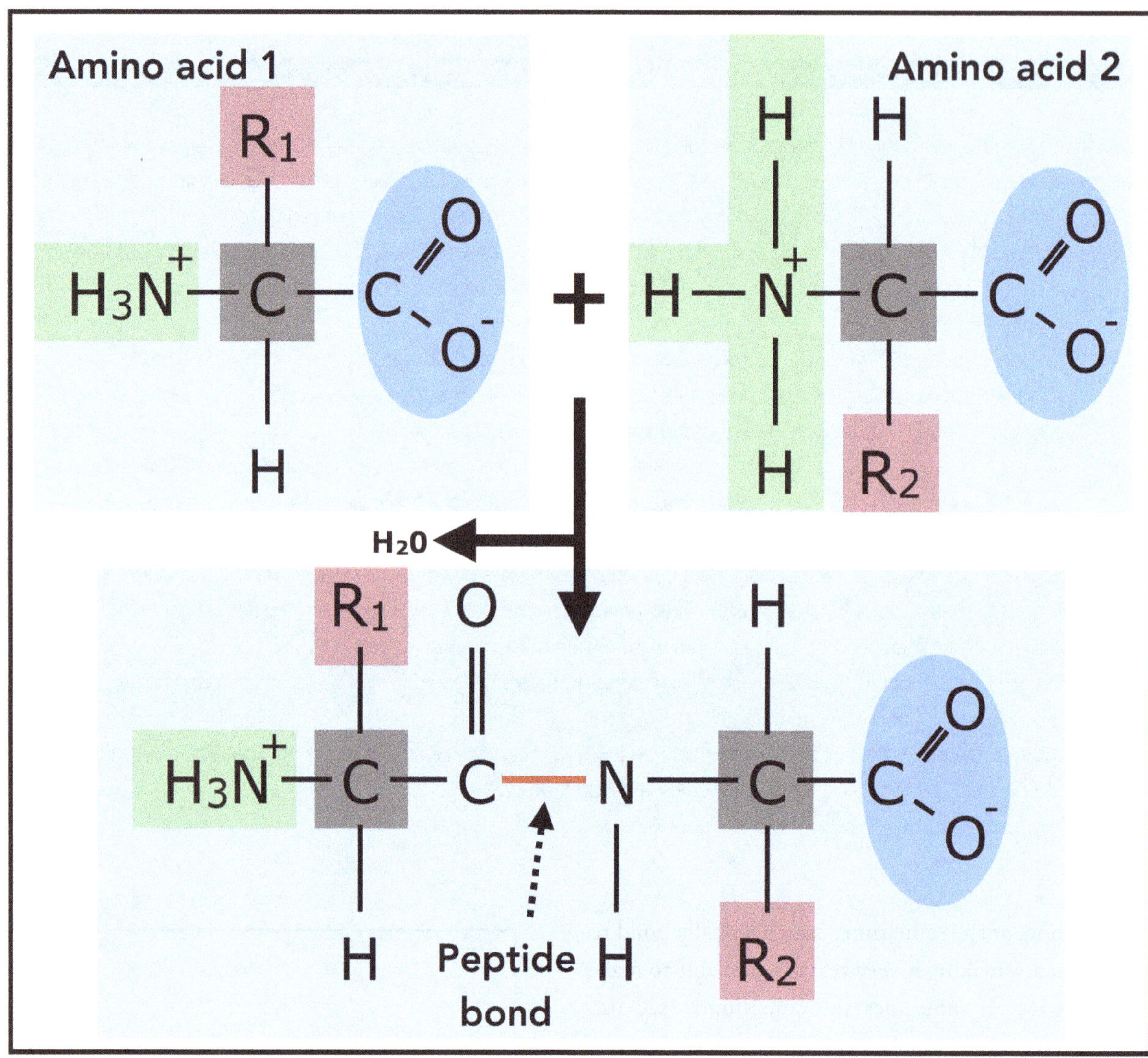

Figure 20.7. Two amino acids are linked via a peptide bond (red line), releasing a water molecule. A long chain of amino acids linked by peptide bonds forms proteins. Note that the nitrogen and carbon are responsible for forming a peptide bond.

THE MOLECULES OF LIFE

The ability of carbon to create millions of diverse molecules with different structures and functions leads to formation of molecules that are the basis of life as we know it. These molecules are relatively large and complex, *polymers,* formed from a combination of simpler molecules, *monomers.* This can create a large number of molecules with limitless diversity. At a basic level, these molecules take care of the structure of cells (formation of a wall to separate the inside of a cell from its environment) and their functions (the storage and transmission of genetic material and storage and utilization of energy as I will explain in Chapter 22).

The molecules that constitute the main molecular components of life, are divided into four categories: *carbohydrates*, *lipids*, *proteins*, and *nucleic acids*. Here I briefly explain each.

Carbohydrates are the source of food and energy. In addition to providing the energy to the cells, they are also responsible for the cellular structures. For example, a carbohydrate called *cellulose* is the main building block of wood.

Lipids can also store energy in the form of fat and are the principle component *forming* the membrane (the structure of the cells). This function of lipids plays an important role in the origin of life (see Chapter 22). The membranes enclose other organic molecules and as such, keep them close to each other to facilitate the chemical reactions necessary to form complex compounds.

Proteins have a large range of responsibilities (see Box 20.5). They are present in all the organisms and perform different functions. They act as structural components in cells or are catalysts accelerating the rate of chemical reactions (in this capacity they are called *enzymes*). Proteins are large molecules built from long chains of smaller molecules called *amino acids* connected through covalent bonds. In other words, they are polymers of amino acids. To study the structure and function of proteins, we therefore need to examine amino acids.

Each amino acid consists of a central carbon atom, called α (*alpha*) *carbon*, which is covalently bound to four groups: a *carboxyl group* (*COOH*), an *amino group* (*NH₂*), a hydrogen atom (*H*), and an *R group*, the side chain, which differs from one amino acid to the next. The type of amino acid is determined by the composition of this side chain. The general structure of an amino acid is shown in figure 20.6. Amino acids are connected to form proteins. The carbon atom in the carboxyl group of one amino acid is connected by a covalent bond to the nitrogen atom in the amino group of the next (figure 20.7). The bond between different amino acids is called *peptide bond* (figure 20.7). When carbon and nitrogen are linked in a peptide bond, the carbon has to release an oxygen atom and the nitrogen has to release two hydrogen atoms. The one oxygen and two hydrogen atoms then combine and form water (figure 20.7). Therefore, a peptide bond involves loss of a water molecule. A total of 20 amino acids are discovered in living organisms. All the living creatures (plants or animals) build protein from a combination of these amino acids. This involves a number of peptide bonds between many different molecules- called polypeptide bonds-that is responsible for the formation of proteins. The sequence with which the amino acids are ordered in

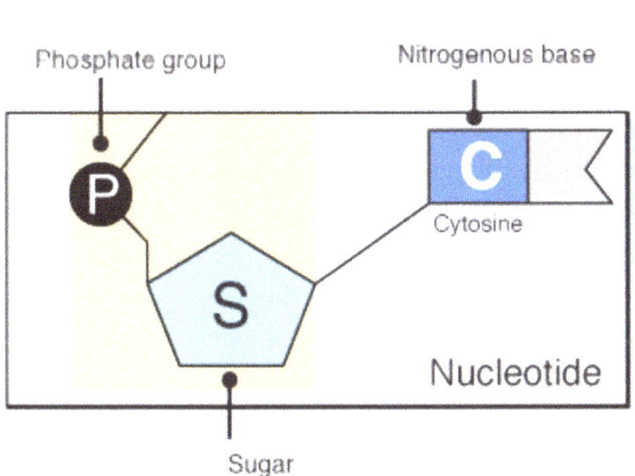

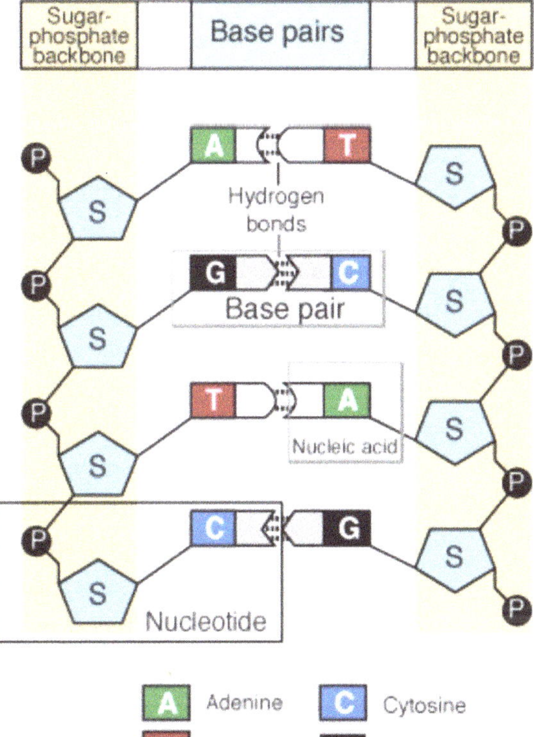

Figure 20.8. General construction of a nucleotide, the units of RNA and DNA (left panel). This consists of a sugar group and a phosphate group that form the backbone of DNA and RNA and a nitrogen group (cytosine in this example) that forms the base.

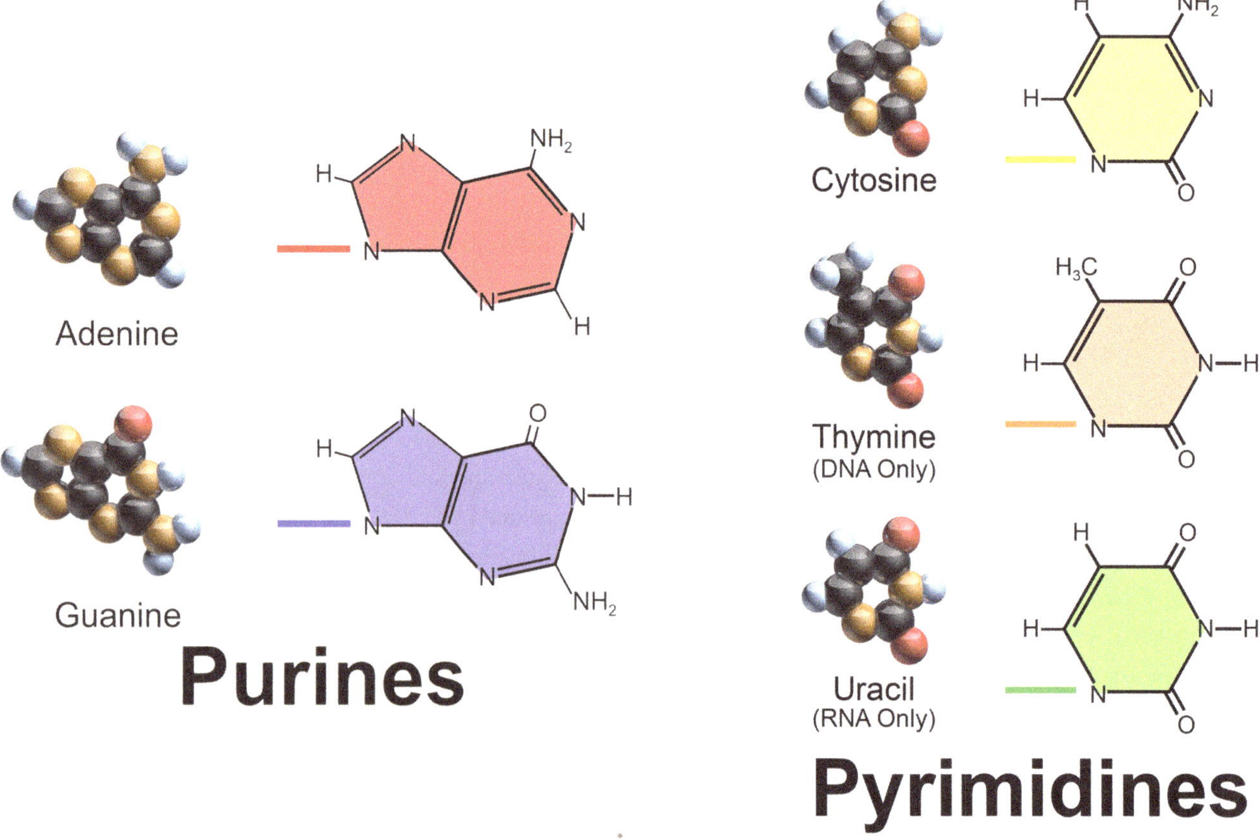

Figure 20.9. Chemical structures of purine and pyrimidine. Purine is formed by two rings and consists of adenine and guanine, while pyrimidine has only one ring, forming cytosine, thymine, and uracil nucleic acids.

a polypeptide chain forming proteins, determines how it folds into a three dimensional structure that in turn, indicates the function of the protein. The fact that living organisms all use the same set of amino acids confirms that they all started from a common ancestor (Morris et al. 2013).

Nucleic acids are formed from a sequence of molecules called nucleotides (see below) and encode the genetic information, the basic hereditary material of life. The nucleic acid *deoxyribonucleic acid* (*DNA*) is the material responsible for transmitting genetic information from parents to offspring. It contains the information for amino acid sequence of all the proteins synthesized in an organism. Changes to DNA modify the inherent characteristic of an organism and allows new species to appear. The nucleic acid *ribonucleic acid* (*RNA*) is responsible for protein synthesis and transport of the genetic material to the sites of protein synthesis. (I will revisit these in later chapters).

DNA and RNA molecules consist of two parts: the backbone and the base (figure 20.8). The backbone connects to the base with covalent bonds, with the base consisting of nucleotides linked together through hydrogen bonds (figure 20.8). The chemical structure of the backbones consist of two parts: a carbon sugar (either ribose or deoxyribose), and at least one phosphate group (figure 20.8, left panel). In RNA, the sugar is *ribose* (with a hydroxyl (*OH*) group connected to carbon), and in DNA, it is *deoxyribose* (with a hydrogen (*H*) connected to carbon). The oxygen is dropped from the sugar (ribose) molecule in DNA and therefore, the name—*deoxyribose*. RNA has the OH in its sugar molecule—*ribose*. The base is constructed from rings containing nitrogen (figure 20.9). They are called *nucleotides* and are divided into two types. Those that contain a single ring are called *primidine bases* that include. *thymine* (*T*), *cytosine* (*C*), *and uracil* (*U*). Bases with two rings are called *purine bases*, that include *adenine* (*A*) *and guanine* (*G*). Figure 20.9 shows the chemical formula for these bases. DNA contains the bases A, T, G, and

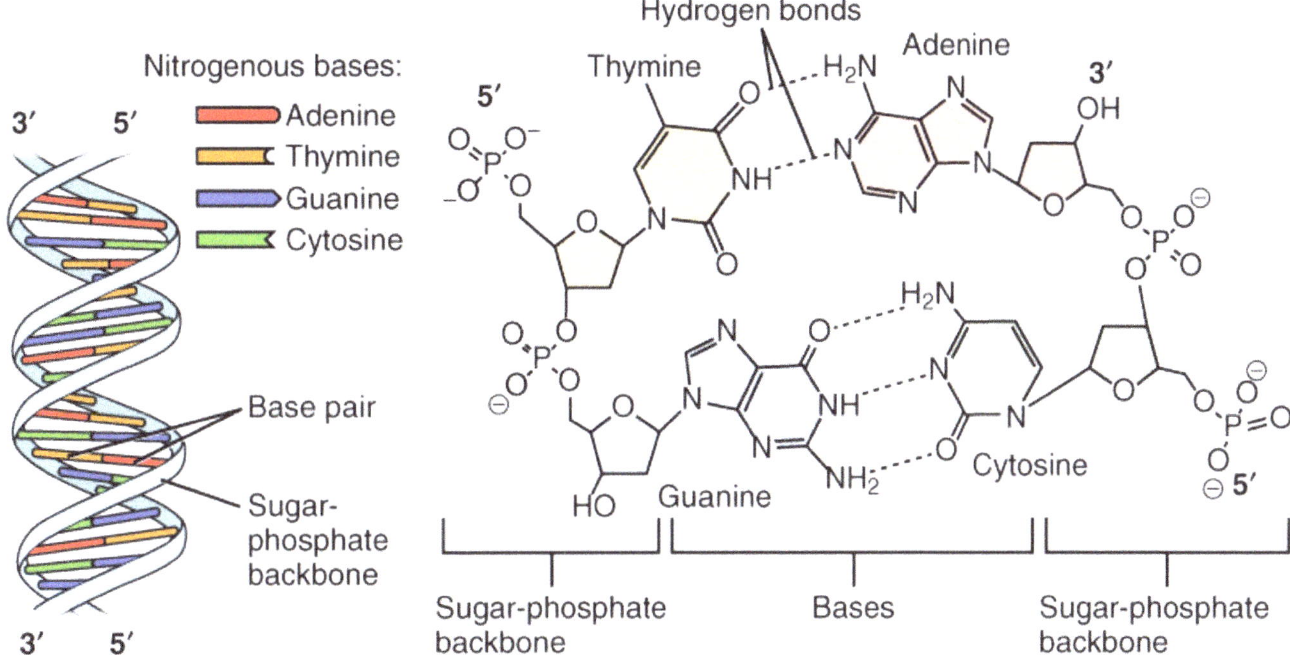

Figure 20.10. The backbone of RNA and DNA forms by connecting a phosphate to a sugar (ribose). This is the sugar-phosphate backbone. The base is formed by hydrogen bonds between nucleotides.

C while RNA contains the bases A, U, G, and C. In RNA, thymine is replaced by uracil molecule. The sequence of nucleotides determines the information in DNA and RNA molecules, similar to the order of amino acids that fixes the type of proteins (Morris et al. 2013).

THE STRUCTURES OF DNA AND RNA

As mentioned in the previous section, DNA and RNA consist of two main parts: the backbone and the base, with the nucleotides being the base units (Figure 20.8). The sugar-phosphate backbones are formed when a phosphate group in one molecule covalently binds to a sugar group in another molecule. The nucleotides in the bases are connected by hydrogen bonds (Figure 20.10). In DNA, adenine always pairs with thymine (through two hydrogen bonds between O—H_2N and NH—N; Figure 20.10) while guanine always pairs with cytosine (through three hydrogen bonds O—H_2N, NH—N and NH_2—O; Figure 20.10), forming the double strand that is the famous structural representation of DNA (Figure 20.11). In RNA, adenine and uracil are paired together, while guanine and cytosine are linked.

I recall that a hydrogen bond between two molecules is the result of non-uniform distribution of electric charge between them, one being slightly positive and the other slightly negative. The hydrogen bonds are responsible for connecting the nucleotides in DNA (*A-T*, *T-A*, *C-G*, and *G-C*) and in RNA (*A-U*, *U-A*, *C-G*, and *G-C*). The chemical structures of the nucleotides indicate that the N—H bonds in the amine group $(-NH_2)$ have non-uniform distribution of electric charge. The central nitrogen atoms in $-NH$ and $-NH_2$ attract electrons to themselves and away from the hydrogen, making the associated hydrogen slightly positively charged. Now consider the carboxyl group $(-C=O)$, the oxygen atom attracts electrons, becoming slightly negatively charged in the process. The positively charged hydrogen atom in the amine group then binds with the slightly negatively charged oxygen in the carboxyl group, forming a hydrogen bond (Figure 20.10). This is why the two pairs *T*—*A* and *G*—*C* bind to form the base in DNA and the *A*—*U* pair binds in RNA. In these bonds, the oxygen and

There are three marked differences between DNA and RNA:

1. They differ in their sugar groups (which, along with a phosphate group, construct the backbone). In DNA the pentose sugar is deoxyribose, which differs from the one found in RNA by the absence of an oxygen atom (Figure B20.3).

2. Deoxyribose provides the sugar base of DNA and comes in four forms: adenine (*A*), cytosine (*C*), guanine (*G*), and thymine (*T*). RNA is also made of four bases, with the difference being that its nucleotide includes uracil (*U*) rather than thymine. The thymine has a methyl (*CH₃*) attached to its central carbon while uracil does not have the methyl group. Apart from these, the two compounds have the same chemical composition and structure (Figure B20.3).

3. DNA molecules have a double stranded structure whereas RNA consists of a single strand (Figure B20.3).

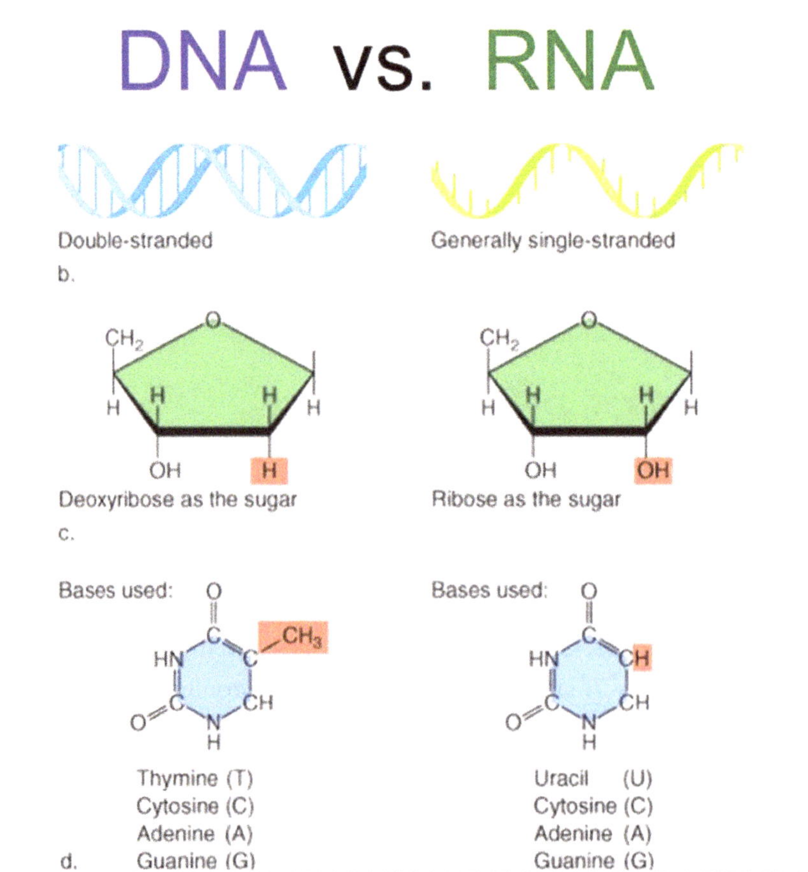

Figure B20.3. The comparison between DNA and RNA molecules. RNA has a hydroxyl group (OH) attached to its central carbon. This is replaced in DNA by only a hydrogen atom (H). The missing oxygen results in a deoxyribose molecule. Also, the nucleotide thymine in DNA is replaced by uracil in RNA. The difference between the two nucleotides is the addition of a methyl group CH₃ to thymine and hydrogen (H) to uracil.

nitrogen (that are slightly negative) from one base can enter into hydrogen bonds with the hydrogen (that is slightly positive) in another base (Figure 20.10). The hydrogen bonds are very weak and could easily break with a small amount of energy. The breaking of the hydrogen bonds in nucleic acids are essential for their role in living organisms and are the reason why DNA and RNA can replicate (figure 20.12). Once broken, DNA could form again by its bases finding their pairs and binding to them. Because of this, DNA has the basic property of life—it is able to reproduce.

Why do only specific pairs of the nucleotides bind, forming the structures of RNA and DNA (*A—T* and *C—G* for DNA; *A—U* and *C—G* for RNA)? Consider adenine and cytosine-one hydrogen bond could be formed between the amine group of adenine and the carboxyl group of cytosine. No other bonds can be formed between these two molecules, making this pair very unstable. This is the reason that, for example, adenine and cytosine do not pair up together, or if they do, the link would be unstable and soon broken.

The RNAs form self-folding structures in which part of the RNA is single stranded and partly binds to itself, forming duplexed regions. The folding in the duplex regions takes place over very short lengths. This allows

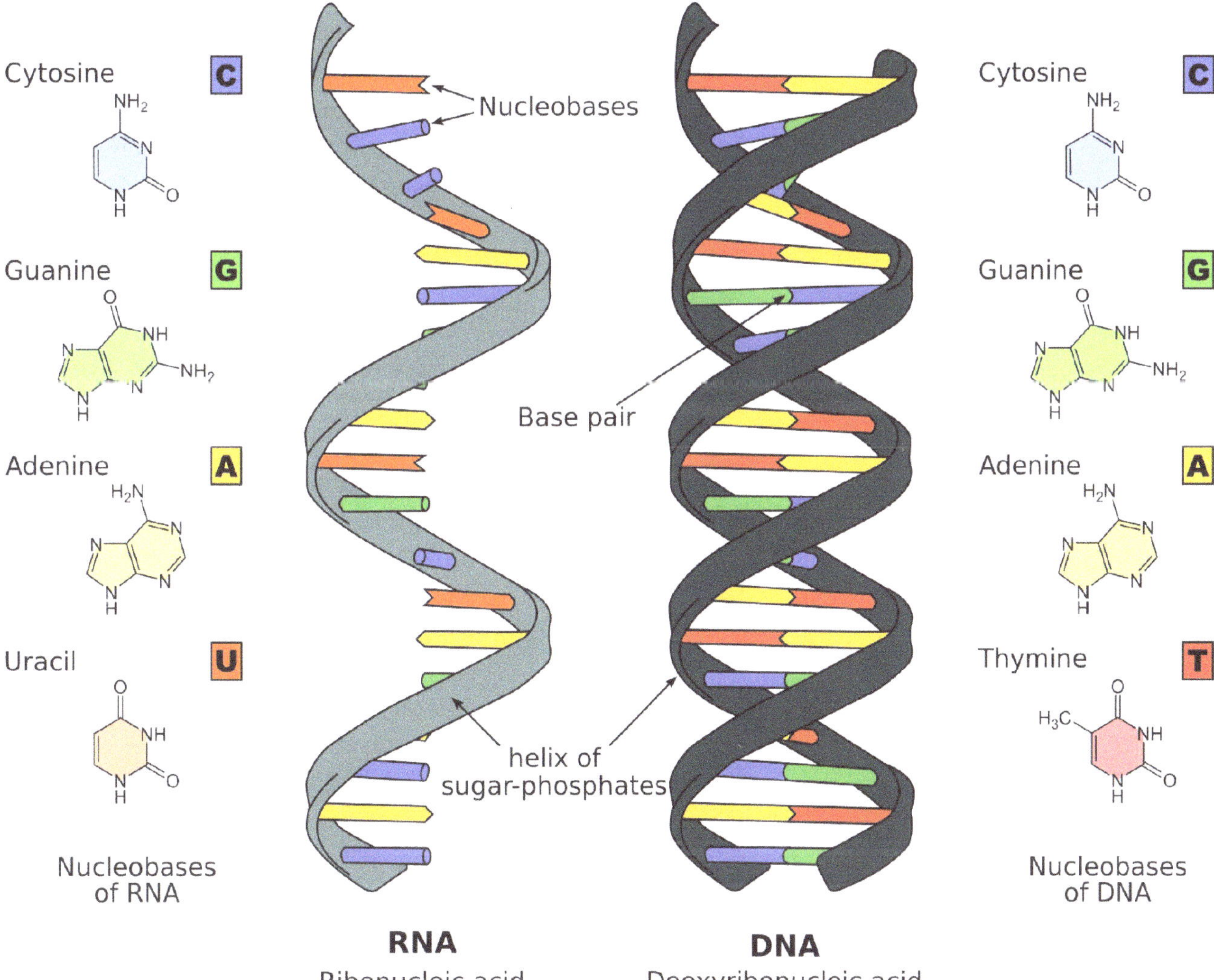

Figure 20.11. RNA has a single strand with bases containing C, G, A, and U, with C—G and A—U pairing. DNA molecule has a double strand with bases C, G, A, and T and the coupling C—G and A—T.

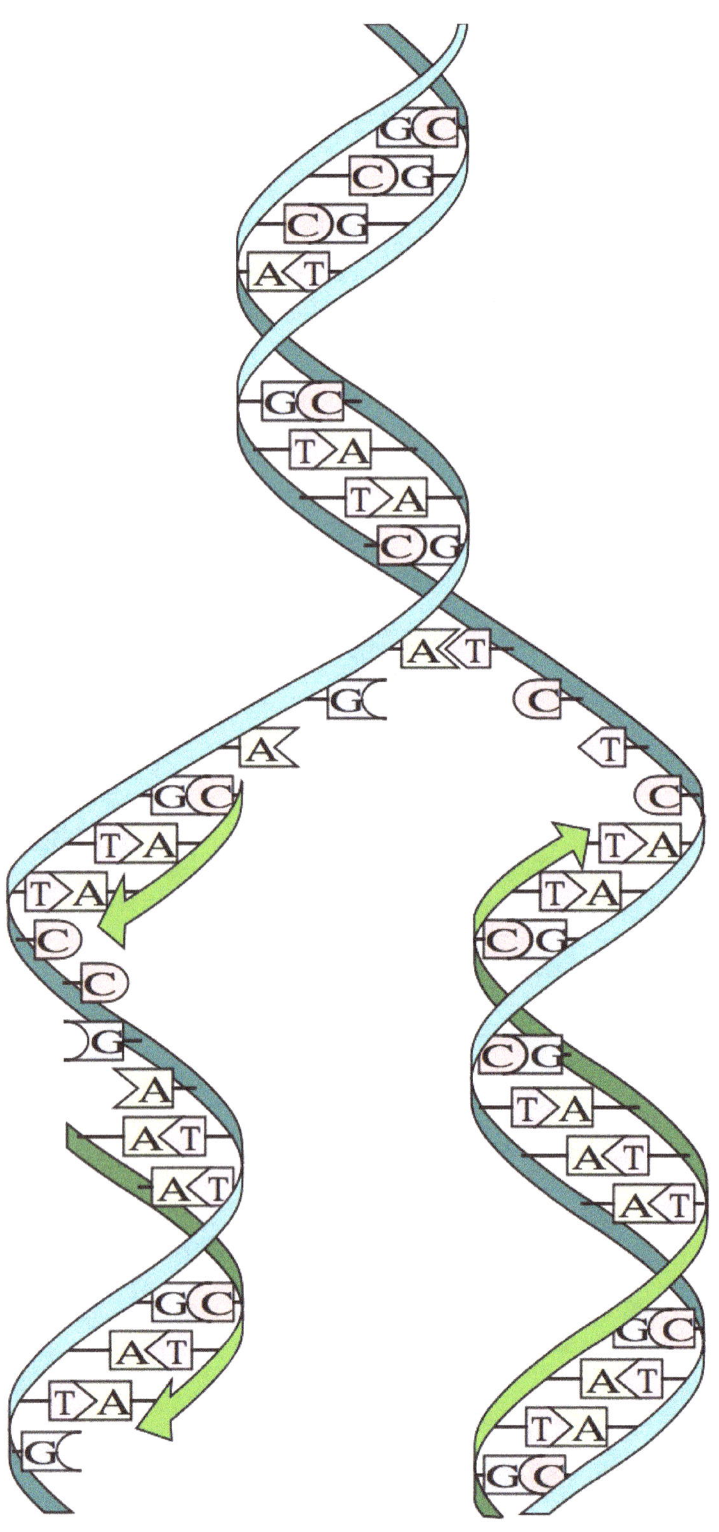

Figure 20.12. An enzyme breaks the bonds between two nucleotide chains of a parent DNA molecule, producing two parent chains. Another enzyme pairs the new nucleotide with those in the two parental chains. This results in the formation of two daughter-DNA molecules, each with one new chain (green). This process is called DNA replication.

the existing base pairing of the RNAs and is responsible for their stability. On the other hand, the single-stranded regions contain the reactive *OH* groups to bind to proteins, creating RNA-protein complexes that have critical roles in processes such as DNA and protein synthesis (see Box 20.3). The RNA structure does not contain long, double helices (like DNA) but a collection of short helices packed together in three dimensions. Such "folding" structures are responsible for the catalytic property of RNA (acting like enzymes). This is confirmed by the study of the structure of the enzymes that catalyze peptide bonds, showing that their active site is entirely composed of RNA.

To summarize, the ribose-phosphate compound forms the "backbone" structure of both RNA and DNA. The way the electric charge is distributed results on the observed geometry causing the polymer to bend into a spiral with the "basic" units (the nucleotides) facing inwards towards the axis (Figure 20.11). This spiral shape is called a *helix*.

THE FUNCTIONS OF DNA AND RNA

The genetic information is carried by DNA and is stored in its base rather than its structure. The information is contained in the sequence of the nucleotides that appear along the DNA molecule. They can occur in any order with an unlimited combination of nucleotide sequences, making DNA a very efficient carrier of genetic information (Morris et al., 2013). For example, the information in the sequence *TCATG* is different from the sequence *AGTGC*. DNA transmits information (the genetic code) in two ways:

1. DNA can reproduce an exact copy of itself using an existing strand as the template. This process is called DNA *replication* (Figure 20.12).

2. A DNA sequence can be copied to RNA in a process called *transcription*. The nucleotides in RNA can then be used to specify a sequence of amino acids in a polypeptide chain. This process is called *translation*. The combined process of transcription and translation is called *gene expression*.

The base-pairing mechanism plays an important role in DNA replication and transcription. Therefore, when a DNA is broken (Figure 20.12), the bases look for their pairs to match with. I recall that the base pairs are *A—T* and *G—C* (for DNA) and *A—U* and *G—C* (for RNA). Also, DNA replication involves the entire DNA molecule. A DNA replication process must be fully and accurately completed so that each of the resulting organisms could have the same set of DNA as its parent. The complete set of DNA in an organism is called the *genome*. The sequences of DNA that are transcribed into RNA are called *genes*. Each individual DNA molecule forms a *chromosome*.

RNA transcription takes place when a region of DNA unwinds and one strand is used as a template for the synthesis of an RNA transcript. This is complementary in sequence to the template and follows the base pairing with the exception that the transcript contains uracil (*U*), while the template (replicated from DNA) contains thymine (*T*). The RNA transcript that was produced by the template DNA contains the genetic information of the gene that was transcribed. This is the information needed to direct ribosome (site of protein synthesis in the cell) to produce the protein corresponding to the gene.

The process of information transfer (genetic code) from DNA to protein (gene expression) needs three types of RNAs to function (Box 20.4), as I briefly describe below:

Messenger RNA (mRNA): When a gene (a sequence of nucleotides) is expressed, one of the two DNA strands in the gene is transcribed, producing an RNA strand. The other DNA strand is *noncoding*. The transcribed RNA then copies the genetic code (the sequence) from DNA and produces *messenger RNA* (mRNA) because it carries the information from DNA to the sites of protein synthesis. In eukaryotic cells, the mRNA travels from the nucleus to the cytoplasm where it is translated into a polypeptide. The coding sequence of the mRNA determines the amino acid (the constituents of proteins) sequence in the protein that is being synthesized in ribosomes [the place in the cell where protein synthesis takes place (Chapter 22)]. Translation of the genetic code carried by mRNA needs enzymes and a source of chemical energy as well as two other kinds of RNA: ribosomal RNA and transfer RNA.

Transfer RNA (tRNA): This mediates between mRNA and nucleic acids (proteins). tRNA can both bind to a specific amino acid as well as read and recognize specific sequences of nucleotides in mRNA. The nucleotides are ordered in three-letter sequences called *codons* (see next section). tRNA converts the three-letter words of nucleic acids to a one-letter word of amino acid and by implication, to protein. A supply of amino acids is available in the cytoplasm of a cell, generated from food or other chemicals. tRNAs carry amino acids from the cytoplasm and match them using base pairing to appropriate codons to form a section of the polypeptide chain. This is performed by their complementary set of codons, called *anti-codons*. In order to do this task, tRNA molecules

BOX 20.4: THE ROLE OF RNA IN THE EARLY LIFE

It is widely believed that in early life, before DNA and proteins existed, RNA alone performed both information storage and catalytic functions, making life so dependent on nucleic acids.

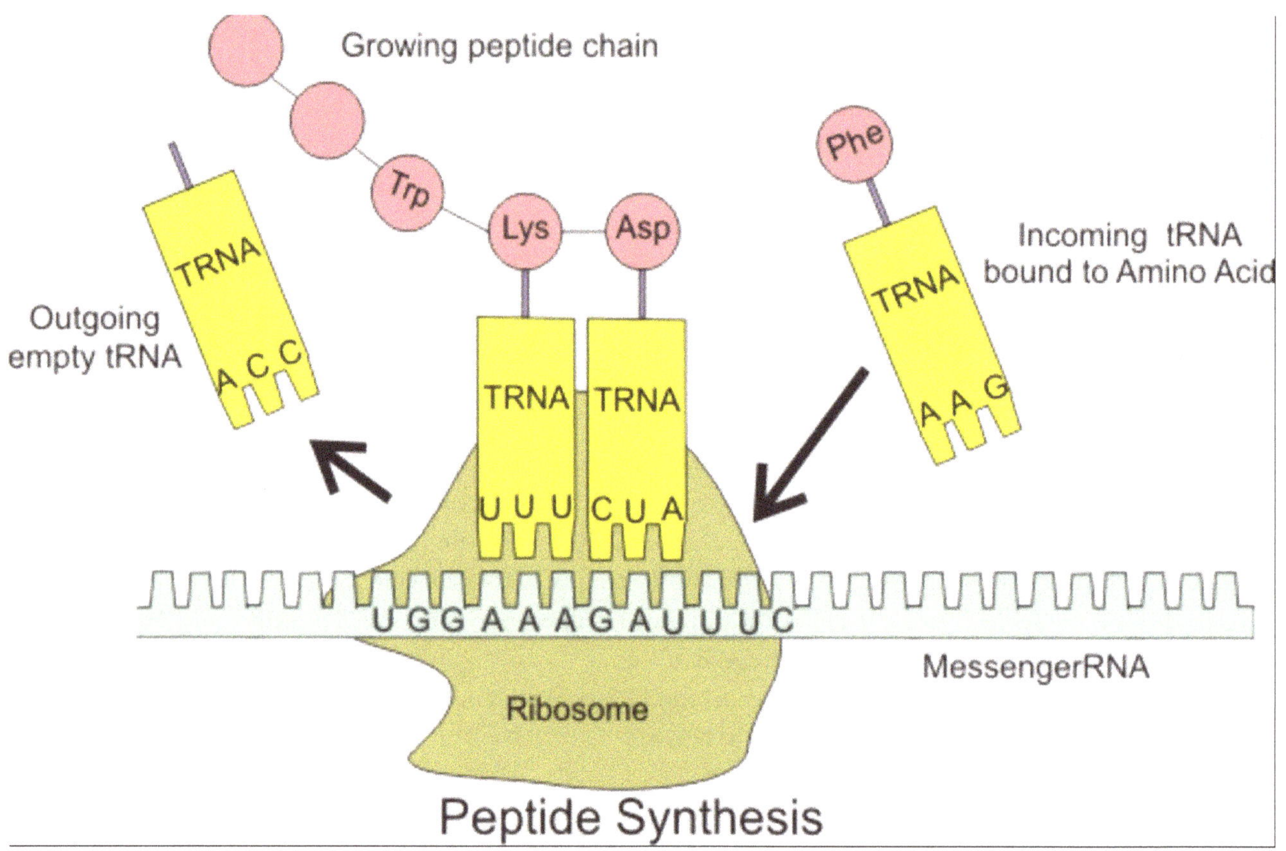

Figure 20.13. A ribosome keeps one molecule of mRNA (for providing the genetic code for the polypeptide-protein) and two tRNA molecules. One tRNA is for generating the polypeptide chain and the other is for bringing a new amino acid. The red circles are different amino acids. The sequence of the amino acids (red circles) form the protein so synthesized.

first pick up the amino acids and then find a base pair, which binds to the appropriate codons in the mRNA (Figure 20.13).

Ribosomal RNA (rRNA): The ribosomes are the protein-producing factory of the cell, and consist of proteins and several rRNAs. The rRNA catalyzes peptide bond formation between amino acids, forming polypeptide chains and hence, proteins. To do this, they coordinate the functions of mRNA and tRNA. A ribosome has two subunits: one is the binding site for mRNA and the other a binding site for tRNA. On each RNA base, the anticodon pairs with a codon on mRNA. The subunits of the ribosome keep the tRNA and mRNA molecules close to one another (Figure 20.13). The ribosome keeps one molecule of mRNA and two molecules of tRNA. One tRNA carries the growing polypeptide molecule and the other carries the next single amino acid to be added to the chain (Figure 20.13).

PROTEIN SYNTHESIS

Producing proteins is the most important task of a cell (Box 20.5). In the last section I discussed the overall process of gene expression that involves many delicate details. Here, I bring different steps together to construct the entire process, breaking down different steps and looking into why they happen the way they do, eventually leading to the production of proteins. The protein synthesis process takes two steps: The first is transcription, in which the

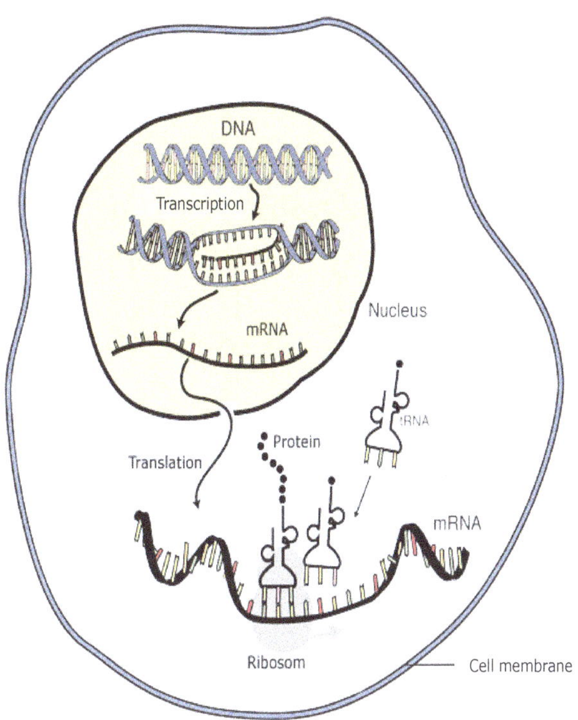

Figure 20.14. Different steps in the process of protein synthesis: 1) DNA in a eukaryotic cell is replicated; 2) mRNA is produced (transcription) then subsequently moves out of the cell nucleus and into the cytoplasm; 3) mRNA is read by the ribosome and matched to tRNAs that carry nucleic acids for protein synthesis, and 4) the nucleic acids are assembled into proteins by tRNAs.

information in DNA is encoded in mRNA, which subsequently leaves the cell and heads out to the cytoplasm. The second is translation, during which the mRNA works with tRNA and ribosomes to synthesize protein (figure 20.14).

The process of protein synthesis starts with the two strands of DNA separating through the replication process (Figure 20.12). One strand of the duplex will be used as a template to code RNA with a sequence of nucleotides complementary to the template itself (transferring the genetic material from DNA to RNA). This task is performed according to the base-pairing rules with the exception that the RNA contains uracil (U), while the template has thymine (T) (Figure 20.11—This is the transcription). The enzyme responsible for this task is *RNA polymerase*, which acts by adding successive nucleotides to the end of a growing transcript. Only the template strain of DNA is transcribed. The transcription starts when RNA polymerase encounters a sequence called *promoter*, which consists of a few hundred base pairs in which the enzyme and associated proteins bind to the DNA template. The transcription ends when it encounters a different sequence called *terminator*.

The RNA polymerase enzyme has the appropriate structural information to separate DNA strands, allow an RNA-DNA duplex by pair matching, increase the length of the transcript by adding new nucleotides, release the finished transcript, and restore the original DNA double helix. The result of this RNA-DNA interaction is an mRNA that contains the genetic information of the transcribed gene. The mRNA carries the genetic information to the site of protein synthesis. This is the information needed for ribosomes to produce proteins (see Chapter 22) with specifications directed by that gene (Figure 20.14).

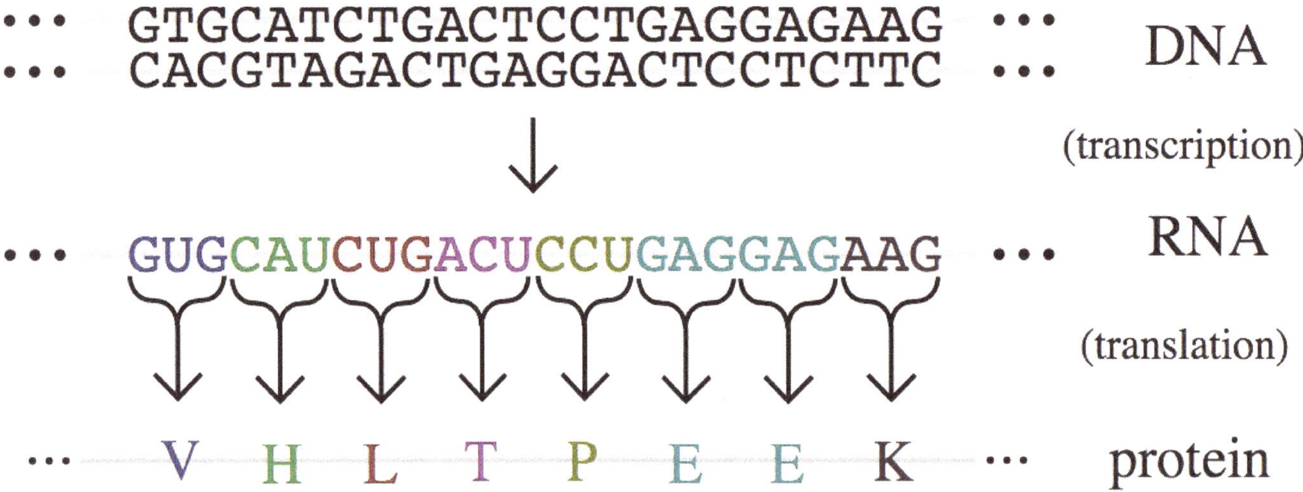

Figure 20.15. The process of transcription and translation leading to the synthesis of protein in ribosomes. The nucleotides are read in groups of three, called codons. The order of the nucleotides indicates the assigned function of the proteins. Note the nucleotides in DNA that are transcribed to RNA (T's in DNA are replaced by U's in RNA).

For eukaryotic cells, the process of transcription takes place in the cell nucleus. Once this is completed, mRNA moves out of the cell nucleus (through the pores) and into the cytoplasm where it combines with ribosomes to synthesize protein. In the case of prokaryotic cells, both the transcription and translation processes are coupled, happening in the same place (since prokaryotic cells lack nucleus).

The step after transcription is the storage and copying of the information in the *genetic code*. The genetic codes are transferred from the genes (DNA) to mRNA and to the ribosomes that make the proteins. The genetic information in an mRNA molecule is a series of sequential, nonoverlapping three-letter words (see next section). These letters are the three adjacent nucleotide bases in the mRNA that carry the genetic code, called a *codon* (Figure 20.15). The three words specify a particular amino acid (amino acids are the building blocks of proteins). The genetic code relates codons to their specific amino acids (figure 20.15).

The next step after the genetic code transfer to mRNA is translation, in which the information contained in mRNA codons are linked with specific amino acids, from which proteins are made (Figure 20.13). This process takes place in ribosomes. While mRNA moves to a ribosome, tRNAs carry specific amino acids from cytoplasm to ribosomes (Figure 20.14). Each tRNA covalently binds to a particular amino acid with the help of a specific enzyme. On the tRNA polynucleotide chain, there is a triplet of bases called the *anticodon*, which complements the mRNA codon for the specific amino acid carried by tRNA (Figure 20.13). For example, the mRNA codon for arginine is CGG, with its anticodon being GCC. tRNA interacts with both mRNA and ribosomes. The

BOX 20.5: WHAT ARE PROTEINS?

Proteins do most of the work in cells and are required for the structure, function, and regulation of the body's tissues and organs. They consist of a chain of many thousands of amino acids attached to one another in a long chain. There are different types of proteins with various functions (fixed by their three-dimensional shape). Examples are: *Collagen*: provides structural support to cells and strengthens bones; *Enzymes*: catalyze interactions (e.g., amylase); *Hormones*: the chemical messengers between cells (e.g., insulin), and *Antibodies*: help to prevent infection by fighting foreign agents.

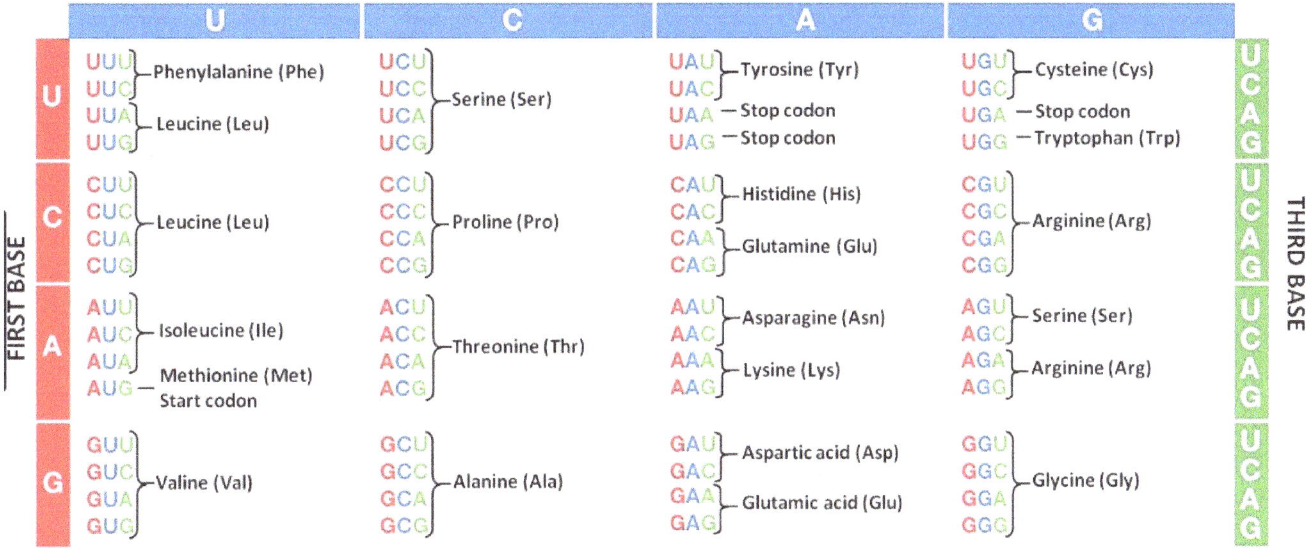

Table 20.1. The list of amino acids and the nucleotide combinations (codons) that produce them. A combination of these amino acids (shown by three-letter codes) produce specialized proteins.

structure of the ribosome is optimized in such a way to keep mRNA and tRNA in the correct positions, allowing polypeptide molecules of amino acids to assemble based on the genetic instructions transferred from DNA to mRNA. The tRNA assembles the proteins as they are read by the ribosome one at a time (Figure 20.13). Protein assembly continues until the ribosome encounters the *"Stop"* codon, which is a combination of three nucleotides that represent an amino acid (Table 20.1).

Finally, tRNA carries the amino acids out of the ribosome and assembles them into protein (Figure 20.14). There are thousands of ribosomes in any given cell, with all taking part in the protein synthesis. To ensure that the protein made through this process is the one specified by mRNA, two conditions must be met: 1) The tRNAs must read mRNA codons correctly and 2) the tRNAs must deliver the amino acids that correspond to each mRNA codon. There is one tRNA molecule for every one of the 20 amino acids.

THE GENETIC CODE

The genetic code uses combinations of the four bases (*A*, *U*, *G*, and *C*) to produce the 20 amino acids (building blocks of proteins). One-letter codes could only produce four combinations (codons) while two-letter codes could only produce $4 \times 4 = 16$ unambiguous combinations. This stops short of generating all the 20 amino acids. A triple code, based on three-letter codons, can produce $4 \times 4 \times 4 = 64$ codons, much more than what is needed to produce the necessary number of amino acids. Therefore, a three-letter code can satisfy this, as shown in the complete genetic code listed in Table 20.1. There are many combinations of the three letters that are redundant and produce the same amino acid. For example, *CGU*, *CGC*, *CGA*, and *CGG* all represent the amino acid arginine. Therefore, there is a lot of degeneracy in the three-letter codes, with not all of the 64 codons representing individual amino acids. The same genetic code is used for all of the species. The codon *AUG* that is the code for methionine is the start codon, i.e., the initiation signal for translation. The codons *UAA*, *UAG*, and *UGA* are stop codons. When one of these codons is reached, translation stops and the polypeptide chain is released and assembled by tRNA.

How does the changing of the order of the letters in the codons affect the resulting amino acids? Changes of the first base of codons produce chemically similar amino acids. For example, consider a Leucine codon set with a *CUX* codon (Table 20.1) where *X* stands for one of *U, C, A,* or *G* nucleotides. In this case, mutation (changes in the nucleotide) of the third nucleotide does not affect the type of the codon (silent mutation). However, mutation of the first nucleotide (*AUX* where *X* again stands for *U, C, A,* or *G*) changes this to Isoleucine or Methionine (*AUG*), which are both similar to Leucine, i.e., medium-sized hydrophobic (water-hating) nuclei acids. Also, only the middle nucleotide is able to fix the properties of the resulting amino acid. The codons with structure *XUX* are all hydrophobic amino acids (Phenylalnine, Leucine, Isoleucine, Methionine, and Valine). However, if we change the middle nucleotide to *A* (e.g., an *XAX* codon), all the resulting amino acids become hydrophilic. Therefore, the second position in the codon indicates whether the nucleic acid is hydrophobic, while the first position identifies the type of the hydrophobic amino acid.

WHY DNA IS THE CARRIER OF THE GENETIC MATERIAL

The RNA has a hydroxyl group (OH) that is highly reactive. This could connect to the phosphate group in the RNA "backbone," dividing it into two. It could also enter into polymerization and bond with hydrogen and other reactants; therefore, RNA molecules are unstable and susceptible to breaking. On the other hand, DNA does not contain the hydroxyl group, making it much more stable. The relative stability of the DNA is a strong reason for making it a better carrier of genetic information. Furthermore, the nucleotide base of the DNA can more easily repair damaged genetic material. RNA contains uracil (*U*) while DNA has a methylated form of this base ($R-CH_3$), known as thymine (*T*), (Box 20.3). In RNA, cytosine (*C*) spontaneously changes to uracil when interacting with water (a process called hydrolysis, i.e., the breaking down of a chemical compound via interaction with water). The presence of the methyl (CH_3) group in thymine then identifies the uracil that was spontaneously created in DNA, restores cytosine, and as a result, repairs DNA. This process is all because of the presence of thymine that makes DNA such a faithful guardian of the genetic code.

ENERGY PRODUCTION PROCESS

The reason that cells could, by themselves, perform so many different and complex tasks is because they can produce their own energy needs. Living cells use the molecule adenosine triphosphate (ATP) to store and release energy. The ATP molecule consists of adenosine, which is made up of the base adenine and the five-carbon sugar ribose attached to three phosphate groups (Figure 20.16, top panel). This has a similar structure to the molecule that constitutes the backbone of DNA and RNA (Figure 20.8). This indicates that the combination of sugar (ribose) + phosphorus + nucleotides is an essential element of life.

The energy in ATP is in the chemical bonds connecting the phosphate groups that are released when the bonds are broken. Each time a cell extracts energy from ATP, it converts it to adenosine diphosphate (ADP) by losing one phosphate group (Figure 20.16, bottom panel). This can then be converted to an ATP molecule by bonding again to a phosphate group. The ATP cycle therefore continues within the cell.

All living things (plants, animals, fungi, and microbes) release energy through a process called *cellular respiration*. This is a set of complicated chemical reactions that convert the chemical potential energy stored in organic molecules (like glucose) into a chemical form that can be used by cells to perform their tasks. This resulting energy source is ATP, which is the energy currency for all cells. For animals, the organelle responsible for energy generation (cellular respiration) is mitochondria, a rod-shaped organelle consisting of two membranes: an outer membrane and a highly folded inner membrane (Figure 22.1). During this process, oxygen is used and carbon dioxide is released. The organelle responsible for energy production in plants is the chloroplast. It captures sunlight and synthesizes sugar through the process of photosynthesis, resulting in the release of oxygen as the waste product. Chloroplasts also consist of outer and inner membranes, inside which there are light-collecting molecules called *chlorophylls*, which are responsible for the green color of plants and for capturing energy from sunlight. By using the energy from sunlight and carbon dioxide, they produce carbohydrates and release oxygen. The mitochondria and chloroplasts contain their own genomes and work independently from other organelles. There are similarities between the DNA in mitochondria and chloroplast with some bacteria. This implies that these organelles originated as bacteria that were captured by eukaryotic cells, and over time evolved to acquire their present functions (this will be discussed in Chapter 22).

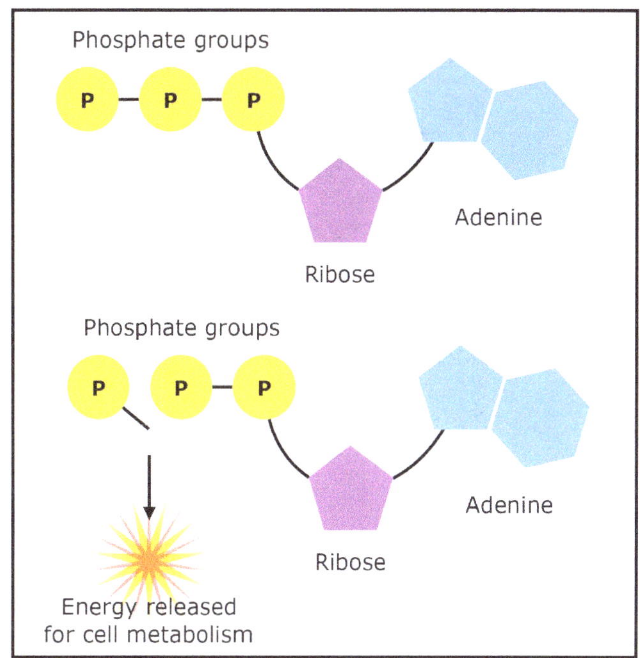

Figure 20.16. Cells producing energy by converting adenosine triphosphate (ATP) to adenosine diphosphate (ADP).

The cellular respiration could take place in the presence of oxygen (called aerobic respiration) or in the absence of oxygen (called anaerobic respiration). Here, I briefly describe the process for aerobic respiration that is the basis of energy production in modern plants and animals. This takes place in four steps, as described below and shown in Figure 20.17.

Step 1: Glucose is partially broken down producing *pyruvate* and a small amount of ATP through a process called *glycolysis*. This process takes place in cytoplasm.

Step 2: Pyruvate is converted to another molecule called acetyl-coenzyme A (acetyl-CoA) with carbon dioxide released. This process takes place in mitochondria.

Step 3: Acetyl-CoA is broken down in the citric acid cycle (the Krebs cycle) with a small amount of ATP and more carbon dioxide produced.

During steps 1–3, chemical energy is transferred to both ATP and electron carriers, which are both energy-storing molecules. Electron carriers are molecules that store and transfer energy in the form of high energy or excited electrons.

Step 4: In a series of reactions, electron carriers donate their high-energy electrons to a final electron acceptor along a series of membrane-associated proteins. The energy of these electrons is then used to generate large amounts of ATP through the process of *oxidative phosphorylation*. The energy produced by electron transfer from an electron carrier to electron transport chain leads to the synthesis of ATP from ADP and inorganic phosphate. In aerobic respiration, oxygen is the final electron acceptor. In this process, oxygen is used and water is produced. The elements responsible for the electron transport are the proteins associated with the inner mitochondrial membrane.

During the steps illustrated in Figure 20.17, the energy stored in glucose, carbohydrates, or lipids are converted to energy in the ATP, which stores large amounts of energy in its phosphate bonds (Figure 20.16). This is the energy currency the cells are able to use to function. The energy generation is a redox process (see the next chapter) with electrons

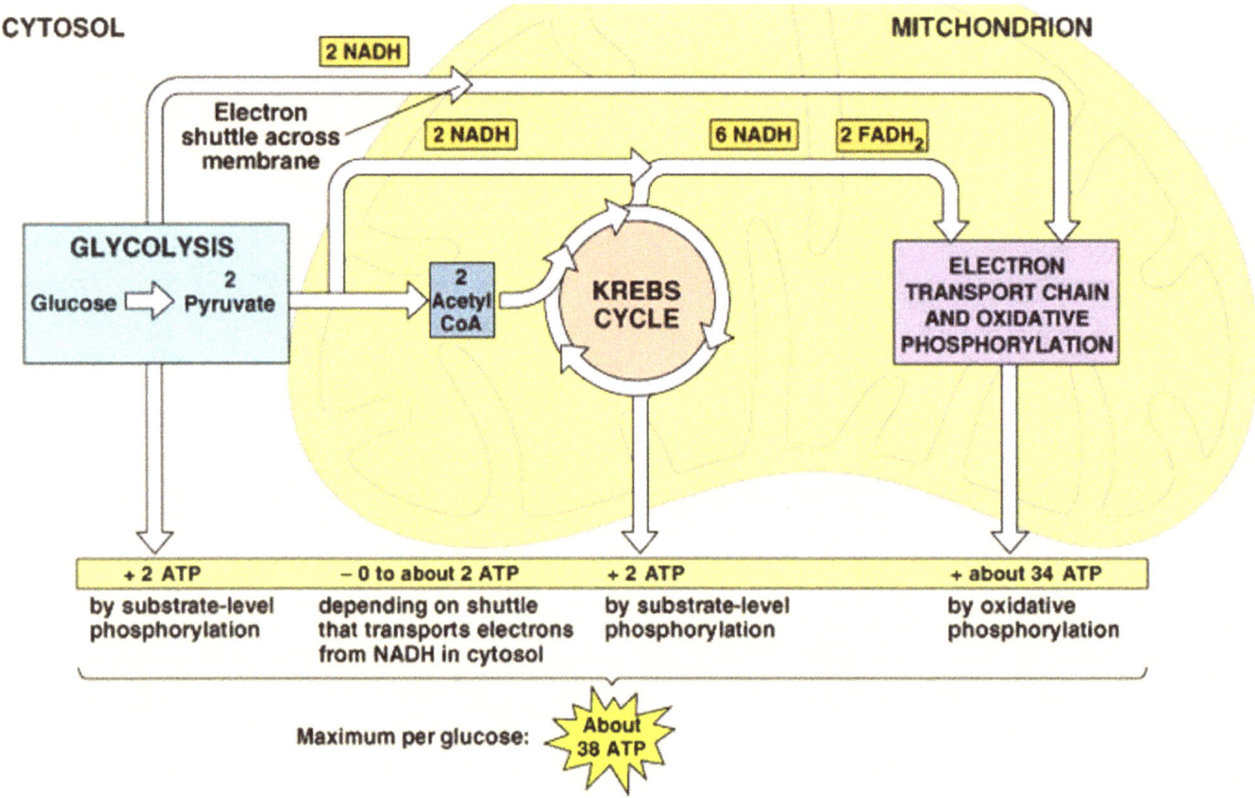

Figure 20.17. The step-by-step process of cellular respiration, which is a source of ATP production. Using glucose ($C_6H_{12}O_6$) as the starting product and oxygen (O_2), CO_2 and large amounts of ATP are released. The electrons are carried by NADH. FADH₂ (Flavin Adenine Dinucleotide-FADH₂), is a redox compound produced during the Krebs cycle that also carry electrons.

carried from one reaction to other by NADH which is a reduced (accepting electrons) form of Nicotinamide Adenine Dinucleotide (NAD^+), generated by the reaction: $NAD^+ + 2H \rightarrow NADH + H^+$ (Figure 20.17).

SUMMARY AND OUTSTANDING QUESTIONS

In this chapter I present a review of the ingredients essential for life, their structure and functions and the basic chemical and biological laws they follow to perform their tasks. Taking a living organism, I reduce it to its parts and study how its different components work as independent entities. The aim in this chapter is to provide the necessary background material to explore scenarios for the origin of life, explained in the next chapter.

The carbon atom, with its special structure and configuration, is the most essential element needed for life. It can simultaneously enter into reactions with four different elements, building very complex molecules through polypeptide bonds. Furthermore, the 3-D shape of composite carbon molecules, when oriented, could generate new configurations (molecules) with different properties than before (isomers). This allows carbon to enter an unlimited number of reactions with other elements- namely hydrogen, nitrogen, oxygen and phosphorous- to produce the organic molecules essential for life. This results the compounds that are in common in all living things and are needed to support and sustain all forms of life. There are four such compounds: Carbohydrate (that provides the source of energy), lipid (used for energy storage), proteins (performs numerous tasks from building structures to working as enzymes—see Box 20.5), and nucleic acids (encodes genetic information and basic hereditary material).

The study of the structure and function of the two most basic molecules- DNA and RNA- shows why they are so essential for life, as we know it. The DNA molecule is reproducible (could generate a copy of itself through the replication process) and is responsible for transfer of genetic material (by RNA) to the sites of protein synthesis. In other words, DNA instructs proteins to take up the tasks they do. They constitute of two parts, the backbone and the base. The backbone consists of a carbon sugar (ribose) connected to a nitrogen base (nucleotide) and a phosphate group. The nitrogen base consists of four nucleotides: adenine, cytosine, thymine and guanine. Because of the chemical compositions of these nucleotides, cytosine always pairs with guanine and thymine always pairs with adenine. This bonding is responsible for the DNA to be reproducible and for the familiar double helix structure of it. Through the process of replication, DNA can produce a copy of itself. Also, DNA can act as a template, copying a sequence of nucleotides to RNA in a process called transcription. This information is then transported to the protein synthesis sites, producing proteins from polypeptide bonds between amino acids. A significant difference between the DNA and RNA molecules is in their backbone where RNA has a sugar base that contains ribose (OH) while DNA molecule is deoxyribose (lacks the oxygen and only containing hydrogen). The OH molecule is highly reactive, making RNA to immediately enter into reactions with other molecules. As a result, DNA is more stable and this is the reason it is the main carrier of genetic information. Other differences between DNA and RNA is in their base where thymine in DNA is replaced by uracil in RNA and the double strand in DNA compared to the single strand for RNA.

There are a total of 20 amino acids in nature, forming the genetic code. The nucleotides combine in groups of three to form amino acids. Many millions of combinations of the amino acids could then be constructed, forming proteins. Changing the order of the nucleotides affects properties of the resulting amino acid causing, for example, the amino acid to be hydrophilic or hydrophobic. The sequence of the amino acids is dictated by the instructions from the DNA, which generates the type of the proteins they produce.

All these chemical reactions need energy to take place. The cells generate their own energy through a molecule called adenosine triphosphate (ATP) consisting of a phosphate group, ribose, and adenine. By losing one phosphate, the stored chemical energy in that bond is released, resulting in adenosine diphosphate (ADP). The phosphorous released then combines with ADP and generates ATP which, in turn, breaks again and releases energy. The initial source of energy is carbohydrate and lipid.

Many of the compound molecules essential for life have similar structures and are composed from the same material. For example, the molecule that builds the backbone of DNA and RNA has a similar composition and structure as the one that generates energy in cells (a nitrogen base, a carbon sugar, and one or more phosphate groups). The question then is: could a combination of the same chemical elements with different configurations be responsible for planting the seeds of first life billions of years ago? Are there another variation of these compounds that could lead to perhaps a different form of life yet to be found?

Why are there only 20 amino acids where we could produce a lot more, given the nucleotides we have? This is likely to have been fixed at an optimum by natural selection. Properties of the amino acids depend on the sequence of the nucleotides forming them, with many of these combinations leading to the same amino acids. Therefore, larger number of amino acids leads to significant redundancy. With the combination of the amino acids, each with its own property, nature has been able to produce the proteins needed to function and sustain life.

Finally, our definition of life is based on the one and the only life we know of- our own. Therefore, any search based on this definition will miss other forms of life, if they exist. Could a version of the scenario we have deciphered on Earth lead to some kind of primitive life somewhere else in the universe, modified based on the environment and available ingredients? This is a hard question to answer given the lack of data but one needs to have this in mind when studying the biochemistry of molecules needed to support life.

REVIEW QUESTIONS

1. What are the main common characteristics that all living organisms have?
2. Explain why a single definition of life is misleading.
3. Why can carbon easily enter into chemical reaction with other elements?
4. Briefly explain chemical bonds.
5. Why is carbon unique in forming the basis of life on Earth?
6. Describe the different parts of an amino acid molecule
7. Describe the process of protein synthesis.
8. What are the differences between deoxyribonucleic acid (DNA) and ribonucleic acid (RNA)? Which one is more stable and why?
9. Briefly explain the structure of DNA and RNA.
10. What are nucleotides?
11. What bonds take place to construct the base of the DNA double helix?
12. How can DNA transfer information to sites of protein synthesis?
13. What are genes?
14. Explain the role of enzymes in DNA replication.
15. Explain how ATP molecule produces the energy needed for the cells.

CHAPTER 20 REFERENCES

Bennett, J., and S. Shostak. 2007. *Life in the Universe*. 2nd ed. Boston: Pearson.

Cleland, C.E., and C.F. Chyba. 2002. "Defining Life." *Origins of Life and Evolution of the Biosphere* 32 (4):387–93.

Morris, J., D. Hartl, A. Knoll, and R. Lue. 2013. *How Life Works*. New York: Freeman.

Sadava, D., D. Hillis, C. Heller, and M. Berenbaum. 2014. *Life: The Science of Biology*. 10th ed. Sunderland, MA: Sinauer.

FIGURE CREDITS

- Fig. 20.5a: Source: https://commons.wikimedia.org/wiki/File:Methane-CRC-MW-3D-balls.png.
- Fig. 20.5b: Source: https://commons.wikimedia.org/wiki/File:Ethane-A-3D-balls.png.
- Fig. 20.8: Source: https://unlockinglifescode.org/media/details/441.
- Fig. 20.9a: Copyright © Bruce Blaus (CC by 3.0) at https://en.wikipedia.org/wiki/File:Blausen_0323_DNA_Purines.png.
- Fig. 20.9b: Copyright © BruceBlaus (CC by 3.0) at https://en.wikipedia.org/wiki/File:Blausen_0324_DNA_Pyrimidines.png.
- Fig. 20.10: Copyright © OpenStax College (CC BY-SA 3.0) at https://commons.wikimedia.org/wiki/File:DNA_Nucleotides.jpg.
- Fig. B20.3: Copyright © Sponk (CC BY-SA 3.0) at https://commons.wikimedia.org/wiki/File:Difference_DNA_RNA-EN.svg.
- Fig. 20.11: Copyright © Sponk (CC BY-SA 3.0) at https://commons.wikimedia.org/wiki/File:Difference_DNA_RNA-EN.svg.
- Fig. 20.12: Copyright © Madprime (CC BY-SA 3.0) at https://en.wikipedia.org/wiki/File:DNA_replication_split.svg.
- Fig. 20.13: Copyright © Boumphreyfr (CC BY-SA 3.0) at https://commons.wikimedia.org/wiki/File:Peptide_syn.png.
- Fig. 20.14: Adapted from https://commons.wikimedia.org/wiki/File:MRNA-interaction.png?uselang=da.
- Fig. 20.15: Copyright © Madprime (CC BY-SA 3.0) at https://en.wikipedia.org/wiki/File:Genetic_code.svg.
- Fig. 20.17: Copyright © by Pearson Education, Inc.

THE ORIGIN OF LIFE

CHAPTER LEARNING OBJECTIVES

This chapter will cover:

- Early experiments to study the origin of life
- Different scenarios for the origin of life
- Origin of protein synthesis
- Origin and evolution of DNA
- Origin of energy production process
- Origin of the genetic code
- Origin of chirality

Among all the theories and hypotheses discussed in previous chapters, some may turn out to be right and some wrong. These will be tested by observations of natural phenomena. However, there is one indisputable fact—that we are here, living on this planet, and are alive. Life on our planet has taken different forms, from simple prokaryotes to eukaryotes, complex multicellular organisms, plants, animals, and finally, the thoughtful creatures like us, able to decipher the secrets of the universe. Life has evolved over billions of years from a primitive form to intelligent beings we are today. With observations of natural phenomena, we acquire knowledge, and by using objective thinking and analytic reasoning, search for our own origin. Digging deep into these observations, we are now at a stage to address one of the most fundamental questions humankind has ever asked: what is the origin of life? The obvious way to explore the origin of life on Earth is by the study of fossils. A fossil of a particular age will indicate that life existed at that time. While geological records reveal much of the history of life on Earth, the theory of evolution will study changes from the time life started to the present. However, these will not elucidate how life started in the first place. The problem is that geological records of the very beginning have become inaccurate and sparse. These records have not survived the harsh conditions during the first hundreds of millions of years after earth was formed. As a result, we do not know exactly when the first life existed.

"Men go abroad to wonder at the heights of mountains, at the huge waves of the sea, at the long courses of the rivers, at the vast compass of the ocean, at the circular motions of the stars, and they pass by themselves without wondering"

— SAINT AUGUSTINE

"An honest man, armed with all the knowledge available to us now, could only state that in some sense, the origin of life appears at the moment to be almost a miracle, so many are the conditions which would have had to have been satisfied to get it going."

— FRANCIS CRICK

The stromatolites (Greek for "rock beds"), rocks structured in layers of sediments mixed with different types of microbes, show evidence for the first life around 3.5 billion years ago. These have microbes near the top that produce energy through the photosynthetic process (as they are exposed to the sun light) with the microbes beneath, using organic compounds to generate their energy. The similarity between the stromatolites and modern layered sedimentary structures supports the fact that stromatolites are likely the first fossil remnants of early life (Bennett and Shostak 2005).

One way to look for the time the primitive life started is by looking for traces of organic carbon in microscopic fossils. This will set a time for when the first cells came to existence. This is however, a very delicate process as the microfossils may have been contaminated by other geological events through the history of Earth, affecting the original signatures. Nevertheless, these constrain the age of the first fossils to 3.5 billion years. If these records are correct, life must have been widespread around 3.5 billion years ago so that it could leave its fossil signatures. This means that life very likely started on Earth well before that, perhaps around 3.8 billion to 4 billion years ago. Given the uncertainties and complications regarding interpretation of geological records, there are problems in determining the age of the first fossils. What is clear is that life is not continuously restarting on Earth or, in other words, is not spontaneously generated but resulted from the life that existed before. The previous chapter used known laws of physics and chemistry to uncover the origin of the ingredients needed for life. Considering a world where the required conditions for life are satisfied and the ingredients for life are available, the question then is: how did all these components come together to start the life, as we know it today?

This chapter extends the limit of our knowledge to the very beginning to find how life started in its most primitive form. It studies the origin of the simplest organic molecules. The chapter presents different competing scenarios for the origin of life. It then studies the origin of different ingredients needed to initiate and sustain life.

THE ORIGIN OF THE SMALL MOLECULES OF LIFE

Looking at our distant past, it becomes clear that all forms of life on Earth resulted from a life that already existed. In other words, spontaneous generation of life does not happen or, life does not originate from nonliving matter.

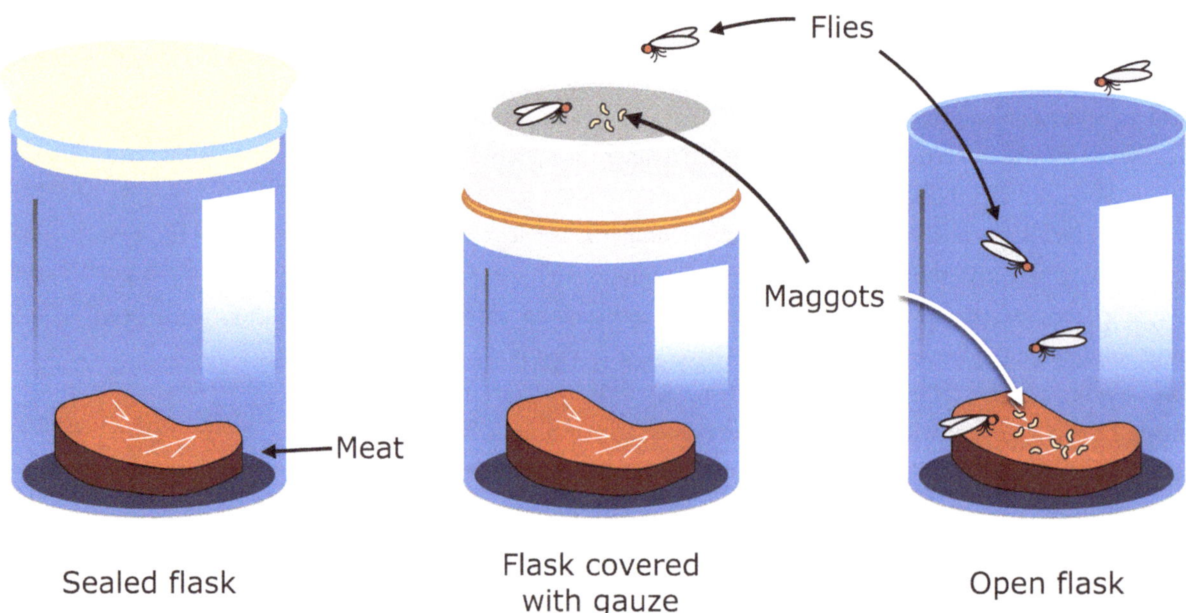

Figure 21.1. An experiment with meat and maggots confirmed that the first life was not spontaneous. In other words, life can only be created from living things.

An experiment performed in 1668 by Italian physician Francesco Redi (1626–1697) examined this hypothesis. He used three jars containing meat and prepared them as follow (figure 21.1):

- The first jar was exposed to both air and flies.
- The second jar was exposed to air but not flies.
- The third jar was exposed to neither air nor flies.

Redi noticed the presence of maggots (and subsequently new flies) in the first jar but not in the others. This indicated that maggots only appear if flies were already around. This simple experiment confirmed that life is not created spontaneously from nonliving material (i.e. meat). In other words, all living things come from preexisting life.

THE ORIGIN OF SIMPLE ORGANIC MOLECULES

The previous chapter used observations to find out about the features that all living organisms share. Now, we study the origin of organic molecules that are responsible for life and in particular, how organic compounds resulted from inorganic material. By breaking down the processes leading to the first living things, it is reasonable to assume that life started through chemical reactions. It is hard to imagine this happening in the oxygen-rich atmosphere today. This is because oxygen is a highly reactive element and will enter into reactions with other elements, breaking organic molecules before they combine to form the complex molecules that are responsible for life on Earth. This implies that early Earth was likely free of oxygen, with only inorganic compounds present. The question now is whether organic compounds can be produced from inorganic material.

In 1953 Stanley Miller and Harold Urey performed a famous experiment that became known as the Miller-Urey experiment. They used a glass flask filled with methane and ammonia, the gases believed to dominate the early atmosphere of Earth (figure 21.2). They filled another flask with water to mimic the oceans. Heating the flask containing water, water vapor was produced and was mixed with the methane and ammonia gas to generate conditions similar to the early Earth atmosphere. This was then subjected to electric discharge, providing the required energy for chemical reactions to start. The

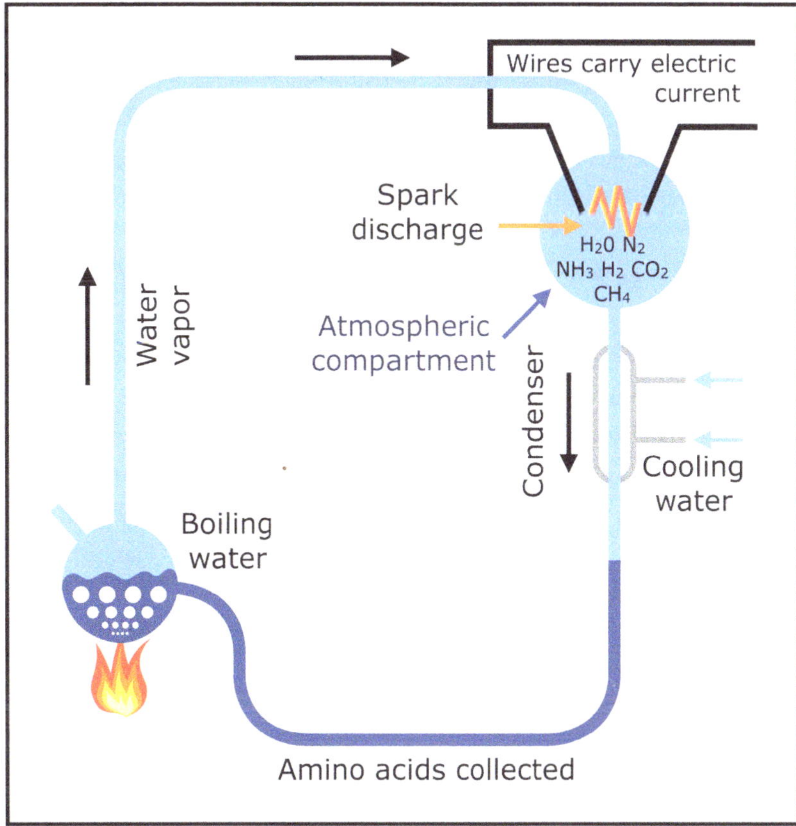

Figure 21.2. The Miller-Urey experiment demonstrates that simple organic molecules can be generated from the initial inorganic molecules present in the early Earth atmosphere.

gas was then cooled and condensed to produce rain, and this was cycled back into the flask containing water (figure 21.2). After letting the experiment run for a week, they analyzed the final product and discovered traces of amino acids and organic molecules. The conclusion from the Miller-Urey experiment was that organic molecules were likely generated from inorganic material in the presence of a strong energy source.

However, results of this experiment depend on the amount of hydrogen in the early Earth atmosphere. Without hydrogen, organic material could not be produced. In the absence of hydrogen, the oxygen present in carbon dioxide would reduce the amount of organic material much lower than predicted by Miller-Urey experiment. The latest studies show hints that hydrogen constituted as much as 30 percent of the early Earth atmosphere.

STAGES OF LIFE

For different components of life to come together and to initiate and sustain living things, they need to go through the following four stages:

Stage 1: The synthesis of small organic molecules such as amino acids and nucleotide monomers.

Stage 2: The joining of these simple molecules into more complex polymers and chains of nucleic acids, producing proteins.

Stage 3: The generation of self-replicating molecules making genetic transfer from parent to daughter cell possible.

Stage 4: The formation of protocells by packaging these molecules into enclosed structures surrounded by membranes that make these different from the surrounding environment.

However, how the "right" elements were generated in the first place, is a different story that I will come back to later in this chapter.

CHEMICAL REACTIONS NEEDED TO START LIFE

There are a number of chemical reactions that are responsible for life on Earth. Chemical reactions will end up in an equilibrium state that is a balance between the reacting atoms and molecules and the product atoms and molecules. If somehow the equilibrium is disturbed (for example by adding more reactants in the process), a state of disequilibrium is reached. This changes the reaction rate to bring it back to the equilibrium again (reactions always work toward equilibrium). The reactions in disequilibrium, moving back to equilibrium, release chemical energy that is used by life to support metabolism (Plaxo and Gross 2006).

The chemical reactions essential to start life on Earth are called *redox* reactions. These involve exchange of electric charge (resulting from movement of electrons) between the reacting atoms and molecules. For example, the process producing water is a redox reaction. First, a hydrogen molecule is converted to two hydrogen nuclei (positively charged protons) and two electrons (negatively charged): $H_2 \rightarrow 2\,H^+ + 2\,e^-$. Second, the two protons and two electrons are combined with an oxygen atom to produce water ($\frac{1}{2}\,O_2 + 2\,H^+ + 2\,e^- \rightarrow H_2O$). Here oxygen accepts an electron becoming more negatively charged—called *reduced*—with the charge of hydrogen increased or hydrogen becomes oxidized (because it combines with oxygen; the word *redox* comes from the combined first letters in *reduced* and *oxidized*). A redox reaction involves transfer of one or more electrons from an electron donor (oxidized) to an electron acceptor (reduced). This transfer produces energy that can then drive biochemical reactions responsible for life.

The process responsible for respiration in animals involves reaction of sugar, such as glucose ($C_6H_{12}O_6$), with oxygen, making carbon dioxide and water and releasing energy:

$$C_6H_{12}O_6 + 6\,O_2 \rightarrow 6\,CO_2 + 6\,H_2O + energy.$$

In this reaction glucose donates electrons while oxygen accepts electrons; therefore, it is a redox. A chain of such redox reactions generate the energy needed to sustain life. In the photosynthesis process the chain begins when chlorophyll absorbs sunlight, creating disequilibrium in the cell, which in turn produces the energy that the cell uses through redox reactions (Plaxo and Gross 2006).

LARGE ORGANIC MOLECULES AND THE ORIGIN OF FIRST LIVING ORGANISMS

Any scenario for the origin of living organisms must be able to address the following three questions: How the first organic molecules essential for life came about? How and in what order these combined, forming the main elements of life? Where did the energy needed for producing the first complex molecules come from?

Even if all of the amino acids needed for life existed on Earth, the probability of these self-organizing by themselves in a single step is extremely low, given the complexity of life. The probability of this is minimal, even for the simplest of the living organisms. Here, I point out some of the complications involved. We know that DNA is the agent responsible for transferring hereditary information, performing this function because of its ability to replicate. For the DNA to assemble, replicate, and perform its functions, proteins are needed to act as catalysts to speed up organic reactions. On the other hand, we know that proteins cannot form, function, and assume their responsibility without instructions from DNA. The question therefore is: Which one came first, DNA or proteins? Assuming that early life followed the same principles as life does today, it for sure needed a self-replicating agent like DNA. However, with its shape and structure, DNA is far too complex to be responsible for replication at this early time in the history of life. On the other hand, without DNA there is no known mechanism for protein synthesis.

It is not yet clear how the early components of life came together. In the following I discuss the competing scenarios that attempt to explain the origin of first living organisms.

1. METABOLISM FIRST

The metabolic-first hypothesis proposes that the first life was formed from a network of self-sustaining chemical reactions of simple monomeric (a unit that chemically binds to other molecules to form more complex molecules called *polymers*) organic molecules. Genetic molecules were then incorporated as this network of reactions evolved in complexity, leading to the development of metabolic life. Based on this hypothesis first developed by German chemist Günter Wächtershäuser in the 1980s, the first metabolic reactions began on the surface of minerals such as iron sulfide and nickel sulfide around hydrothermal vents deep in oceans. These minerals act as catalysts producing carbon from the existing inorganic material like carbon monoxide (CO) existing in the vents or carbon dioxide (CO_2) generated by volcanic activities and finding their way to ocean floors through ridges at the floor of the oceans. The process of carbon release from these compounds requires hydrogen. The hydrogen will be supplied from hydrogen sulfide (H_2S) in the vents or from ocean water.

Using the reducing power of hydrogen sulfide (the ability to release electrons), iron sulfide and nickel sulfide catalyzed the reduction of carbon dioxide to small organic molecules, accelerating conversion of inorganic to organic molecules. The basic substrate material for this reaction is iron sulfide in the surface of the rocks.

Furthermore, the iron sulfide donated electrons to carbon monoxides, converting them into acetic acid and starting reactions that led to the formation of large chains of amino acids that formed proteins (Plaxo and Gross, 2006). The energy needed for these reactions came from the redox process (see previous section) because of the chemical disequilibrium when the hot water from the hydrothermal vents merged into the colder ocean water (see Box 21.1).

The prevailing metabolic pathway in living organisms is the citric acid cycle (Krebs cycle). Aerobic organisms use the citric acid cycle to oxidize acetate, $C_2H_3O_2^-$ (from food consisting of carbohydrates, fat, and protein) and produce chemicals needed for the synthesis of amino acids and the energy-producing molecules, ATP, needed for the cells to function (Figure 20.17). Carbon dioxide is a waste product in this cycle. Wächtershäuser argued that by running this process in reverse, carbon (needed for the synthesis of organic molecules) could be generated from carbon dioxide, producing acetate as the waste product. The catalyst in this process is sulfide mineral, namely troilite (*FeS*).

The metabolism-first scenario argues that the presence of hydrogen and carbon dioxide in hydrothermal vents, combined with an iron-nickel-sulfur catalyst, yield a number of chemical substances that assemble and react to conduct the process of reverse Krebs cycle. The problem here is that the Krebs (citric acid) cycle generates the energy-producing molecule ATP (see Figure 20.17). However, while running it in reverse, it needs a source of energy, probably in the form of the ATP organic molecule. At that time, none of the complex organic molecules existed and therefore, ATP could not be produced (see Box 21.1). This is one of the challenges for this scenario. The production of ATP will lead to a self-sustaining cycle forming organic compounds. Somehow, the initial energy requirement must be supplied (Plaxo and Gross, 2006).

In summary, it seems that biochemistry-like reductive reactions can take place in the presence of mineral sulfides. The surface of iron sulfide would constrain the distribution of the products from each reaction and support a complex, self-sustaining sequence of metabolic reactions, leading to the formation of new and more complex catalysts and metabolic pathways. Because of its dependence on iron sulfide, the metabolic-first scenario is also known as *Iron-Sulfur World Hypothesis*. Hydrothermal vents are the locations where these reactions take place. These are placed deep in the oceans where mineral-rich water heated by geothermal energy come out of the openings in the seafloor.

There are a number of complications with this scenario. For example, it is not clear how a set of very different chemical reactions could lead to a metabolic network that could spontaneously self-organize. Furthermore, gene-free networks are resistant to evolutionary changes since this requires simultaneous mutations that are difficult to explain in this framework. It is therefore difficult to explain how this process led to sustainable life in the absence of evolution and natural selection (Plaxo and Gross, 2006).

The carbon dioxide in the atmosphere is dissolved in ocean water producing carbonic acid (H_2CO_3) (Figure 19.2). This gives up protons (in the form of hydrogen atom) to produce bicarbonate (HCO_3^-), which is then converted to carbonate ion (CO_3^{2-}) as well as hydrogen ion (H^+). This increases proton concentration in ocean waters, making this environment more acidic. On the other hand, the fluid coming out of the vents is alkaline (low in proton concentration). The difference in the proton concentrations between the ocean water and the fluid coming out of the vents produces the energy needed (in the absence of ATP molecules) to run the reverse citric acid (Krebs) cycle and to conduct the first chemical reactions leading to production of the first organic compounds.

The first living organisms in this scenario are *genes*. These are small molecules that contain information and can replicate, acting as catalysts themselves. The fact that they could catalyze their own formation and replication suggests that the molecules could evolve (if the changes occurred could be passed through the generations). This is the significant advantage of the gene-first hypothesis over the metabolism-first hypothesis.

As mentioned earlier in this chapter, chances that all of the available amino acids assembled and spontaneously formed the molecules of life are extremely small. The abundance of the fundamental building blocks is too small and the speed with which they could assemble is too slow to allow this to happen without a catalyst to help the process. Several inorganic minerals are likely to facilitate the assembly of these complex organic molecules. The kind of mineral called *clay* is likely to have played that role in the assembly of the genes and the origin of life. The oldest known terrestrial material, *zircon grains*, indicates that clays existed in great abundance on Earth 4.4 billion years ago, implying that clays were common at the time we believe life started (Bennett and Shostak, 2004). Structurally, the clay minerals consist of layers of molecules to which other molecules can attach (see Box 21.2). As a result, when organic molecules combine with clay, they are kept in close proximity to each other, allowing them to react with one another, creating a long chain of molecules. Different layers of clay consist of different chemical elements and compounds. This variability between different layers of clay could make them function like genes. For example, the charged ions in one layer of clay could act as templates and catalyze the formation of the next layer of clay. Any errors in replication (mutation), which is the packing of alumina and silica that occur during the copying process, would be reflected in all of the next layers. If these mistakes increase the efficiency of the replicating process, they will generate a selective advantage by speeding up the replication and production of new chemical compounds that would satisfy our definition of life (Plaxo and Gross, 2006). Also, ions in clay layers could act as catalysts to speed up the organic reactions and polymerization of the RNA. Given this scenario, it is likely that clays in Earth's oceans may be responsible for generating the first organic molecules able to replicate themselves (hence fulfilling one of the definitions of a living system). This is the so-called *clay world hypothesis*.

BOX 21.2: COULD CLAY BE THE CATALYST FOR THE FIRST LIVING ORGANISMS

Clay consists of layers of minerals containing charged alumina and silicates packed in alternate layers with sodium and calcium. It has the formula $(Na, Ca)_{1/3} (Al, Mg)_2 (SiO_4O_{10}) (OH)_2 \, nH_2O$ representing a layered structure. Positively charged sodium and calcium can replace one another at some negatively charged sites on the surface of the alumina and silicate layers. Therefore, while the total number of sodium plus calcium ions in the mineral are conserved, the sodium to calcium ratios differ. Similarly, while the total number of alumina and magnesium is conserved, their ratio changes in the alumina layer. This leads to an infinite number of molecules, each differing in the ratio of sodium to calcium and alumina to magnesium. This forms a layered structure for clays separated by layers of water (nH_2O). This variability of clay from layer-to-layer resembles the genes because each layer of charged ions. Each layer of charged ions in clays can now act as a template to catalyze the formation of a complementary new layer. Sometimes, irregularities take place during the copying process, leading to a *mutation* that influences the new layers. If the mutation increases the efficiency of the copying process, it will provide a selective advantage, resembling the living molecules. The ions in clay could also act as catalysts speeding up organic reactions and hence help formation of the RNA. Therefore, the clay could be considered as a catalyst to bring together inorganic molecules and after many millions of years form a kind of primitive life (Plaxo and Gross 2006).

However, there are some problems with the clay world. No trace of clay-based metabolism has been found in the living organisms known today. Furthermore, no known chemistry in a laboratory has ever produced any molecules. It is possible that clays were not good catalysts and were overtaken by better catalysts as soon as new and more efficient organic-based catalysts were created by the evolution (Plaxo and Gross, 2006).

3. RNA WORLD HYPOTHESIS

Here I re-state again the question I raised in the introduction to this section. All cells have DNA that they pass to descendant cells. These cells use the information in the DNA to synthesize proteins. Some of these proteins are enzymes that synthesize new DNA which, in turn, is passed to descendant cells as this process continues. Therefore, the protein synthesis depends on DNA, which itself is built by proteins (acting as enzymes). The question therefore is: Where did this cycle start?

Biologists have proposed the possibility that RNA could play a major role here. It has been found that in some viruses (e.g., the retroviruses, such as HIV, the virus that causes AIDS), RNA can carry the genetic material. Thomas Cech at the University of Colorado, discovered in 1981 that RNA has the ability to act as an enzyme. RNA molecules able to act as enzymes (i.e., catalyzing biochemical reactions similar to protein enzymes) are called *ribozymes* (ribonucleic acid enzymes).

Because of these properties, being able to store information in its sequence of nucleic acids, like DNA, and acting as enzyme facilitating chemical reactions, it is likely that RNA may have been the first molecule to encode genetic information as well as catalyze its own production and replication. Therefore, it is possible that the first living things on earth used RNA as both genetic material and for catalytic activity, which includes the replication of genetic material. This concept is known as *RNA World hypothesis*.

Like other polymers, RNAs form from the polymerization of simpler precursors by the formation of a specific polymer sequence based on the sequence of another, in a process called *template directed polymerization*. In RNA World hypothesis, the first living things consisted of three parts: a ribozyme with RNA polymerize activity (the formation of a complex RNA molecule by combining smaller and simpler molecules), a template RNA to direct polymerization, and a physical container (membrane) (Plaxo and Gross, 2006). Two RNA molecules (not just one) are needed to start this process and for the ribozymes to catalyze reactions. This is because ribozymes need to fold into complex structures in order to perform their functions. Therefore, given this complex shape, it is very unlikely that any self-replicating molecule can act as a template to replicate itself. In order for an RNA molecule to serve as a template to synthesize a new RNA molecule, a molecule must be unfolded and exposed to the monomer (simpler molecule) that would polymerize on it. In this scenario, the formation of two ribozymes is essential. A container is also needed to keep together the genetic material and the molecules it encodes. Without this, the material would diffuse away and not interact.

To summarize, in the RNA World hypothesis, our very first ancestors started life as two RNA molecules (a self-replicating ribozyme) within a lipid membrane. Nucleotide monomers then "leaked" into the membrane and polymerized into new copies of ribozymes. This process continued, generating more RNA molecules and leading to increased molecular weight within the lipid membrane. The membrane that resembled a protocell at this time then split under the weight of the ribozymes trapped in it. The lipid membrane then evolved into cells with more efficient lipids surrounding it. The process of diffusing simple molecules (monomers) through the lipid membrane becomes more efficient from protein-based pores that are more selective and only allow in the small molecules that are needed by the cell. Therefore, the RNA World hypothesis naturally leads to the first replicating molecules as well as the first cells (Plaxo and Gross, 2006).

The question is, if RNA played such a significant role in the origin of information storage and transfer, why do cells use DNA to store information and proteins to perform cellular processes? As discussed in the last chapter, this is primarily due to RNA being less stable than DNA because it carries a highly reactive hydroxyl group (OH) that easily enters into reactions with other molecules. Furthermore, single-strand RNA breaks more easily than double-stranded DNA and mutates more often. Given this, a change from RNA to DNA makes synthesizing larger and more stable genomes possible (Plaxo and Gross, 2006).

There are two serious shortfalls with the RNA World hypothesis. First, there is a very small likelihood of spontaneously generating a sequence that could replicate itself through random polymerization of RNA monomers. The situation becomes even more difficult because one needs two RNA molecules to start this process. Secondly, the ribose in RNA (sugar), amino acids, and organic compounds are *chiral*. This means they cannot be superimposed on their mirror images (see later in this chapter for a discussion of chirality). For example, proteins are found to be always composed of left-handed amino acids (this is called *homochirality*). A mixture of left-handed and right-handed amino acids cannot fold into three-dimensional structures that are needed for these complex molecules to perform their functions. The problem is that early-Earth chemistry produced an equal number of left- and right-handed molecules. It is therefore very unlikely that random processes in chemistry could produce molecules that are one-handed.

ORIGIN OF THE BASIC INGREDIENTS NEEDED TO SUSTAIN LIFE

In chapter 20 I discussed the basic ingredients that are essential to support and sustain life. Once all these ingredients are available, with the information carrying units and the necessary energy sources in place, the process of life starts. It would then sustain itself, adapt to the environment and evolve. Looking around us today, every living thing is a result of this very complicated and amazing chain of events. I continued the discussion in the present chapter by studying different scenarios for the origin of life and how these different components came together to create the kind of life we see around us today. However, we still don't know how the basic ingredients originated. In the following sections I continue our discussion in chapter 20 and take it to a deeper level by studying the origin of DNA, the origin of the genetic code, the origin of protein synthesis, the origin of the energy generation process and chirality.

THE ORIGIN OF DNA

All cells have double-stranded DNAs. Studying the origin of DNA is therefore essential to our understanding of early life evolution. This is particularly important because it is believed that DNA predates the earliest life on the planet. It is likely that DNA originated from RNA in an RNA/protein world. This hypothesis is supported by the evidence showing that DNA is a modified form of RNA, with the *ribose* sugar in RNA (OH) reducing to *deoxyribose* in DNA, and the base uracil (*U*) into thymidine (*T*) through the *methylation process* [addition of a *methyl R-CH$_3$* group (Figure 20.11 and Box 20.3)]. A complication here is that DNA synthesis requires protein as a catalyst and proteins cannot be synthesized without DNA to fix the order and sequence of the nucleotides.

The first step in the synthesis of DNA was the formation of *U-DNA* (DNA containing uracil). This took place through a chemical reaction converting *deoxyuridine Triphosphate (dUTP)* to *deoxyuridine Monophosphate (dUMP)*[1] – *dUTP + H$_2$O → dUMP + diphosphate* – catalyzed by the enzyme *dUTP diphosphatase*. This enzyme

---

[1] Nucleotides are commonly abbreviated with three letters. The first letter indicates the identity of the nitrogenous base (e.g., A for adenine, G for guanine), the second letter indicates the number of phosphates (mono, di, tri), and the third letter is P, standing for phosphate.

has two functions: first it removes *dUTP* from the deoxynucleotides, reducing the likelihood of this base being incorporated into DNA and hence DNA containing uracil, and second, it produces *deoxythymidine triphosphate (dTTP)* that is one of the four nucleotide triphosphate molecules (these contain a nitrogenous base, a sugar ribose, or deoxyribose and phosphate groups bound to the sugar) that are used in the synthesis of protein. This is how uracil in RNA is replaced by thymine in DNA. The molecule *dTTP* is produced in the cells by the modification of *dUMP* into *dTMP* (and *dTTP*) by the enzyme *thymidylate synthases* (followed by phosphorylation). The above explains formation of DNA from RNA.

There are some known viruses (noncellular particles that reproduce inside a cell) that have RNA rather than DNA as their genetic material. With its nucleotide sequence, RNA (rather than DNA) can perform the task of information carrier and be expressed as a protein. Some viruses like the human immunodeficiency virus (HIV) have this property. After infecting a host cell, such a virus makes a DNA copy of its genome, becoming incorporated into the host's genome. This virus relies on the host transcription machinery to make more RNA. This RNA can be translated to produce proteins or incorporated as a genome into the virus. The synthesis of DNA from RNA is called *reverse transcription*, with such viruses referred to as *retroviruses*.

There is also the possibility that the initial responsibility of DNA was to store phosphate, with the genetic duty evolving later on. Phosphates are needed in the cells, not only to produce the system responsible for genetic material like ribosomes, but are needed for building the molecule ATP—the energy powerhouse of the cells—fatty membrane molecules, and a lot more.

As I described earlier, DNA replaced RNA as genetic material because it is more stable and can be repaired more easily because of its molecular structure. Replacement of RNA by DNA as genetic material has thus opened the way to the formation of large genomes, a prerequisite for the evolution of modern cells. Once the first DNA molecules were manufactured, Darwinian selection took over and cell populations with DNA genomes finally eliminated cells with RNA genomes.

An important step after formation of RNA is the sequencing of base units (the order the nucleotides follow in the DNA) needed to code for different amino acids in order to synthesize proteins (the genetic code). This sequencing is an optimization that resulted after millions of years of evolution. To decipher the origin of the genetic code, scientists look for similarities and differences in the composition, shape, and structure of the chemical molecules responsible for protein synthesis. By tracing this back in time, they reduce this to a smaller number of diverse molecules that were in place at the very beginning. The near universality of the basis of the genetic code implies that all of the components (RNAs and the enzymes required for them to perform their tasks) were already in place when the first signs of life were detected.

The molecules responsible for the transfer of genetic codes are tRNAs. While these molecules cover a range of properties, they are very similar in terms of their sequence. This observation is used to conclude that they all evolved from a smaller set of molecules. There are 20 different tRNA enzymes (one for each amino acids) that have common features and similar chemical structures, which indicates that they all came from two original RNAs at the earliest time of protein synthesis. This dates back to the time before the last universal common ancestor (LUCA) since those are known to have had a much larger number of enzymes. This indicates that molecular composition was in existence during the time before LUCA. Therefore, the history of protein synthesis before LUCA is revealed in the sequence diversity of tRNAs and their enzymes.

In Chapter 20 I discussed that molecules *A*, *U*, *G*, and *C*, from which amino acids are made, combine in groups of three letters called codons (e.g., *AUG*). This is the basis of genetic information for all forms of life. It was also discussed that most of the information are embedded in the first two letters of a codon fixing the properties of the amino acid they produce. The question therefore is: Why does one need three letters rather than two to build the genetic code? This is likely because of the selection pressure to produce a large number of amino acids to allow the combinations able to synthesize protein. This process is a result of optimization after millions of years of evolution. The other question is: Why did it stop at 20 amino acids? This is likely the optimum number to allow a large enough variation and combination of amino acids; any larger than that would increase the likelihood of errors when the transfer of genetic material to protein is made, with natural selection working against that.

THE ORIGIN OF PROTEIN SYNTHESIS

The first protein synthesis process is likely to have started with the association of an RNA molecule with amino acids. To efficiently execute this process, a catalyst (often in the form of protein) is needed. However, there were no proteins at the time that the very first protein synthesis took place. The role of RNA as a catalyst is therefore essential for initiating this process.

It is known that RNAs acquired their functionality from amino acids. All of the RNA molecules have two components in common, a phosphate unit and a ribose group, with a functional group (consisting a combination of A, U, G, and C) that is different depending on specific RNAs. This phosphate-ribose unit forms the "backbone" of the structure of the RNA with the nucleic acids forming the base (Figure 20.11). Because of the chemical composition of the RNA molecule, they can add more amino acids to themselves to enhance their functionality. The formation of the first RNA-amino acid molecule provided a selection advantage by making more RNAs with effective functionality. Such RNA-amino acid systems could be the basis of the first tRNA.

The RNA-amino acid molecules are the major components of protein synthesis taking place in ribozymes where polymers of amino acids (proteins) are manufactured (but note that at the time the first protein synthesis took place, there were no cells or ribosomes). Because of the role these polymers play in forming the structure of the cells, there is significant selective pressure to form such complex molecules; the ribozymes evolved to generate complex polypeptide molecules using tRNAs associated with amino acids.

Another essential component needed to synthesize proteins is the sequencing of polymers. This is known to have been produced by mRNA acting as a template to use amino acids to make sequenced polypeptides. The ribosomes use these templates to direct protein synthesis by bonding its primitive tRNA to another RNA, forming the first mRNA. The sequence of this mRNA dictates the sequence of amino acids that is incorporated into the resulting polymers (proteins). Natural selection has played a significant role here by giving selective advantage to the sequences that can sustain and be used in protein synthesis. The result is proteins with the same sequence as in the mRNA. Once the first proteins were produced, they catalyzed a more efficient production of proteins with difference sequencing.

ORIGIN OF ENERGY PRODUCTION IN CELLS

How is the ATP molecule synthesized? Organic molecules like carbohydrates, lipids, and proteins are good sources of energy. The sugar produced in animal cells and the carbohydrates produced in plant cells are the starting molecules for the cellular respiration process and ATP synthesis.

Now, what is the source of energy in glucose, carbohydrates or lipids? They contain energy because these molecules are able to take part in an oxidation-reduction (redox) process (Box 21.3). Oxidation is the loss of electrons and reduction is the gain of electrons. The loss and gain of electrons always takes place together in a single redox reaction. Electrons are transferred from one molecule to another with one molecule losing and the other gaining the electron. Electron transport processes are used to extract energy from molecules such as glucose, or in photosynthesis to extract from sunlight. The movement of electrons in the redox reactions is therefore the main cause of the carrying and transferring energy (Figure 20.17).

However, all these scenarios take place after the organic compounds are formed. The question then is: How are the energy requirements for the first organic molecules produced? One can extract energy from a system in disequilibrium through the redox process. The hot material coming out of hydrothermal vents on the floor of the oceans enters into the exchange of thermal energy with the colder ocean water. This also generates a difference in proton concentration in the oceans (Box 21.1). These combined, initiate a redox process producing the energy needed to start chemical processes that result in the production of the first organic materials.

BOX 21.3: CLASSIFICATION OF ORGANISMS IN TERMS OF THEIR ENERGY AND CARBON SOURCES

The organisms get their required energy from the environment in two different ways: chemical compounds or sunlight. Organisms that acquire their energy from chemical compounds are called *chemotrophs*. Animals are in this group; they obtain organic molecules, such as glucose, that they break down using oxygen, producing carbon dioxide, water, and energy. Organisms that produce their energy from sunlight are called *phototrophs*. Plants are in this group; they use the energy from the Sun to convert carbon dioxide and water to sugar and oxygen. Sugar is then used to synthesize ATP, the energy currency in the cells.

Organisms can also be classified in terms of the source of carbon. Some organisms convert carbon dioxide (an inorganic form of carbon) to glucose (an organic form of carbon). These organisms are called *autotrophs* (self-feeders). Other organisms obtain their carbon directly from organic molecules that were synthesized by other organisms. These are called *heterotrophs* (or other feeders). Some organisms do not fit in either of these groups, those are called *chemoautotrophs* or *photoheterotrophs*.

THE ORIGIN OF CHIRALITY

Chirality (or handedness) is the geometric property of ions and molecules. An ion or molecule is called *chiral* when they cannot be superimposed on their mirror image (the same way the glove for our right hand does not fit our left hand). This is an important concept, as the chemicals needed to start life (amino acids and sugars) are chiral (they are not superimposable on their mirror image). The proteins exclusively composed of left-handed amino acids while nucleic acids (DNA and RNA) are composed of right-handed sugars. Naturally proteins are made up of left-handed amino acids and therefore, are left-handed. Right-handed amino acids make right-handed proteins which are very rare in nature. Having the same chirality (called *homochirality*, being all either left-handed or right-handed) is an essential requirement in forming molecules of life. There are two obvious questions here:

1. Why must homochirality be present? Because polymers of mixed chirality monomers cannot fold and hence cannot perform their functions. This provides a selective advantage for organisms that need to fold their protein and RNA into active forms. Amino acids found in nonliving things have a mixture of left-handed and right-handed chiralities
2. Is there a selective advantage associated with left- and right-handedness? Most likely the answer is no, as a molecule and its mirror image have the same chemistries. Polymers that the chirality of their monomers change randomly do not function. The conclusion is that *homochirality is an essential property of the living things*.

The origin of homochirality is not clear. It is likely that there were a selective pressure for one-handedness. For example, left-handed proteins only bind to left-handed substances. But why "one hand" was selected and not its mirror image, is likely to be a completely random selection. For example, if carbon-based life exists somewhere else in the universe, it is quite possible that it would possess a different chirality than the terrestrial life on Earth. It is also possible if the first amino acids were formed in comets, circularly polarized stellar radiation would selectively destroy one chirality of amino acids, resulting the life on Earth being homochiral. Chirality is also important in the way enzymes connect to their substrates. Enzymes often distinguish between chirality of their substrates. Enzymes will fit better to a substrate of the same chirality (handedness) and poorly to a substrate with opposite chirality.

SUMMARY AND OUTSTANDING QUESTIONS

Results from two simple but classic experiments provided important insights into studies of the origin of life. First, it was found that living things could only come from preexisting life. Second, under the conditions prevailing in the early Earth, it was possible to produce organic materials that are needed for life, from inorganic compounds. There are particular chemical reactions called redox that are essential in generating the energy required for life. This involves exchange of electric charge between the reacting atoms and production of energy in the process. An example is burning of sugar (glucose) with oxygen resulting carbon dioxide, water and energy. Such reactions are needed to initiate and sustain life. Therefore, any attempt for studying the origin of life must allow for these processes.

The combined processes responsible for the origin of life are rather complicated. Even if we have all the amino acids and the ingredients required for life, extreme fine tuning is needed to go from known chemistry to biology and hence life. There are a number of competing scenarios for the origin of life. The metabolism first hypothesis states that life started from a self-sustaining set of chemical reactions on the surface of rocks on ocean floors close to hydrothermal vents. The material present at the vents, carbon monoxide (CO), ammonia, and hydrogen sulfide, interact with iron and nickel sulfide minerals resulting in organic material and metabolic pathways. The main compound that would catalyze this process is iron sulfide. The gene first scenario considers genes as the first living organism. In this scenario the organic materials are kept close together by a catalyst, likely to be the mineral clay. After millions of years of being kept in proximity, they interact and generate long chain of organic molecules. The clay minerals are very old and have a structure and composition that is able to assemble complex organic molecules. The RNA world hypothesis proposes RNA as the starting point. The protein synthesis requires DNA to provide the required instructions while some of these proteins are needed as enzymes for the synthesis of DNA. The question is, where did this cycle start? Since RNA is a simpler molecule than DNA but shares many of its properties, and since it could act as an enzyme, it is likely that RNA was the first molecule responsible for genetic information.

To study the origin of life we need to investigate where the main ingredients responsible for the formation of life came from. This includes the first organic molecules, the initial source of energy and the genetic material for information transfer. It is likely that RNA was the first molecule because of its simple shape (single strand) and multi-task ability. It is versatile and is able to replicate itself and work as a catalyst. The source of energy in a cell is the redox process. Finally, the chemicals that form the backbone of life are chiral (they cannot be superimposed on their mirror images). Proteins compose of left-handed amino acids while nucleic acids (DNA and RNA) contain right-handed sugar. Right-handed protein or left-handed amino acids are not stable in nature. Enzymes enter into interaction with molecules of the same chirality.

Despite significant progress, it is still not clear how or when the first life started. Every one of the scenarios proposed for the origin of life has strengths and weaknesses. These need to be supported by scientific data, in nature or in laboratories. We also note that if life exists in other places in the universe, it would not necessarily follow the same path as the terrestrial life we know on our planet. For this, we need to look for biomarkers and then search for them in other worlds. For sure, we have a lot to do before we know the origin of life.

REVIEW QUESTIONS

1. Based on current estimates from fossil records, when did life start?
2. What is the historic experiment that shows life could not have started spontaneously?
3. What is the significance of the Miller-Urey experiment?
4. Explain redox reactions and why they are essential for generating life. Give an example of a redox reaction.
5. Explain the weaknesses of the metabolism first scenario for the origin of life.
6. What is the basis of the clay world hypothesis for the origin of life? What properties of clay make it a good catalyst?
7. What is the RNA world hypothesis? Explain the role of RNA in this scenario.
8. What is the origin of DNA?
9. What is the definition of autotroph and heterotroph?
10. Why is homochirality an essential property of any molecule responsible for life?

CHAPTER 21 REFERENCES

Bennett, J., and S. Shostak. 2005. *Life in the Universe*. 2nd ed. Boston: Pearson/Addison-Wesley.

Plaxo, K.W., and M. Gross. 2006. *Astrobiology: A Brief Introduction*. Baltimore: Johns Hopkins University Press.

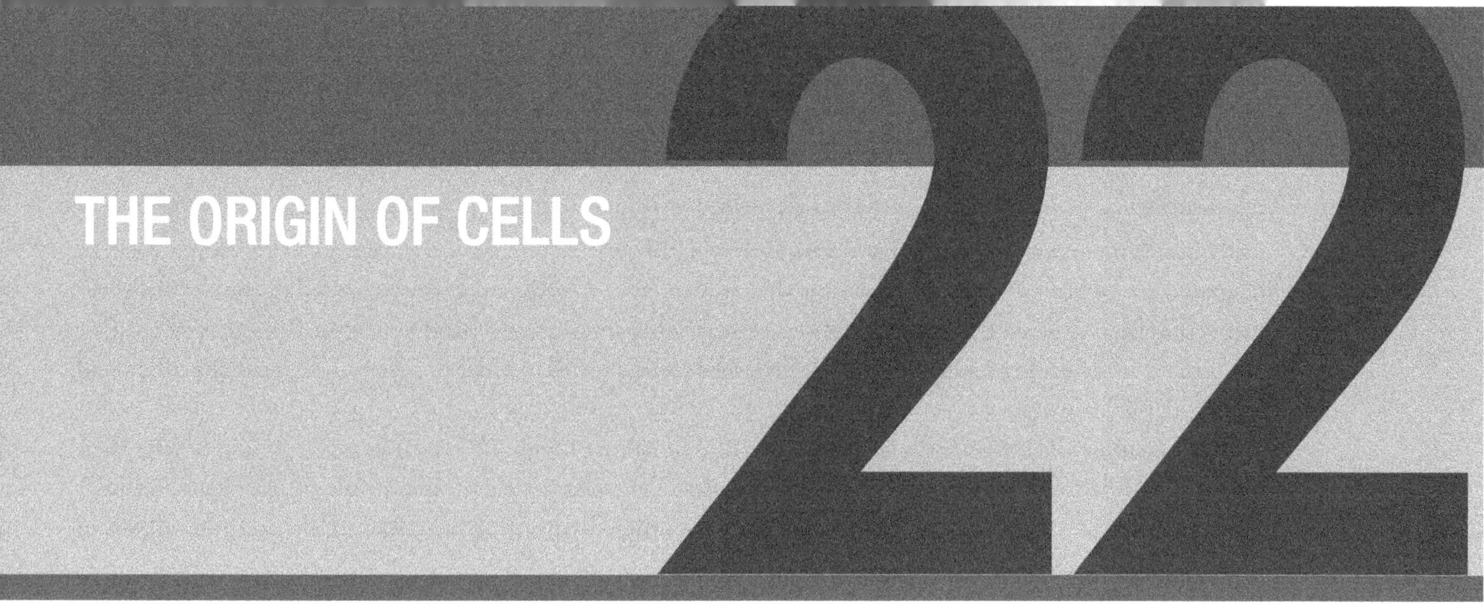

THE ORIGIN OF CELLS

CHAPTER LEARNING OBJECTIVES

This chapter will cover:

- The structure and function of cells
- Protocells and their origin
- The origin of cell organelles
- The endosymbiotic theory for the origin of eukaryotic cells
- The origin of multicellular organisms
- The evolution and diversity of the prokaryotic and eukaryotic cells
- The first living things
- Origin of diversity
- Origin of the nervous system
- Common origin of all life

"If you find me not within you, you will never find me. For I have been with you, from the beginning of me."

- RUMI

"To raise new questions, new possibilities, to regard old problems from a new angle, requires creative imagination and marks real advance in science"

- ALBERT EINSTEIN

Cells are the smallest units of life and the most basic elements responsible for it. Understanding these most basic units is essential in order to study the origin and development of life. The chemical processes essential for metabolism, energy production, and protein synthesis, all basic requirements of life, cannot take place in diffuse environments of water or air. It would simply be too unlikely for the chemical reactants and enzymes to come together, to collide and enter into reactions if they are freely floated in their respective environment. A compartment is required to bring together all these different components and to concentrate all the molecules in an enclosure, allowing them to combine and initiate reactions that would eventually lead to the complex molecules of life. Cells create such protective environments, separated from their surrounding by membranes. There are about 60 trillion cells in an adult's body, all working the same way. Each cell contains around ten thousand different molecules. They use these molecules to perform their functions.

Cells divide to other cells. Daughter cells are produced from their parent cells with the sequence going back indefinitely to the first cells. But how did the first cells form? What were the conditions under which the cell formed? How did different components of a cell (called organelles) form and inherit their functions? How did different organelles coordinate their tasks to produce an integrated entity like a living cell? Given the delicate structure of cells, any process that led to the first cells was surely complicated, being the result of billions of years of natural selection and evolution. The story of the first cells is the story of the beginning of life. We need to know about this in order to better understand the origin and evolution of life.

In this chapter I first present the basic structure and function of cells. The scenarios for the origin of cells, their function and their different organelles will be discussed. The chapter then explores the origin of multicellular systems followed by a study of the evolution of the cells. I then discuss the origin of diversity, domains of life and the common origin of life.

STRUCTURE AND FUNCTION OF CELLS

Cells are highly complex and organized units with numerous internal structures (Figure 22.1). They are sites of protein synthesis, storage of genetic material, and places where energy is produced (Sadava et al., 2014) (Box 22.1). Animal and plant cells contain many common components, with each performing a specific task. These specialized parts of the cells are called *organelles*, i.e., analogous to organs. Cells are separated from external environment by cell walls or *membranes*. The main organelles of a cell and their functions include:

Membranes: These are thin layers composed of *phospholipid* molecules and proteins that separate internal organelles from outside (Figure 22.1). The shape of the membranes is maintained through the interaction of these molecules with the outside environment. The phospholipids are polar molecules with one end soluble in water

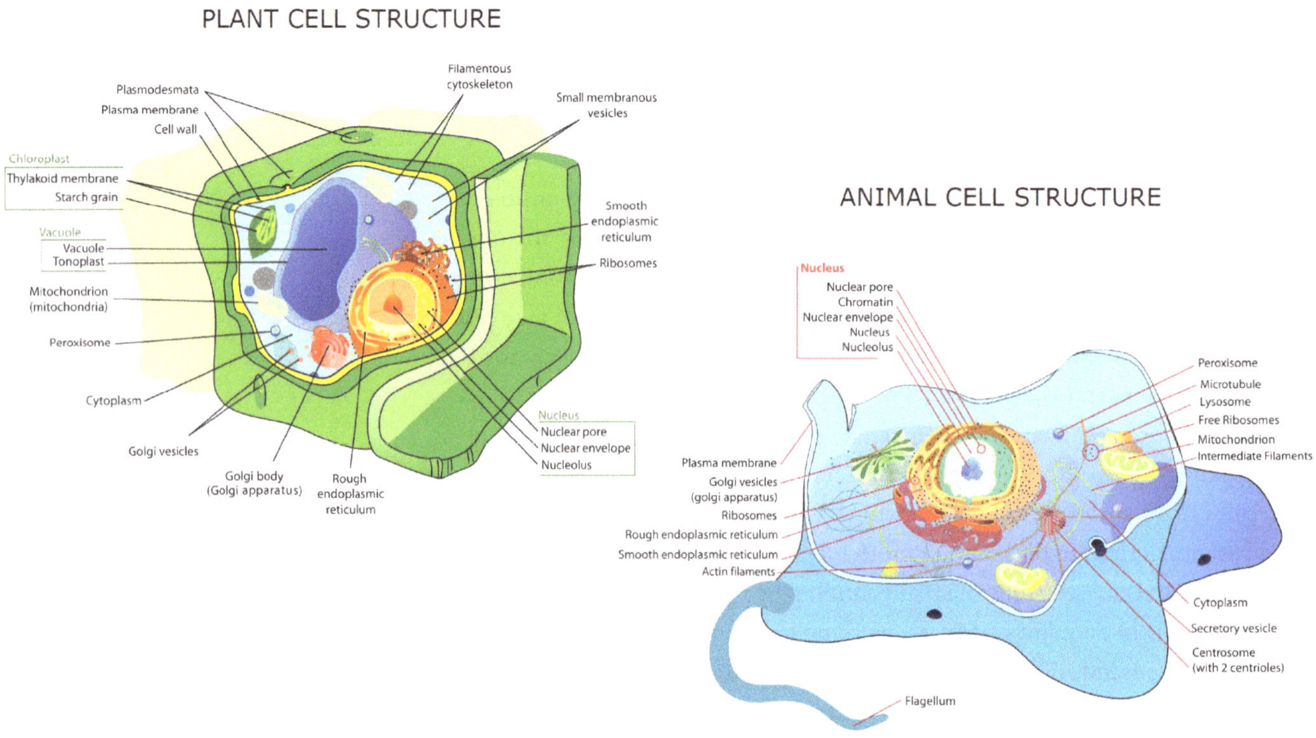

Figure 22.1. Shows different organelles in the animal and plant cells. There is significant similarity between the structures of two cells and their internal organelles implying that they likely share a common origin.

(the *glycerol*) and the other not water-soluble (*fatty acid*). The protein components of cellular membranes are either on the surface of the membrane or inside the membrane.

Nucleus: This is the largest organelle in a cell. It is the site for the genetic material (DNA) of the organism. It is surrounded by a nuclear membrane that separates the content of the nucleus from the rest of the cell (Figure 22.1).

Mitochondrion: This generates the energy needed by the cell and only exists in animal cells. The number of mitochondria per cell depends on the cell tissue type and the associated organism.

Chloroplast: This exists in both plant and animal cells and conducts the task of making energy from the sunlight (Figure 22.1).

Cytoplasm: This is the fluid that fills the cell and is responsible for the shape of the cell. Organelles are suspended in cytoplasm. It also stores DNA and chemical substances.

Endoplasmic reticulum: This is the site of protein synthesis and surrounds the nucleus (Figure 22.1).

Ribosome: This is the site of protein synthesis where mRNA is used to dictate the type of the proteins to be produced.

WHY ARE CELLS SO TINY?

For a cell to function efficiently, it has to have a small size. The size of a cell is limited by the ratio of its surface area to its volume. For any object, its surface-to-volume ratio decreases as its size increases. This is because as an object increases its volume, its surface also increases but not as rapidly. This relationship is of extreme importance for biological systems like cells, for two reasons:

1.   The volume of a cell is an indication of the amount of metabolic activity it is able to perform per unit of time.
2.   The surface area of a cell indicates the rate of substances that can enter into the cell or the waste that can exit from the cell.

As a cell becomes larger, its metabolic activity and therefore, its rate of waste production grow faster than its surface area. It cannot get rid of its waste efficiently, given its small surface area. Therefore, a cell under this condition will not survive. Also, substances must be able to move from one location to other in a given cell. This is more easily accomplished for smaller cells. Given these requirements, a large area-to-volume ratio is required for cells to perform their functions and to survive (Hillis et al. 2012). The size of a cell is therefore optimized by evolution and natural selection to make the cell functions most efficient.

THE ORIGIN OF THE CELL MEMBRANE

Cells are separated from one another and from their external environment by membranes. The membranes are also responsible for what goes in or out of a cell. *Lipids* are the main constituents of cell membranes. To understand the secrets of cells, we need to know about the biochemistry of lipids and how they form the enclosed structure of cells.

The types of the lipids found in cell membrane are *phospholipids* (figure 22.2). These have two parts—*hydrophilic* (water loving) and *hydrophobic* (water fearing) molecules. The phosphate "head" group is hydrophilic, polar, and enters into interaction with water, while its two long fatty acid "tails" are hydrophobic, nonpolar, and noninteracting with water (figure 22.2). Therefore, the head enters into interaction with water through hydrogen bonds whereas the tails do not. Molecules with both hydrophilic and hydrophobic regions are called *amphipathic*.

When positioned in water, the lipid molecules orient to have their "heads" grouped on the outside, interacting with water, while their hydrophobic (nonpolar) tails are inside, away from water. The fatty acid therefore forms a structure isolating the inside and outside regions. If some water is trapped inside this structure where the hydrophobic fatty acids are, an unstable situation arises (as the hydrophobic part of the fatty acid does not interact with water). To stabilize this structure, the fatty acids form a second layer, called *lipid bilayer*. This is a two-layered structure in which the polar heads of the fatty acids face both outward and inward as they are attracted to the polar water molecules in each side of the double layer (figure 22.2- left panel). The nonpolar tails extend to the interior of the structure where there is no water. This is called *liposomes* (figure 22.3). These structures, surrounded by lipid bilayer membrane resembling cells, are *protocells*.

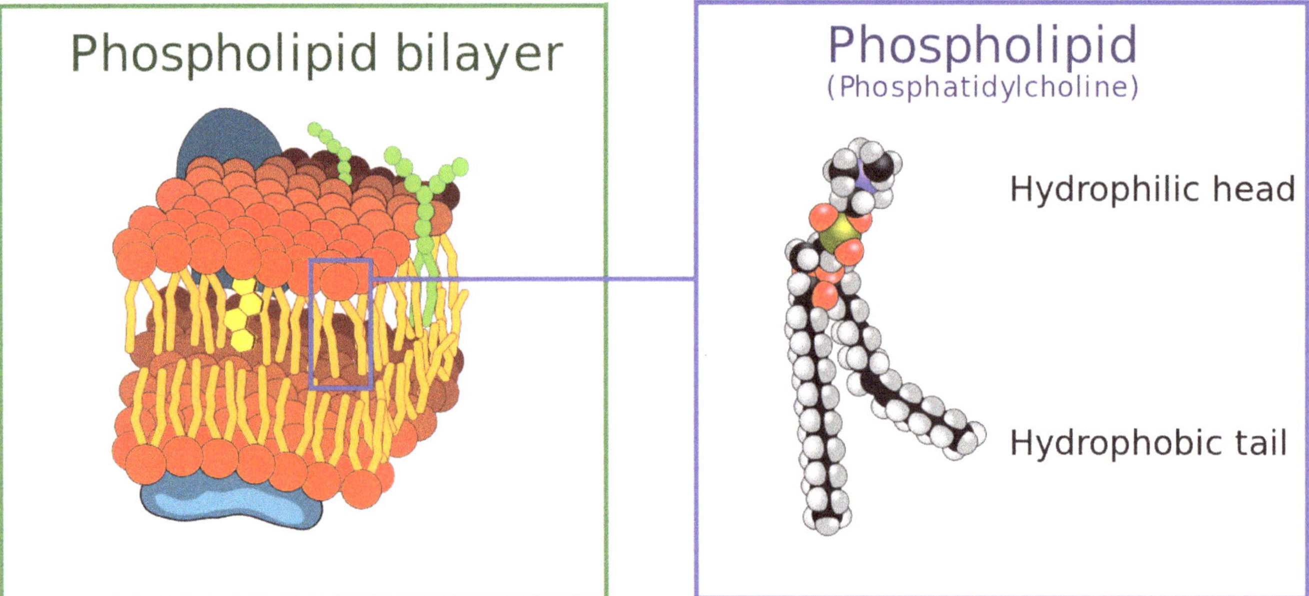

Figure 22.2. A phospholipid molecule with the hydrophilic and hydrophobic ends marked (right). A phospholipid bilayer formed from phospholipid molecules is oriented so that the hydrophilic ends are away from the aquatic environment (left).

Large DNA and RNA molecules cannot pass through these lipid bilayers but smaller molecules like sugar and nucleotides can. This is important as the nucleotides penetrate through the protocell membrane and integrate with the nucleic acid inside, forming polynucleotide chains. This replication takes place without the presence of protein (enzymes) and could be the first step toward cell production.

These protocells have still a long way to go to form the modern cells, the smallest units for life. They cannot perform all the metabolic activities required for a cell. However, this simple lipid bilayer model has some basic functionalities of a real cell: (1) It produces the interior and exterior environments that are separated from one another; (2) it has the organized system a cell needs to have, with the amino acids interacting; and (3) it can replicate itself (Morris et al. 2013).

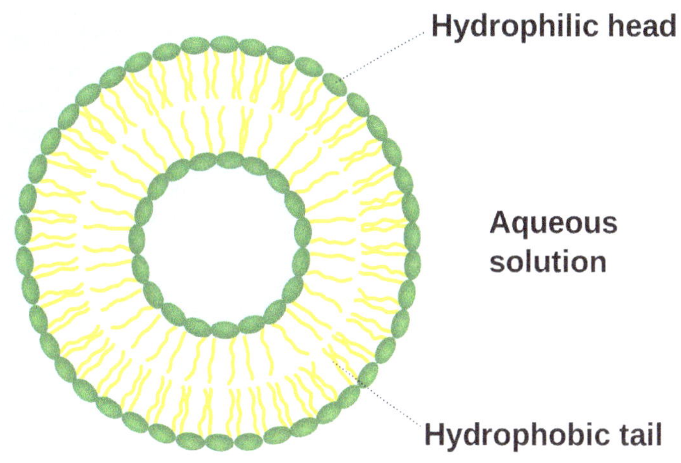

Figure 22.3. The structure of a liposome with hydrophobic tails oriented away from the aqueous environment.

THE FIRST CELLS

For the self-replicating molecules to perform chemical reactions and conduct metabolism, they should be confined in an enclosure, called a membrane. The membranes separate different parts of the cells as well as their inside and external environments. This makes the chemicals inside of the cells have a higher concentration compared to those outside, allowing the biochemical reactions that are essential for life to flourish. Therefore, the coexistence of a self-replicating molecule and chemicals (needed for metabolism to function) within a closed membrane is the first step in generating the first cells and hence, the life.

How the first cells formed? It is likely that a mixture of phospholipids and water produced spontaneous circles that resembled cells. The self-replicating material (e.g., RNA) likely penetrated through the surface of the circle and introduced the genetic material into the structure, which formed the first cell. The essential fact here is the compartmentalization of the genetic material and the chemicals needed to conduct metabolism. The earliest cell was in the form of bacteria and appeared approximately 3.5 billion years ago.

CELL TYPES

There are two cell types: *prokaryotes* and *eukaryotes*. Cells with no nucleus and no internal compartments are called prokaryotes (figure 22.4- right panel). They are in the form of single-cell bacteria. The first identifiable fossils that show chemical fingerprints of prokaryotic cells are from 3.8 billion years ago found on sedimentary rocks while the oldest prokaryotic fossils, found in Western Australia, have an estimated age of 3.46 billion years. The features existing in the prokaryotic cells are the ones present in the common ancestors of all organisms (because these were the first living things on Earth), with their DNA not surrounded by any membranes. Organisms with a nucleus and internal components separated by membranes are eukaryotes (Figure 22.4- left panel). The membranes separate the two main parts of the cell, the nucleus and the cytoplasm, as well as separating the cells (both prokaryotes and eukaryotes) from their surroundings. Eukaryotes come in simple, single-cell organisms

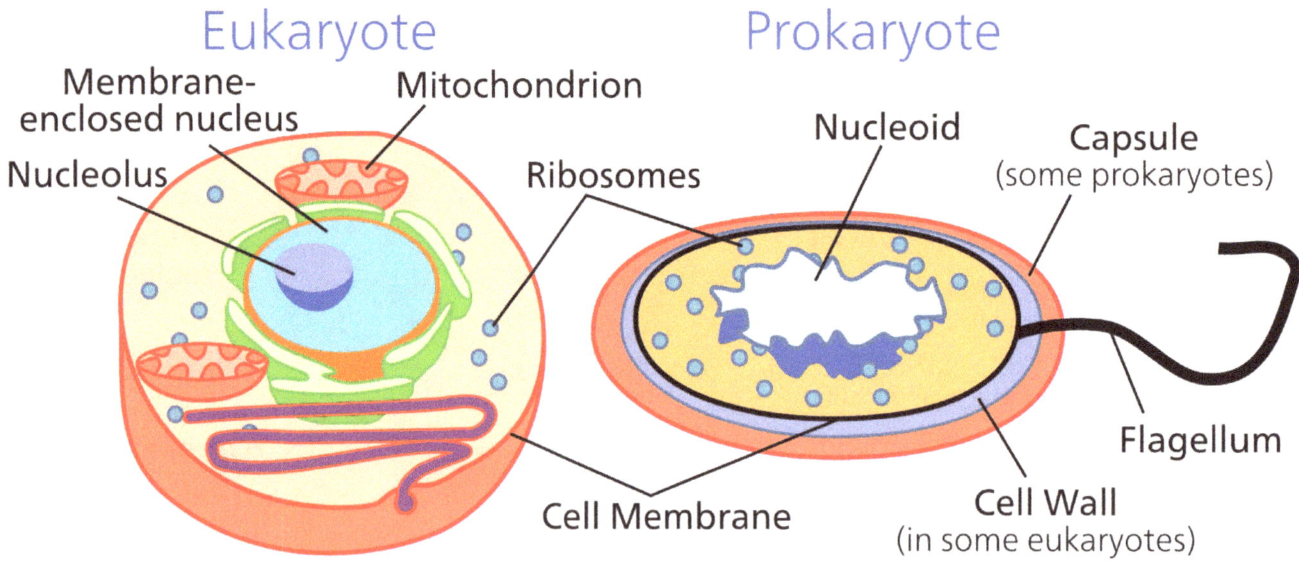

Figure 22.4. Shows the structure and different membranes of a prokaryote and a eukaryote cell.

or complex multicellular forms and are much more complex than prokaryotes (Sadava et al., 2014). They were developed years after prokaryotes, roughly 1.2 billion years ago.

The prokaryote cells are divided to two broad domains: *bacteria* and *archaea*. The bacterial cells have very simple structures with their DNA floating in the cytoplasm and their mRNA immediately translated to proteins by ribosomes. Like bacteria, the archaea have no membrane-bound nucleus. Their membranes are made from lipids, different from the fatty acids in bacteria and eukaryotic membranes. Like eukaryotic cells, the DNA transcription in archaea is made through RNA polymerase, different from that in the bacterial cells.

The bacteria and archaea cells diverged around 3.5 billion years ago (about the time the first prokaryotic cells were dated). The difference between these two domains lies in their cell walls. The cell walls of bacteria contain a polymer of amino sugars that results in a structure around the cell. No such material exists in the cell walls of archaea. Soon after their divergence, some of these bacteria started to use light energy in photosynthetic pathways that did not produce oxygen. Two of the most important bacteria that have tremendous influence over the environment and life are: *cyanobacteria* and *proteobacteria* (Box 22.4).

The cyanobacteria are prokaryotes that live through photosynthesis, similar to plants (figure 22.5). They started life in oceans since at that time (around 3.4 billion years ago) there were no lands or continents and the atmosphere had not yet developed its protective ozone shield. Therefore, water provided a protective layer against intense UV radiation from the sun. The cyanobacteria removed CO_2 and added oxygen to the atmosphere, contributing to the oxygen build up in the atmosphere. The increase in oxygen level led to formation of the ozone shield against the ultraviolet radiation from the sun, allowing living organisms to move from sea to the land. Over many years, due to cell growth and sediment deposition, the cyanobacteria resulted in a dome-shaped structure called *stromatolites*. The cyanobacteria can be seen in the outer surface of stromatolites, found today in shallow waters of Australia's western coast. Archaea live in extreme conditions in deep-sea hydrothermal vents (with a temperature exceeding 100 deg C). Recently, they have been found in less extreme conditions, e.g., soil, lakes, and seas.

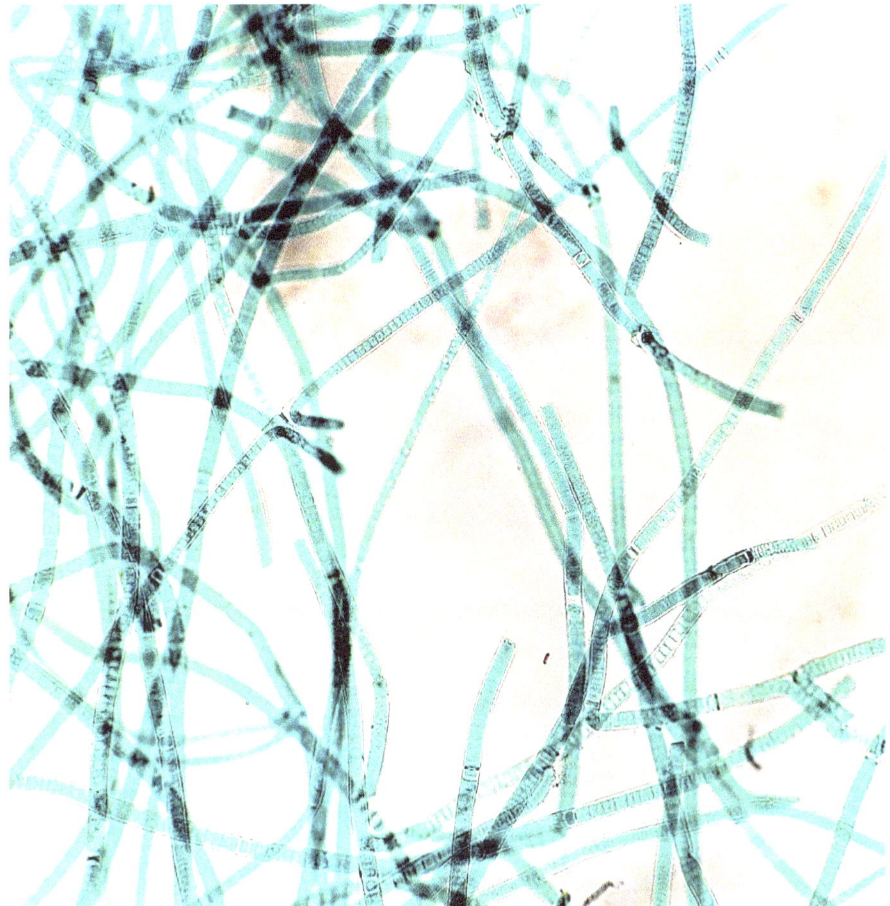

Figure 22.5. An example of 3.8-billion-year-old Cyanobacteria.

There is evidence that eukaryotic cells contain archaea genes. This gives clues toward formation of eukaryotic cells. Starting from a prokaryotic cell, growing in size and infolding by having part of the outer membrane folding inside (a process called invagination), the structures inside the cell were formed when the infolding eventually separated from the outer membrane of the cell, forming the cell nucleus. The DNA of the prokaryotic cell then moves to the nucleus. At some point an aerobic (oxygen-loving) organism, expected to be proteobacteria, becomes symbiotically involved with the initial cell and is moved inside the cell, forming mitochondria. This allows the cell to thrive in an oxygen-rich environment, forming the first eukaryotic animal cells with mitochondria as the main energy agent. Sometimes, the "folded" prokaryotic cells symbiotically engulf cyanobacteria that are biochemically similar to chloroplasts (see next section). This is the reason that chloroplasts are also found in eukaryotic animal cells. Therefore, the prokaryote tree of life is composed of two branches: bacteria and a separate branch leading to archaea and eukarya.

The eukarya were separated from the prokaryotic cells (namely archaea) around 2.5 billion years ago when symbiotic bacteria became incorporated in an ancestor of the eukaryotes, producing mitochondria. Around 1.5 billion years ago, the bacteria and eukarya joined through the symbiosis process, forming chloroplasts (a more detailed discussion will follow in the next section). Diversification of eukaryotes started around the same time when they were divided into *plants, animals* and *fungi*. Any other organism with eukaryotic cell falls in the group called *protists*. These are all microscopic organisms and although classified in the same family, have widely different properties. The unique characteristics of eukaryotic cells imply that they all come from a single eukaryotic ancestor

BOX 22.2: DIFFERENCES BETWEEN THE ORIGIN OF PLANT AND ANIMAL CELLS

If a prokaryotic cell engulfs a proteobacteria, it will evolve to have mitochondria and hence, will turn to an animal cell.
If a cyanobacteria is engulfed, it will evolve to have chloroplast and hence, become a plant cell (se also Box 22.4).

diversified to different protists as well as plants, animals, and fungi. Eukaryotes are more closely related to archaea than to bacteria (because of the similarity of their genes) but their chloroplasts and mitochondria resemble the bacterial family (Sadava et al. 2012)- (Box 22.2).

THE ORIGIN OF CHLOROPLASTS AND MITOCHONDRIA IN CELLS

Earlier in this chapter we studied the mechanism for the formation of cell membranes separating a cell from its environment. However, there are other organelles in a cell, each responsible for a different function that complement each other's tasks and create today's living cells. Knowledge of these organelles and their origin is needed in order to understand the origin of cells themselves. The first requirement for a cell to function is for their energy production organelles to be in place. These are the *mitochondria* and *chloroplasts*, independent organelles found in animal and plant cells respectively (although mitochondria are present in both animal and plant cells).

The chloroplasts in cells closely resemble the *cyanobacteria*, a type of photosynthesis bacteria from the prokaryotic cell family believed to be among the first living organisms on the planet (this is explained later in this chapter). The process of photosynthesis is similar between the chloroplasts and cyanobacteria (Box 22.2). Both use photosystems to acquire energy and the same reaction to reduce carbon dioxide into organic material, releasing oxygen. Furthermore, it was found that there are some organisms that host algae within their tissues, implying that at some point the organism engulfed another simpler system. These observations combined, led to the development of the *endosymbiotic theory*. This postulates that mitochondria and chloroplasts are the result of endocytosis of bacteria and algae (transport of molecules into the cell by engulfing them), followed by years of evolution. In other words, these were not swallowed by cells but became symbiotic (by invagination of its membrane) instead (figure 22.6). Endocytosis is when a material enters a cell without passing through its cell membrane. The membrane in the cell invaginates and keeps the external substance inside (figure 22.6). More evidence for this scenario came from the observations that the chloroplasts in algae were separated from the cytoplast by two membranes. This is only possible if cyanobacteria are symbiotically engulfed by a eukaryotic cell. The inner membrane is that for the cyanobacteria while the outer membrane is that of the engulfing cell (figure 22.6). Another major discovery was the evidence that chloroplasts have their own DNA organized in the form of circular chromosomes, similar to that of the bacteria but different from other eukaryotic cells. The conclusion from the above observations is that the photosynthetic chloroplasts in cells were originated from cyanobacteria being engulfed by an existing eukaryotic cell (Sadava et al. 2012).

Mitochondria have close properties to chloroplasts. They have similar biochemistry to free-living bacteria, have DNA close to another form of bacteria, *proteobacteria*, and were originated as endosymbiotic bacteria. While most eukaryotic cells contain mitochondria, there are some found in oxygen-free environments that do not (Box 22.2). Every mitochondria cell examined today has remnants of mitochondrial genes in its nuclear genome, indicating that these cells at some point had mitochondria but lost them. The eukaryotic cells that do not have mitochondria

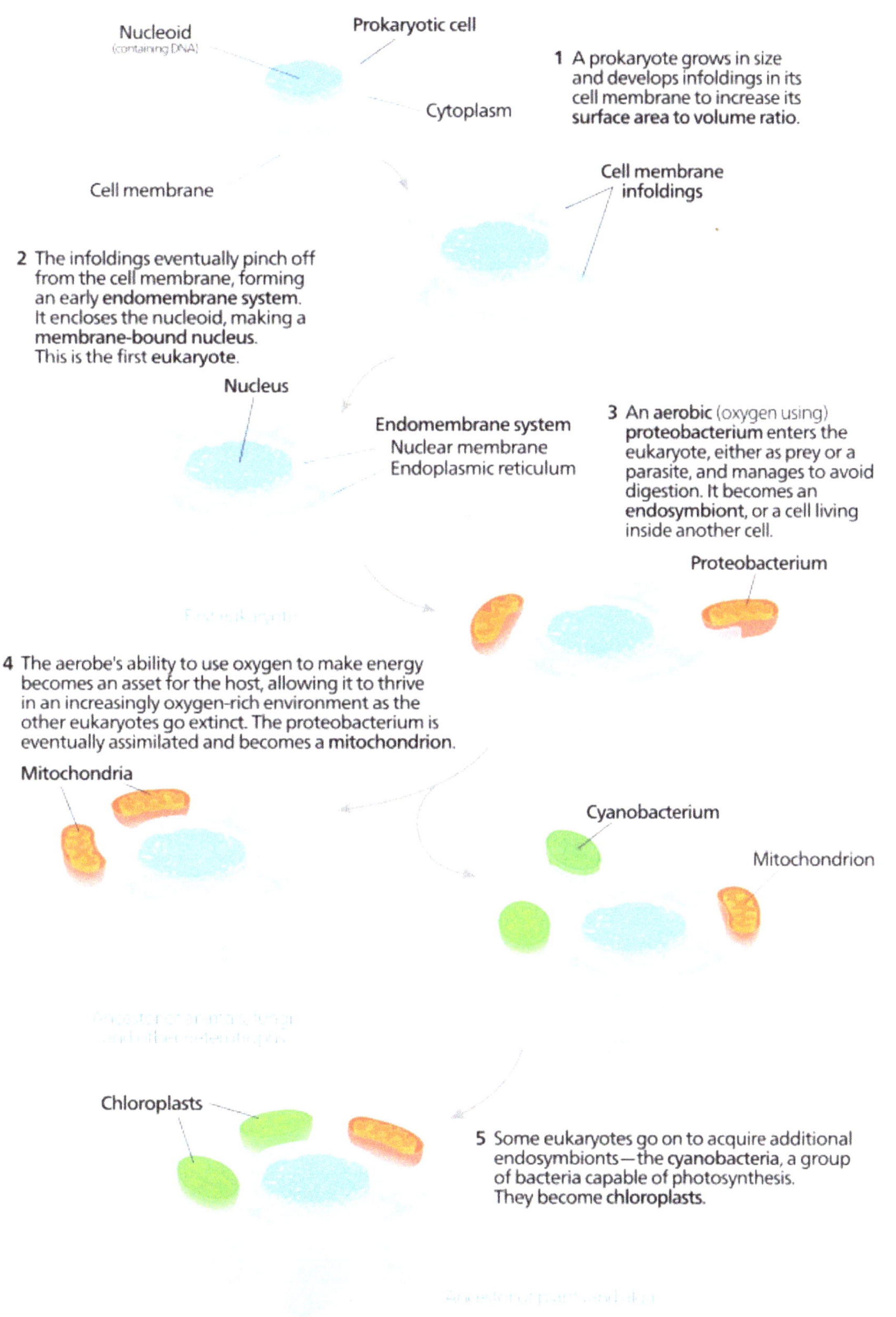

Figure 22.6. The endosymbiotic process for the origin of eukaryotic cells. The figure shows the formation of mitochondria and chloroplasts for animal and plant cells respectively. It explains the origin of different structures in the eukaryotic cells through a prokaryotic cell folding in itself. This eventually forms the nucleus of the eukaryotic cell.

have other small organelles called *hydrogenosomes* that generate energy for the cell. The hydrogenosomes are therefore altered mitochondria, tuned for oxygen-poor environments (Sadava et al. 2012).

THE ORIGIN OF MULTICELLULAR ORGANISMS

So far we have discussed the origin of single cells, their structure, and their function. However, humans, animals, and plants consist of trillions of cells all communicating and working together to produce the complex multicellular systems around us. These came to the world stage over three billion years ago after first microorganisms appeared and then followed by a long evolutionary history. Only eukaryotic cells have the ability to evolve to multicellular organisms and to do this, they need to satisfy the following common characteristics (Morris et al. 2013):

- **Adhesion between cells:** Cells must be able to stick together and have the property to develop a relationship with each other for the organism to function. Molecular bonds between cells, by means of proteins, are responsible for this.
- **Communication between cells:** Cells communicate by molecular signals. The signaling molecules, in the form of proteins, are synthesized by one cell and binds with a receptor molecule (another protein) to another cell. This generates a molecular switch that activates or represses gene expression in the receptor cell's nucleus. All cells have receptors that respond to signals from the environment. Complex multicellular organisms develop cellular pathways for movement of molecules from one cell to another.
- **Network of regulatory genes:** Cells have different functions depending on what genes (Box 22.3) are switched off or on by the molecular signals they receive. Each signal alters the production of proteins, regulating the genes. In a three-dimensional multicellular system, exterior and interior cells are exposed to different environments. There is a change in the oxygen and nutrition level from the outer boundary of the multicellular systems to the inside, causing differentiation of cells within the multicellular organism. These conditions in oxygen or nutrition starve interior cells, resulting in the expression or repression of certain genes. The increased genetic control of cellular responses to this signaling difference leads to gene regulation within multicellular organisms.

Study of cell adhesion, communication, and gene regulation in plants and animals has revealed that multicellular systems were evolved independently. In other words, the common ancestor of plants and animals was not multicellular.

BOX 22.3: UNPACKING THE GENOME

Genome in an organism is its complete set of DNA. In eukaryotes, this information resides in the nucleus of the cell.

Chromosomes are one or more unique pieces of DNA, and make the complete genome of an organism. They vary in length and can consist of millions of base pairs. Humans have 23 unique chromosomes. There are two copies of each; one from our mother and the other from our father, a total of 46 chromosomes.

Genes are a specific sequence of DNA about 3,000 bases long. They contain the information necessary to produce all or part of a protein.

For complex multicellular systems to grow, they require an oxygen-rich environment, as large and active animals can only live in such environments. With large concentration of oxygen in the atmosphere, these could diffuse through the interior cells of large organisms, allowing them to generate their energy and to function. Because of this, deep in the oceans, when the oxygen level is only 10 percent of the surface, only small animals could live. The oxygen in the atmosphere reached its present level just over 500 million years ago. Interestingly, this is the time that fossils of first multicellular organisms were found.

USING CELLS TO MONITOR EVOLUTION

The organelle responsible for protein synthesis is ribosomal RNA (rRNA) that are the main constituent of ribosomes in cells. While the messenger RNA (mRNA) carries instructions to make specific proteins, the rRNA has a catalytic role, which is to establish a chemical bond between two amino acids. The ability of rRNA to catalyze provides significant support for the *RNA world hypothesis* for the origin of life (chapter 21). The genes that encode rRNA evolve with time in a unique way and therefore, provide excellent markers to trace evolutionary history or to study different species. This property of rRNA is extensively used for evolutionary studies of all living things.

By the study of rRNA biologists found convincing evidence concerning the evolution of prokaryotes. By comparing rRNA genes from different organisms, they found evolutionary relationships between them throughout history. The rRNA is particularly useful for studies of living organisms for the following reasons:

Two of the most commonly known organisms in prokaryotic domain are:

Cyanobacteria: These are sometimes called blue-green bacteria because of their pigmentation. They are photosynthesis bacteria that require water, nitrogen, oxygen, mineral elements, light, and carbon dioxide to survive. They use **chlorophyll** for photosynthesis and release oxygen. These bacteria were responsible for enriching earth's atmosphere with oxygen. Also, the chloroplasts of photosynthetic eukaryotes are derived from endosymbiosis of cyanobacteria.

Proteobacteria: These contain the largest group of bacteria. Genetic and morphological evidence indicate that the mitochondria of eukaryotic cells were derived from a proteobacteria through endosymbiosis. Among the proteobacteria are **rhizobium**, which contributes to the global nitrogen and sulfur cycles.

- rRNA was present in the common ancestor of all living things and therefore, existed from the very beginning.
- All organisms contain rRNA and therefore, these can be compared through the tree of life.
- rRNA has evolved slowly so that gene sequences from distantly related species can be analyzed.

ORIGIN OF DIVERSITY

Why is there a diversity of different species in nature today? Why don't we have only one species of every living creature instead of a wide family of them?

After the self-replicating RNA (and later DNA) appeared on Earth, they were subject to natural selection followed by many millions of years of evolution. The RNAs that could replicate faster and more efficiently dominated the population. However, before the natural selection and evolution took place, some genes went through a process called *mutation*. This is believed to have been the main cause of the observed diversity today, being the result of an error in the copying of the genes resulting in some variations in the new genes. The most successful of these adapt to their environment and survive, grow, reproduce, and evolve. After many years of natural selection and evolution, these simple organisms turn to today's living cells.

The mutation happens when a gene is damaged or altered, resulting in the change in the message carried by that gene. This is a random change in the sequence of a gene and is caused by mistakes in the DNA replication process or environmental effects that damage DNA. The result of this is permanent alteration in the DNA composition and therefore, the genetic code.

I recall that from a sequence of three nucleotides an amino acid is made (Chapter 20). These bind together through polypeptide bonds to produce proteins. This is how the genetic message is transcribed from DNA to mRNA. As an example, consider the sequence of nucleotides in Figure 22.7. The sequence *CAG* produces the amino acid Glutamine (*Gln*). If one of the nucleotides (for the reasons mentioned above) randomly changes to another type—*C* changing to *T*—the new sequence, *TAG*, transfers to *UAG* in mRNA (remember that Thymine in DNA replaces Uracil in mRNA). This is the stop codon and causes the sequence to stop. As a result, the protein synthesis stops at that point due to mutation, causing an incomplete amino acid sequence and dysfunctional protein (Figure 22.7, top panel). Another nucleotide sequence in Figure 22.7 consists of a *CAT* sequence that produces amino acid Histidine (*His*). If the nucleotide *A* in one of the sequences converts to *C*, it converts *CAT* to *CCT* codon that is a different amino acid—Proline (*Pro*). This changes the sequence resulting in a different, mutated protein. In some cases, mutations are harmful or could not survive. In other cases, they could successfully go through the natural selection process and Darwinian evolution and lead to new species and hence, diversity of the living things.

THE THREE DOMAINS OF LIFE

By studying the sequences of nucleic acids, it has now become possible to classify different organisms into groups with similar internal compositions without regard to the external morphology that is often misleading. In 1977, Carl Woese (1923–2012) postulated that the more similar the genetic sequences between two species, the more related they were. To look at the similarities of all the organisms existing on Earth today, one needs a molecule that exists in all the living organisms (Box 22.5). We could then look at the degrees to which this changes from organism to organism. As discussed above, the molecule ribosomal RNA (rRNA) satisfies this requirement. The task of rRNA is to translate genes into proteins; it has the same function in all the organisms on Earth, very likely because it is generated from a common ancestor of all organisms. The genetic structure of the rRNA changes over millions of years of mutation and evolution. Studying these changes allows us to reconstruct the process of diversification. Woese studied the sequence of rRNA between the bacteria (organisms with no nucleus) and compared their similarities and differences. Based on this and other independent studies, the living things were divided into three domains: *bacteria, archaea*, and *eukarya* (Figure 22.8).

Bacteria are the oldest of the three domains and have developed various metabolic activities using organic molecules as the source of energy; these are called *heterotrophs*. They cannot make their own food supply by producing carbon from inorganic material, but will get this by eating other plants and animals. Some of them use anaerobic and some use aerobic respirations. Also, some bacteria like cyanobacteria perform photosynthesis, taking energy

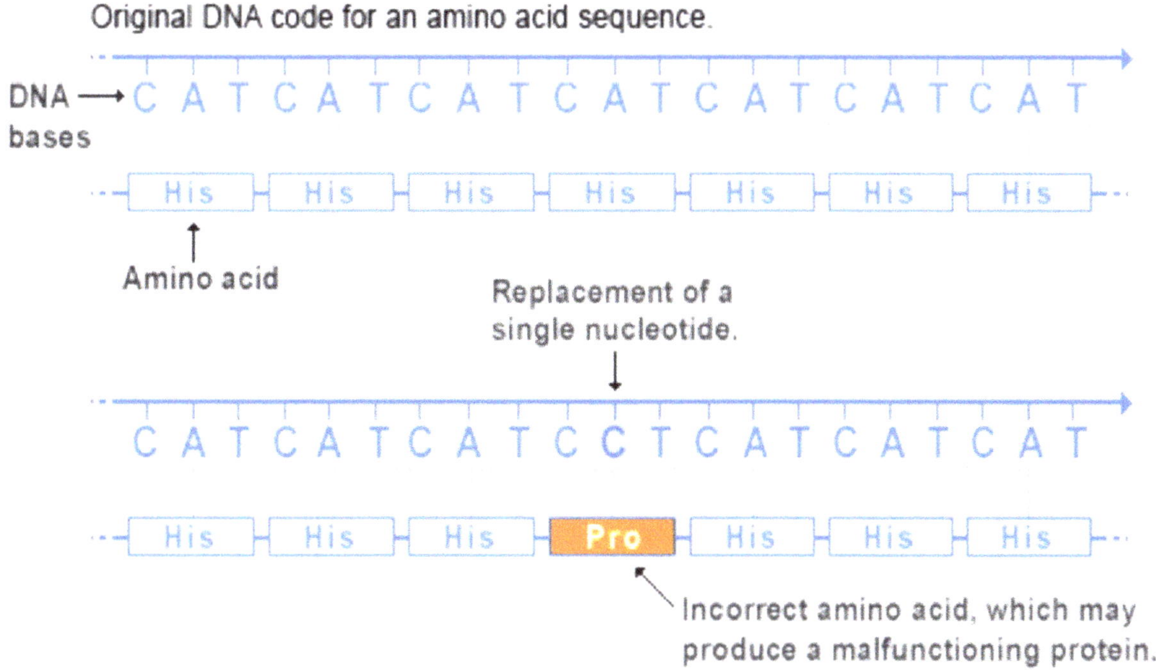

Figure 22.7. The mutation process when a nucleotide changes to another, causing a change in the sequence of amino acids. This leads to either premature synthesis of protein or generation of a new amino acid sequence. If this could adapt to the environmental conditions, it will evolve to produce a new species.

Phylogenetic Tree of Life

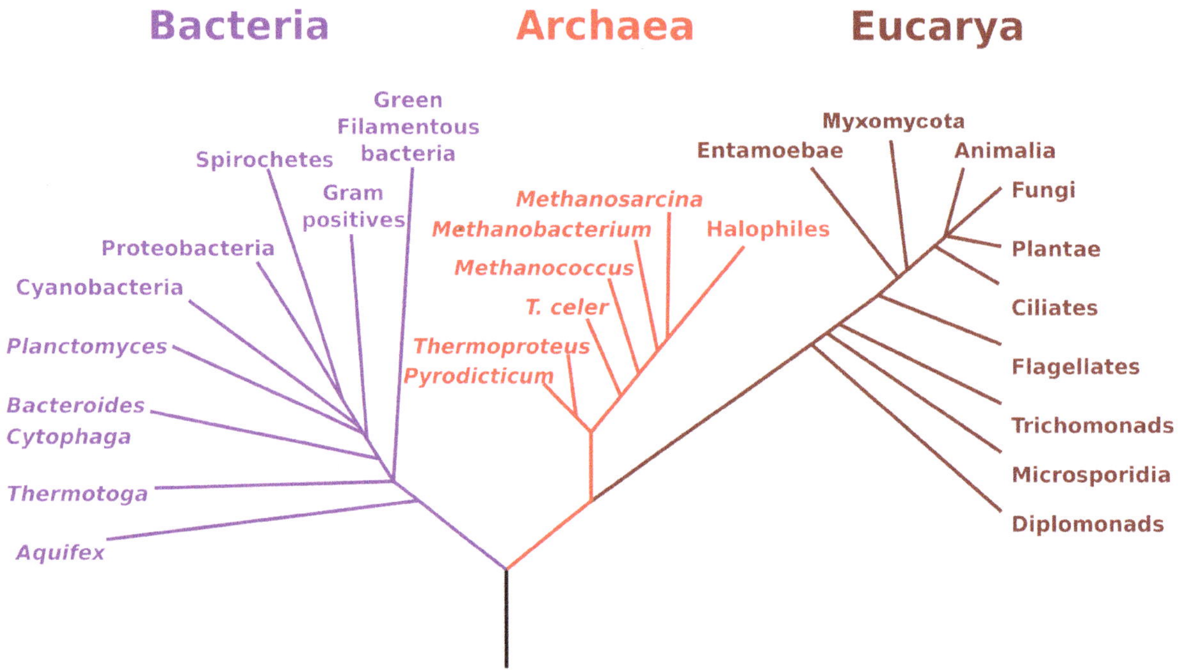

Figure 22.8. The three domains of life: archaea, bacteria, and eukarya. Each of these domains is divided into different kingdoms.

BOX 22.5: THE FIRST LIVING THING

The first identifiable fossils are that of the prokaryotes whose chemical fingerprints are found on sedimentary rocks dated roughly 3.8 billion years ago. The oldest prokaryotic fossils are found in Western Australia. These have an age of 3.46 billion years and are similar to the cyanobacteria, prokaryotes that live through photosynthesis, similar to plants. They were living in oceans since at that time there were no lands or continents and the atmosphere had not yet developed its protective ozone shield. Therefore, water provided a protective layer against intense UV radiation from the sun, allowing the cyanobacteria to thrive. The cynaobacteria can be found in the outer surface of stromatolites, found today in shallow waters of Australia's western coast.

The cyanobacteria removed CO_2 and added oxygen to the atmosphere, contributing to the oxygen build up in the atmosphere. By approximately 2 billion years ago, the presence of oxygen made most environments unsuitable for anaerobic (organisms that do not need oxygen to grow) prokaryotes. As a result, new metabolic pathways evolved from the photosynthetic cyanobacteria and aerobic (organisms that need oxygen to grow) bacteria. The increase in oxygen level led to formation of the ozone shield against the ultraviolet radiation from the sun, allowing the living organisms to move from sea to the land.

from the Sun and carbon dioxide from the enivornmnet, releasing oxygen, while others use *chemosynthesis*, getting energy from inorganic chemicals.

Archaea also have diverse forms of metabolic activity. Some use inorganic chemical reactions to generate the energy they need to make organic matter, a process called *chemoautotrophic*. These reactions produce methane (CH_4) or hydrogen sulfide (H_2S) as wastes. They are often found in extreme environments (hot springs or very salty or acidic environments).

Eukarya consist of animals, plants, fungi, and protozoa. The protozoa consist of the organisms that do not belong to any of the other categories. The eukaryotic cells are more complex than the prokaryotic organisms. Chloroplasts and mitochondria both have structures similar to the bacteria and are found inside the eukaryotic cells.

COMMON ORIGIN OF ALL LIFE

All living things support their structure and function from the same elements: hydrogen, oxygen, carbon, and nitrogen. These are the most abundant elements in the universe and certainly within the atmosphere and oceans of our planet. The previous chapter demonstrated that the chemical elements responsible for life are formed from complicated organic molecules of 20 amino acids. These have the same composition except for a side chain that is different and makes them have different properties (figure 20.6). These common compounds that exist in all forms of life are the fundamental building blocks of life.

The next step is for the amino acids to form the structural units of life, the *proteins*. The proteins in all living things are responsible for the way they are. Again, this is the common feature in all the life on Earth. For the cells to function and synthesize proteins, they need energy. The energy supply of the cells for all living things comes from the complex molecule of adenosine triphosphate (ATP), another common property of all the living things. The instruction for building proteins comes from the molecule's DNA, i.e., the double chain of molecules connected by molecules called base pairs. When a cell reproduces, the DNA divides into two chains. The chains then chemically attract the missing half of each other to form an exact copy as the parent DNA. At the end of this process, two identical DNA molecules are produced that carry exactly the same properties of the parent DNA.

Another common feature between living things is the way they process complex molecules to generate energy

BOX 22.6: LAST UNIVERSAL COMMON ANCESTOR (LUCA)

The characteristics shared between all the living things today are inherited from the Last Universal Common Ancestor (LUCA). The genetic information in LUCA was stored in DNA, which possessed proteins working as enzymes. The transporters and receptors used the same amino acids that are in our biochemistry today and used RNA to convert information in the genes, producing proteins that could function. Some of the characteristics of the living organisms today are the result of millions of years of evolution and therefore did not exist in LUCA.

to support their metabolisms. An example of this is the production of carbohydrate in plants and animals to provide their energy needs. To convert carbohydrates into energy, plants use *chlorophyll* through the process of photosynthesis. This is structurally similar to *hemoglobin*, which transports oxygen in the blood of animals. The only difference between these is that chlorophyll is built up around a magnesium atom while hemoglobin is built

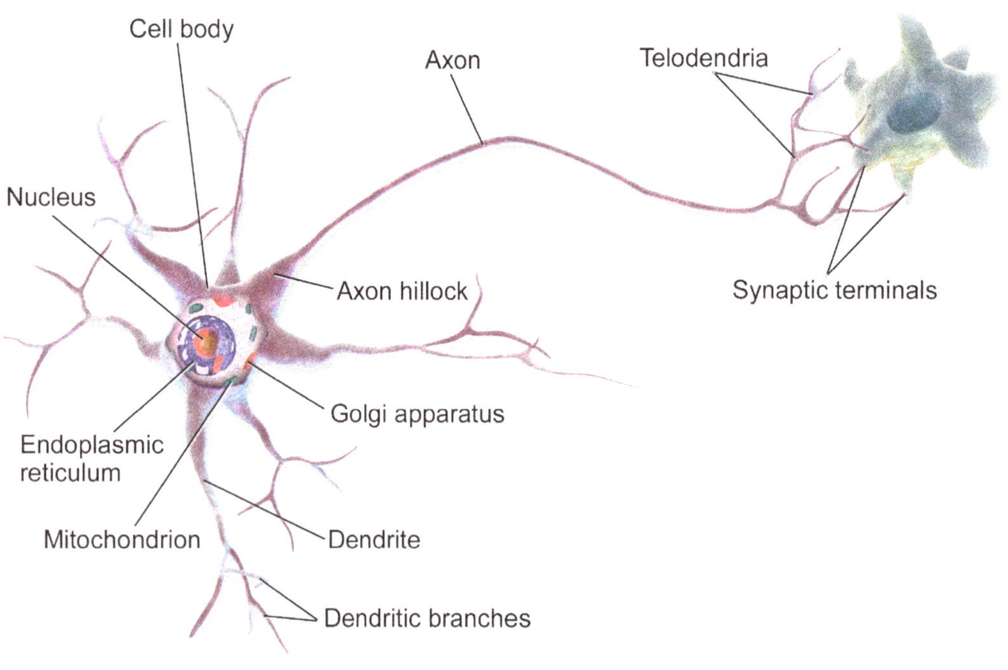

Figure 22.9. Indicates the structure and different components of a neuron. It also shows how two neurons interact with one another through their synaptic terminals by releasing chemical substances that diffuse between two neurons.

around an iron atom. This discussion confirms that all life on Earth started from similar but complex chemical molecules (Box 22.5) and that life, be it plant or animal, has had the same origin (Bennett and Shostak, 2004)-(Box 22.6).

ORIGIN OF THE NERVOUS SYSTEM

Study of the origin and evolution of the nervous system is a complex problem. The basic elements of the nervous system are *neurons*. These are specialized cells with different shapes, types and functions. Our brain contains between 100 billion and 1,000 billion neurons. This is more than the number of stars in our Galaxy or the number of galaxies in the observable universe. Neurons have two branches called *axons* and *dendrites* (Figure 22.9). Each neuron has one axon but many dendrites. Dendrites are like branches of trees. They are thick at the point they connect to the cell with their thickness decreasing as they extend out. In contrast, axons are thinner and much longer with the same thickness all through their length. Input to nerve cells is through dendrites while axons carry cell's output (figure 22.9). The point of contact between neurons is called *synapses*. At synapses, neurons release chemical substances that diffuse to an adjacent neuron that is contacted by that synapse. These are chemicals that may excite, inhibit or modulate the contacted neuron. A feature of axon terminals is the synaptic vesicle. These are small vesicles that store chemicals released at the synapse. When the chemical content of the vesicles is released in the space between the two cells, they interact with the molecules (proteins) present within the membrane of the contacted cell. When activated by synaptic chemical, the molecules activate the cell (Figure 22.9).

Given the above summary of the structure of the nerve cells (neurons), a study of the origin of the nervous system should ideally start with the investigation of the origin of neurons. The neurons share many of the properties of normal cells and therefore, must have somehow attained these characteristics, making them distinct from other cells. One characteristic that all the neurons have in common is that they could communicate with one another or with non-neural cells. Therefore, development of intercellular communication was a turning point in evolution allowing cells to transmit, exchange and integrate information. This is indeed one of the common characteristics of multicellular systems, as I discussed earlier in this chapter.

There are a number of outstanding questions regarding the origin and subsequent evolution of the nervous system. Among them are: What is the origin of neurons? When did animals with nervous systems first appear? How did the first centralized nervous systems come to being? How did neurons acquired their function? Fossil records, although rare, provide useful information regarding the evolution of early nervous system. One way to study these questions is to study the origin of *Metazoans*. These are groups of multicellular animals that have cells differentiated to different functions and into tissues and organs. In the Metazoans, some cells turn out to form neurons. The origin of Metazoan is estimated to be 750-800 MYA, coinciding with the geochemical events that led to the increase in oxygen level in the atmosphere from 0.1% to 3%. The first fossil evidence for Metazoa is from 600 MYA. It is also found that the shape and morphology of Metazoa became more complex and diverse around 542 MYA. This indicates a "missing" 150 million years of early Metazoan evolution. It is likely that early nervous system evolved during that time. Interestingly, this time coincides with major climate and geological events including glaciation.

The mammal's nervous system emerged from millions of years of evolution of lower vertebrates. The neural elements today, are the ones that survived the evolutionary process. For example, the neurons that successfully connected to others and established a network for information transmission are those that survived. Early development of the nervous system in mammalian embryo may give some clues towards understanding the origin and evolution of early nervous system (Box 22.7). As a result, the origin of the human nervous system may be related to the cells that form the skin and the sensory structures close to the body surface (*ectoderm* that is the outer layers of the cell forming the embryo). At that location they could receive stimulation from the environment. In later stages of development the neural part separates from the part that becomes the skin when sensory neurons diverge and become specialized to take on their responsibilities.

The earliest neurons were likely a combined receptor (receiving environmental stimuli) and motor units (producing muscle or gland responses). Such prototypes of neurons still exist in today's sponges. After millions of years of evolution these were separated to specialized cells—receptor and effector (motor) neurons. This developed into greater complexity when forming the nervous system. In this evolutionary scenario the primordial nervous system composed of neurons on the body surface that finally evolved within the surface (skin) to the nervous system that interacts with the external world through the sensory nerves. Study of the fossils from vertebrate remains indicate that in all of them the brains and cranial nerves are built on a similar structure with bilateral symmetry (neural structures being symmetric between the left and right sides of the body). This may indicate a common origin for the nervous system. The presence of proteins responsible for synaptic connections in Metazoans suggests that the process necessary for synaptic transmission started and evolved prior to the synapses (Box 22.7).

For primitive organisms like jellyfish or sea anemones, the nervous system consists of neurons that receive sensory information. Organisms with developed nervous systems show bilateral symmetry (one half of the body is a mirror image of the other) and segmentation (the body is composed of organized parts). More recently evolved spices such as clams, snails and octopuses have clusters of neurons that resemble primitive brains that are the

BOX 22.7: EARLY NERVOUS SYSTEMS

The first neurons were formed around 700 MYA with the first brain forming 250 MYA. A humanlike brain evolved around 6 MYA and our modern human brain has been around for 200,000 years, the time Homo sapiens appeared on the scene.

Australopithecus were the first to show human characteristics 4 MYA with a brain size similar to apes. The oldest fossils of genus Homo (of which modern human is a lineage) were found 2 MYA and called Homo habilis. The size of the brain of early creatures increases from Australopithecus to H. habilis and H. Sapiens. This is found by the increasing size of their skulls. Interestingly, the increase in the size of humankind coincided with climate changes. Other factors proposed for the increase in the brain size include nutrition (animals eating fruits have larger brains than those eating grass), anatomical changes and slow rate of maturation allowing for more time for the brain to develop.

command centers. Although nervous systems of different animals differ, they were built based on the same principles. Similar genes specify segments of the nervous system.

Study of early branching of metazoans will provide important steps towards understanding of the evolution of the neurons. For example comparison between characteristics of bilaterian and non-bilaterian (animals, including humans with two-sided symmetry are called bilaterians) metazoans provides important clues. The presence of neuron-like cells in non-bilaterian metazoans can be used as an argument in favor of a common origin for all the neurons. Furthermore, it is important to study their similarity to various types of neurons in bilaterians when going down the Metazoan tree of life.

SUMMARY AND OUTSTANDING QUESTIONS

There have been significant progress over the last two decade in understanding the properties and functions of the cells. For example, we have come a long way to address some fundamental questions: How was the nucleus of a cell formed? How did other organelles in a cell form? How could all these work together in harmony to sustain a living cell? When the multi-cellular systems developed? biologists have found detailed answers to many of these questions.

Cell membranes are made up of phospholipids that consist of two parts: hydrophilic "head" and hydrophobic "tail." Once positioned in aquatic environment, they orient with their head outside interacting with water and their tail inside away from water. The fatty acid then isolates the outside and inside regions. If water finds its way inside the structure, the fatty acid forms a second structure, a lipid bilayer. The structure surrounded by the lipid bilayer forms a protocell. The sugar molecules and nucleotides are small enough to penetrate inside the cell and integrate with the nucleic acids inside. No proteins exist at this stage of first cell formation, and this process takes place without the need for proteins acting as enzymes.

The mitochondria and chloroplasts (the energy generating part of the cell) are formed by endocytosis of bacteria and algae. This is when a material enters a cell without penetrating through the cell membrane and when two different species could live and function together. The membrane in the cell then keeps the external material inside. Both mitochondria and chloroplast have their own DNA, indicating that they originated from prokaryotic cells. Both organelles use their DNA to produce the protein needed for their functions (Chapter 20). Furthermore, a double membrane surrounds both mitochondria and chloroplast, confirming that they were produced through the invagination process. Endosymbiotic theory explains how a large cell and ingested bacteria could become dependent on each other and, after many million years of evolution, mitochondria and chloroplast become specialized and integrated parts of a cell, not being able to live by themselves outside the cell. The nucleus of eukaryotic cells were also formed through the endosymbiotic process.

The endosymbiotic model is supported by a number of findings. Present-day mitochondria and chloroplasts resemble prokaryotic cells. Both contain small amounts of DNA, RNA and ribosomes that are similar to prokaryotic cells. The organelles transcribe and translate their DNA into polypeptides, contributing to some of their own enzymes. Finally, they replicate their own DNA and reproduce within the cells.

The prokaryotic cells are divided into bacteria and archaea, with the difference being their cell walls. Two of the important prokaryotic cells are proteobacteria and cyanobacteria. Study of the evolution of prokaryotic cells is performed by their ribosomal RNA. The eukaryotic cells are divided into plants, animals, fungi, and protists with widely different properties. The eukaryotic cells originate from a common ancestry.

Multicellular organisms appeared when the oxygen level in the atmosphere reached its present level around 500 million years ago so that it could penetrate to the interior of large cells. These were only formed from eukaryotic cells and evolved independently from one another. Therefore, multicellular organisms did not generate from a common ancestor. The transition from single cell to multicellular organisms is a complex process. Details of biochemical reactions between different cells coming together to form a multicellular system are not well known.

Among the most outstanding challenges is the study of the origin of the nervous system and its early evolution. A logical way to do this, is to study the multicellular animals at the time when their cells were just differentiated to perform different functions and when they were divided into different tissues. This is estimated to have happened around 750-800 MYA. The fossil evidence of these organisms indicate a change in their shape around 542 MYA. The early nervous system must have developed during that time when different parts of a multicellular system (including the nervous system) assumed their exclusive functions. The neural systems that survived and evolved are those that successfully connected to others and established a network of information transmission. Early development of human nervous system were likely consist of combined receptor and motor units. After millions of years of evolution and natural selection, they were separated to specialized units. It is estimated that the first brain formed around 250 MYA with a human-like brain around 6 MYA and a modern human brain no later than 200,000 years ago.

Although significant progress has been made in understanding the origin of the prokaryotic and eukaryotic cells, a number of open questions still remain. The fossil records and the universal presence of mitochondria in eukaryotic cells indicate that they first appeared well after the prokaryotic cells. This is supported by the evidence for cyanobacteria (themselves prokaryotic cells) very early in the history of Earth (over 3 billion years ago) and the fact that mitochondria themselves are formed from another kind of bacteria, the proteobacteria. While there is abundant evidence that all eukaryotic cells came from the same common origin, and symbiosis played a crucial rule in the formation of eukaryotic cell, there are still a number of unknowns regarding the origin of eukaryotic cells. It is not clear what evolutionary process the protoeukaryotic cells went through before the acquisition of mitochondria. How important was the environment in this evolutionary process? Were other bacteria responsible for shaping the nucleus of a eukaryotic cell? What was the metabolic reaction between the mitochondria and its host? How did the nucleus and other organelles of eukaryotic cells evolve? How the cells in multicellular systems were differentiated to assume different functions? How early nervous system developed? These are among some of the outstanding questions to be addressed in future years.

1. Why do cells have small sizes?
2. What is the main constituent of a cell membrane? Explain the properties of these components.
3. What are liposomes?
4. What is the origin of the energy producing organelles in animal and plant cells?
5. Explain endosymbiotic theory.
6. What are the characteristics needed to develop multicellular organisms?
7. What are the four kingdoms of eukaryotic cells?
8. Why oxygen is needed to develop multicellular organisms?
9. Explain the properties of ribosomal RNA (rRNA) and their significance in studying the evolution of cells.
10. What are the characteristics of cyanobacteria and proteobacteria?
11. What physical processes were likely responsible for the first neurons?
12. What is the time sequence from the first neurons to the first nervous system to the first brain?

CHAPTER 22 REFERENCES

Hillis, D.M., D. Sadava, H.C. Heller, and M. Price. 2012. *Principles of Life*. New York: Freeman.

Morris, J., D. Hartl, A. Knoll, and R. Lue. 2013. *How Life Works*. New York: Freeman.

Sadava, D., D. Hillis, C. Heller, and M. Berenbaum. 2012. *Life: The Science of Biology*. 10th ed. Sunderland, MA: Sinauer.

FIGURE CREDITS

- Fig. 22.1a: Source: https://en.wikipedia.org/wiki/File:Plant_cell_structure-en.svg.
- Fig. 22.1b: Source: https://en.wikipedia.org/wiki/File:Animal_cell_structure_en.svg.
- Fig. 22.2: Copyright © Dhatfield (CC BY-SA 3.0) at https://en.wikipedia.org/wiki/File:Cell_membrane_detailed_diagram_4.svg.
- Fig. 22.3: Copyright © SuperManu (CC BY-SA 3.0) at https://en.wikipedia.org/wiki/File:Liposome_scheme-en.svg.
- Fig. 22.4: Source: https://en.wikipedia.org/wiki/File:Celltypes.svg.
- Fig. 22.5: Copyright © Matthewjparker (CC BY-SA 3.0) at https://en.wikipedia.org/wiki/File:Tolypothrix_(Cyanobacteria).JPG.
- Fig. 22.6: Copyright © Kelvinsong (CC BY-SA 3.0) at https://en.wikipedia.org/wiki/File:Serial_endosymbiosis.svg.
- Fig. 22.7a: Source: https://ghr.nlm.nih.gov/primer/illustrations/nonsense.jpg.
- Fig. 22.7b: Source: https://commons.wikimedia.org/wiki/File:Missense_Mutation_Example.jpg.
- Fig. 22.8: Source: https://commons.wikimedia.org/wiki/File:PhylogeneticTree.png.
- Fig. 22.9: Copyright © BruceBlaus (CC by 3.0) at https://commons.wikimedia.org/wiki/File:Blausen_0657_MultipolarNeuron.png.

THE EARLY EVOLUTION OF LIFE ON EARTH

23

CHAPTER LEARNING OBJECTIVES

This chapter will cover:

- The evolution of the first organisms on Earth
- The Cambrian explosion and radiation of life
- The first sea animals
- The first plants
- Migration from sea to land
- The first invertebrates and vertebrates
- Our very first ancestors in the sea
- Evolution in sea and on land
- The first birds
- The age of dinosaurs

"A story should have a beginning, a middle and an end, but not necessarily in that order"

- JEAN-LUC GODARD

"I love to think that animals and humans and plants and fishes and trees and stars and the moon are all connected"

- GLORIA VANDERBILT

The life started deep in the oceans. The oceans provided a protective shield for living things against the intense and harmful ultraviolet radiation from the sun, the extreme environment early life needed to prosper, and the nutrition water could supply. The first living organisms influenced the environment and through that, made it suitable for different types of plants and animals to foster. Indeed, from the very beginning, the living organisms have had significant effect on the development and evolution of life. Through mutation and natural selection, different species were developed. Those species that could adapt to their environment survived and evolved and produced off-springs. This process took billions of years and is still ongoing. Once Earth's atmosphere was formed and the oxygen (produced by cyanobacteria) accumulated in the atmosphere, the ozone layer was formed, providing a shield against the ultraviolet radiation from the sun. This caused migration of life (in the form of plants and animals) from the sea to the land around 488 MYA. A few million years before that, around 530 MYA, most of the ancestral species of plants and animals present in the world today came to the existence during a period known as the *Cambrian explosion*. There were a number of factors responsible for this, which will be discussed in this chapter.

Our planet, Earth witnessed a number of catastrophic mass extinctions during its history. In each event up to 95 percent of the existing species at the time were extinct. Through amazing resilience, life started all over again and continued to the present. This is an indication that once conditions for life are present, and the organic material available, life in some form will appear, either from the preexisting life or through inorganic compounds producing organic material.

The history of life on Earth is revealed by the study of fossils (from the word *fossilis*, in Greek meaning "dug up"). The science concerning the study of fossils is *paleontology* (in Greek: *palaios* meaning "ancient," *ontos* meaning "having existed," and *logy* meaning "study of"). Fossils exist in hard forms (shells and bones) or in the form of soft bodies (footprints and trails). The most common form of fossil is sedimentary rock. This is the result of settling of materials consisting of eroded rocks throughout Earth's history. These were carried by water and eventually settled deep in the seas. Over many millions of years, they form layers of material that vary in size, property, and nature—called *sediment*. The sediments later become *stratum*—a layer of sediments with known properties. Study of the stratum reveals history of the Earth. Since these layers are built upon one another, going back billions of years, they provide a sequence of the history of life on Earth. Each stratum is therefore older than the one above it and younger than the one below it.

This chapter presents a study of the evolution of life during the Precambrian time and after the Cambrian explosion through geological times. The main guide here is the study of fossils. Using methods for estimating ages of the fossils, the chapter connects the information at different times to elucidate the evolution of the living creatures since life began on Earth.[1]

EVOLUTION AND NATURAL SELECTION

Evolution is defined as the change with time in the genetically related characteristics of a *population*. Population is defined as a group of organisms of the same species that are able to interbreed and are therefore, genetically identical. The genes (DNA) determine the characteristics of a population. The mix of genes within a population can change. Therefore, evolution only takes place within populations. Evolution can occur by four different mechanisms:

Mutation is the change in the base-pair sequence of an individual's DNA that results in a change of the structure of its gene passed to subsequent generations. This leads to evolution if it takes place in an individual's gamete-producing cells. Mutation causes permanent alteration of the nucleotide sequence of an organism (Figure 22.7).

Genetic drift is the change in the frequency of a gene variant in a population, caused by a random sampling of the organisms. The genes in the offspring are a subset of those in the parent population.

Migration is the change in the frequency of gene variations in a population caused by individuals moving into or out of a population.

Natural selection is the process that encourages passage of beneficial genes to the next generation and forbids transmission of harmful genes. This is the process that selects which gene will be passed to the next generation and therefore, drives the evolution.

New variants of the genes can only be created by mutation within a population and therefore, mutation generates the variation on which natural selection could act. The theory of natural selection was first developed by Charles

---

[1] In this chapter Million Years Ago is denoted by MYA and Billion Years Ago is denoted by GYA.

Darwin and Alfred Wallace and presented by Darwin in his book *On The Origin of Species by Means of Natural Selection*, published in 1859. Today, there is substantial evidence supporting the concept of evolution and natural selection.

PRECAMBRIAN TIME (3500–550 MYA)

The Precambrian epoch covers more than 87 percent of the history of Earth from 3.5 GYA, when the first sign of primitive life was identified, ending with a rapid explosion of life and formation of different species around 550 MYA, called the *Cambrian explosion*. The first cells were likely prokaryotes (forming around 3.5 GYA). They lived in extreme environments like salty lakes and oxygen free waters. The oldest prokaryotic cells were found in south-

Figure 23.1. Shows stromatolites in shallow waters of Shark Bay, Western Australia. These are the oldest known fossils, appeared around 3 billion years ago. They were created through photosynthesis by cyanobacteria, which are believed to be among the first living cells on the Earth.

western Australia and are dated back to 4.46 GYA. These are similar in structure to today's *cyanobacteria,* the prokaryotes that carry on photosynthesis functions (releasing oxygen into the atmosphere) like plants. During the Proterozoic era (around 2.5 GYA), the oxygen producing pathways of photosynthesis evolved through mutations that changed the existing pathways. The cyanobacteria and other photosynthetic bacteria grew in aquatic environments and became dense. They captured minerals and sediments and over many years, formed outer parts of large dome-shaped structures called *stromatolites*. These lived around shallow waters and are found today in the coasts of Western Australia (figure 23.1). The release of oxygen by cyanobacteria into the atmosphere made life difficult for anaerobic (oxygen hating) cells around 2 GYA. At this stage, a new metabolic pathway for life was developed by the combination of photosynthetic cyanobacteria and aerobic (oxygen loving) cells. This was the first stage of the development of life of the aerobic cells.

The oxygen buildup in the atmosphere continued and around 500 MYA reached to the level it is today. An oxygen-rich atmosphere had a number of major impacts on the development of life (Prothero and Dott 2004):

- Oxygen molecules break apart and rejoin as ozone (Chapter 19). The ozone then accumulates in the upper atmosphere and blocks the harmful solar ultraviolet radiation reaching Earth. Without the protective layer of ozone, life would not have been able to move onto the land. The ultraviolet radiation cannot penetrate water and therefore, lack of an ozone layer did not affect development and evolution of life in the oceans.
- The oxygen reacts with the self-assembly of complex organic compounds.
- The abundance of oxygen provides a more suitable environment for organisms that thrive under aerobic conditions and the species that could not adapt to these conditions extinct.
- Multicellular organisms need oxygen to generate their required energy and to function efficiently. Therefore, the growth of the organisms and efficiency of their food production process directly rely on their oxygen intake (Prothero and Dott 2004).

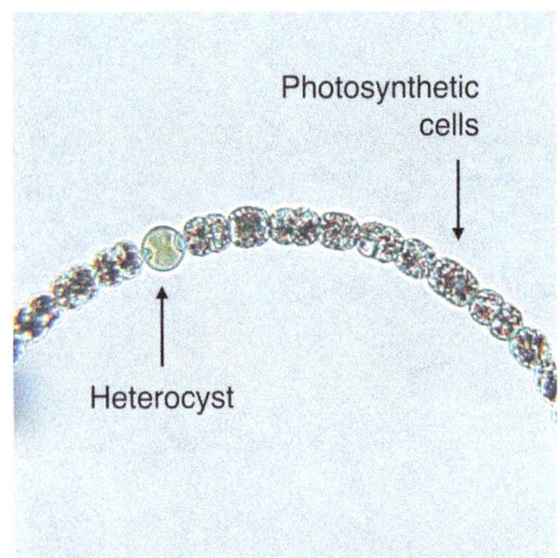

Figure 23.2. The bead-like structure of photosynthetic prokaryotic cells in cyanobacteria. Heterocytes are hollow-looking photosynthetically inactive cells with thick walls that block atmospheric gas from entering. They are specialized cells that contain enzymes for fixing atmospheric nitrogen (nitrogen fixing), providing nitrogen in the form of ammonia (NH$_3$) that plants can use.

Apart from building the oxygen level in the atmosphere, cyanobacteria made other ecologically important contributions. They incorporated carbon from carbon dioxide to other organic compounds as well as converting nitrogen from the air (N_2) into ammonia (NH_3). The cyanobacteria have filamentary structures (figure 22.5) with some chains in the structure containing nitrogen-fixing cells, called *heterocysts* (figure 23.2). These cells contain special enzymes to convert N_2 in the air to nitrogen. Plants need nitrogen but they cannot use the gaseous nitrogen (N_2) as they do not have the enzyme to break them into nitrogen. However, they can take dissolved ammonia (built by cyanobacteria) from the soil (Hillis et al. 2014). The nitrogen is needed for plant and animal growth and to build protein and nucleic acids.

The eukaryotic cells started to develop after oxygen dominated the atmosphere, as they needed oxygen to function. The evidence for the first eukaryotic cells dates back to about 2.7 billion years ago, as lipids in rocks. These acquired energy from cellular metabolism in the presence of oxygen. The origin of mitochondria in eukaryotic cells is from free-living bacteria that synthesized ATP (as discussed in chapter 22), while chloroplasts were free-living photosynthetic prokaryotes (like the cyanobacteria). By 850 MYA single cell eukaryotes reached their maximum diversity. Over the following years, they declined in abundance and diversity, reaching their low point by 675 MYA. The reason for this decline is mainly attributed to a change in the climate to colder temperatures, reduction in atmospheric CO_2, and increase in O_2, affecting the photosynthesis process in micro-algae. The decline sets the scene for the appearance of multicellular life (Prothero and Dott 2004).

The formation of the eukaryotic cells was the first step toward the creation of multicellular organisms. The oldest evidence for multicellular activity is from the fossils dating back to 1.4 billion years ago. These are also the first cells experiencing sexual reproduction (Box 23.1). At this stage of the evolution, some cells were specialized to be the male reproductive cells (gamete cells) and some to female reproductive cells (somatic cells). The emergence of multicellularity and rise of oxygen in the atmosphere allowed the evolution of larger bodies with specialized parts to perform different functions. This was a step toward the evolution of the invertebrates (animals without a vertebrate column) that started to appear around 630 MYA. The first such fossils were discovered in the Ediacara Hills of Australia. This made the scene ready for the Cambrian explosion.

BOX 23.1: THE EMERGENCE OF SEXUAL REPRODUCTION

Evidence for the first sexual reproduction is from 1.4 billion to 1.1 billion years ago during the Proterozoic eon around the time the multicellular life started. This was a major advance in the evolution. Asexual or splitting reproduction produces only identical copies of the parent cell. In this case, new variations would be generated very slowly. Contrary to this, sexual reproduction provides the opportunity for the organisms to exchange and mix genes, resulting to offspring with new characteristics. This allows the evolution to move faster by providing a wider variety of organisms to go through the natural selection process. Sexual reproduction also helps to repair genetic damage. All sexually reproducing eukaryotic organisms are likely to have originated from a single common ancestor.

THE CAMBRIAN EXPLOSION (545–500 MYA)

The fossil record before 600 MYA shows evidence for only stromatolites and microfossils, such as *acritarchs* (single-cell eukaryotes). Evidence for "jellyfish" from 550 MYA was found in the Ediacaran Hills of southern Australia. These were soft-bodied organisms that dominated the world and given the size of their fossils, they were likely multicellular organisms. They lacked any respiratory organ and therefore, needed to increase their surface area for nutrition exchange (by increasing their surface they increased chances of absorbing more nutrition from their environment). They disappeared early in the Cambrian period and were replaced by creatures containing shells, believed to be the ances-

Figure 23.3. An example of a trilobite (Animalia kingdom). They were among the first arthropods on Earth and lived in oceans for 270 million years.

tors of shelled invertebrates found today. The discovery of significant burrowing and trace fossils in sediments indicates the presence of animals capable of burrowing—animals like worms with hydraulically stiffened bodies. At the same time the marine fauna grew richer and large invertebrates with hard skeletons appeared. Later in the Cambrian period marks the first appearance of *trilobites* (meaning "threelobes"; figure 23.3), the unique features of this period. These are one of the first groups of marine *arthropods* appeared by a radiation of life at the beginning of the Cambrian and lived in the oceans for 270 million years. They disappeared in a mass extinction about 250 MYA (the Permian extinction; Prothero and Dott 2004).

Around 545 MYA an explosion of life happened, different spices appeared and life became abundant. This took place in the Paleozoic era and is termed the *Cambrian explosion*. Most of today's plants and animals can be traced back to this era. The animals at this time developed skeletons, the remnants of which can be found today in the form of fossils. The mutation process, leading to the appearance of different species, prominently happened at this time. This was caused by errors appearing during DNA sequencing, producing species that would subsequently adapt to their environment and, with time, grow into new species. The Cambrian period lasted for about 40 million years.

The cause of the Cambrian explosion is not clear but it is likely that a number of independent events all taking place at the same time may have been responsible. Some of these events include:

- The level of the oxygen in the atmosphere reached a critical level to allow larger and more energy-intensive forms of life to develop and prosper.
- The eukaryotes developed more complex genetic variations in their DNA, allowing more diversity in populations. Once the genetic and molecular structure of the organisms become more complex, more variations of the genes became possible.
- The cold climate on Earth came to an end during the Cambrian period. It is likely that the colder climate initiated the process of genetic diversity so that the organisms could adapt and survive. Once the climate eased, the diversification happened.
- Evidence from *strontium isotopes* indicates that there was a rapid increase in nutrients like calcium and phosphate in deep oceans. Also, the rapid increase in carbon abundance at the beginning of the Cambrian period again implies rich resources of nutrients, allowing explosive growth of organisms. Tectonic activities may have

relocated these nutrients into the marine environment, to be used later by shell building organisms (Prothero and Dott 2004).

EVOLUTION OF MARINE LIFE

As discussed in chapter 21, a likely place for the primitive life to have started is the environment around hot springs in the deep sea. These places were protected from cosmic impacts and the ultraviolet radiation from the sun, both posing serious problem to development of life on land. It is therefore reasonable to assume that the first stages of the evolution of life happened deep in the oceans.

The diversity of marine life significantly changed after the Cambrian explosion and during the Ordovician time (488–443 MYA). Only about 150 families of animals were known from the Cambrian, and this increased to 400 families around 443 MYA. One reason for this diversity was greater ecological options, reflected into a more complex food chain. During this time, some of the organisms, for the first time, grew to reach above the seafloor. The most common fossils found in the Ordovician were *brachiopods,* consisting of shells of calcium carbonate (figure 23.4a). These filtered the water to get the food through their shells. After the brachiopods, the next abundant organism during this time was *bryozoans* (figure 23.4b). These were coral animals that mostly lived in warm waters and filtered food from the water. They constituted a very diverse population (Prothero and Dott 2004).

The first coral reefs appeared around 450 MYA, with the earliest examples built by bryozoans (figure 23.5). Some corals were also made by *stromatoporoids*, themselves consisting of cyanobacteria or "sunflower corals"

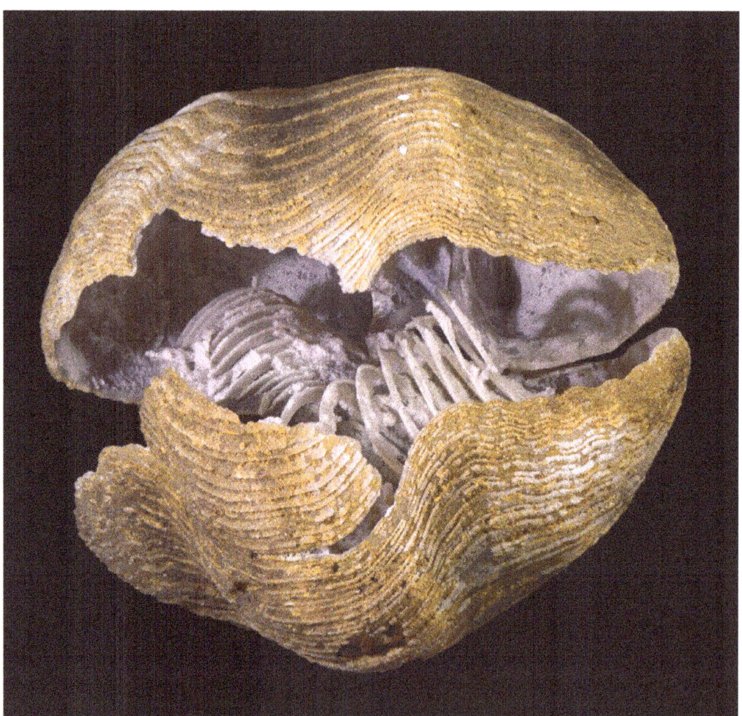

Figure 23.4a. An example of branchiopods, marine animals that have shells on the upper and lower surfaces. The shells are hinged in the back and open for feeding and close for protection.

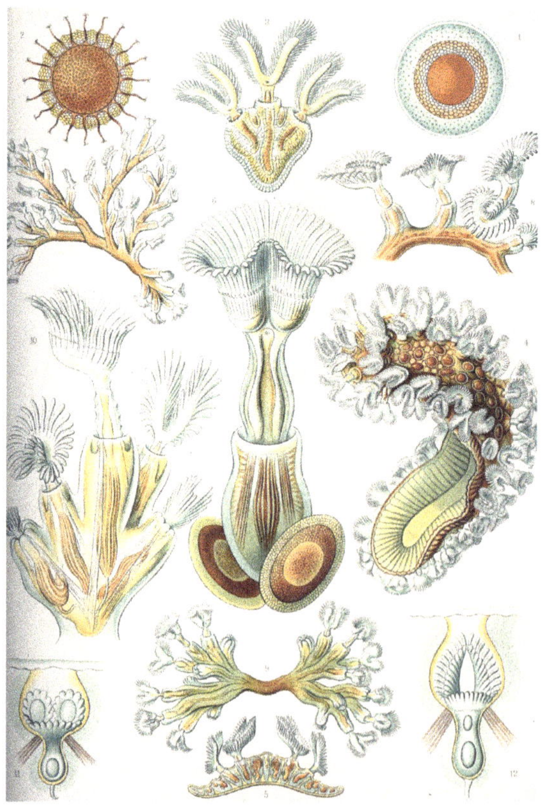

Figure 23.4b. A diverse population of bryozoans, marine animals found in warm waters. They filter seawater for nutrients.

made from algae and plants. The corals were fed by catching small fish and planktons but most acquire their energy from photosynthesis. A significant change during the Ordovician was the presence of predators that affected the abundance of the population. Among these we could name *nautiloids*, a group related to squids and octopus (figure 23.6). Another organism that appeared during the Ordovician period was the *molluscs*, which are invertebrates and highly diverse family constituting over 25 percent of marine life. They evolved to today's snails (figure 23.7; Prothero and Dott 2004).

The earliest ancestors of vertebrates are traced back to Ordovician period (about 480 MYA), in the form of *jawless fish* with a skeleton built by cartilage (figure 23.8). They acquired food from mud at the sea floor or from water. Their fossils are found in the sea, indicating that the life of vertebrates first started from the sea. These are early ancestors of all the vertebrates in the world, including humans.

Figure 23.5. An example of corals, inhabiting tropical waters. They build reefs and secrete calcium carbonate to build a hard skeleton. Individual heads grow by asexual production, but they also breed sexually. Corals of the same species release gametes simultaneously over a period of several nights around a full moon.

The Ordovician was the time of a significant radiation of life, increasing the diversity, quantity, and complexity of marine life (Prothero and Dott 2004). This is because during the transition time from Cambrian to Ordovician, the sea level rose, flooding all the land and providing suitable environment for marine life to develop, and given differences in the environmental conditions, it could diversify. Furthermore, the oxygen level during the Cambrian time increased, reaching close to the present time. A high oxygen level (16 percent of the present level) is needed in order for calcite in the skeleton to develop and to support vertebrate organisms. This level reached during the Ordovician time. This is the reason for the radiation of life resulting in the first vertebrates. Around 443 MYA the world oceans became cold, leading to a major mass extinction, with only the organisms able to adapt to cold conditions surviving.

Figure 23.6. The fossil of a nautiloid from the Ordovician time.

Figure 23.7. An example of a mollusc. About 80 percent of all molluscs are arthropods containing snails and slugs. They are the largest marine organisms and are a highly diverse population (about eighty thousand species). The gastropods first appeared in the Cambrian period (541–485 MYA).

Figure 23.8. An example of a jawless fish from the Ordovician time. These are mostly extinct species with only two species of the jawless fish still present—hagfish and lamprey.

Figure 23.9. An example of a marine plankton, tomopteris. These drift on the water and produce their energy from photosynthesis and by absorbing nutrients from water.

The majority of jawless fish at the time were extinct with only two species remaining until present (Prothero and Dott 2004).

The cooling of the seawater around 443 MYA resulted in only a small family of cold-adapted animals to survive. Their diversity therefore substantially reduced in the following years into the Silurian period (443–416 MYA). Some of the animals from the Ordovician period (like brachiopods) acquired thicker shells and more robust bodies. The predators during this time were again the nautiloids and a new thing called *sea scorpions*. The seawater became warmer during the Silurian period, making the environment appropriate for *planktonic organisms* that thrived during this time (figure 23.9). The planktons are diverse group of organisms drifting in water (plankton means "errant" in Greek or "wandering") with no ability to swim. They are a source of food for marine animals, while they acquire their nutrients from the photosynthesis process by absorbing and processing light from the sun. For this reason they are often floated on the surface of water to receive sunlight. Planktons are divided into two broad groups: *phytoplankton*, the organisms that are like plants (*phyton* or "plant" in Greek) and includes organisms like diatoms cyanobacteria, and *zooplankton*, the organisms that are like animals (*zoon* or "animal" in Greek) and contain the subgroup protozoans. Phytoplankton makes significant contribution to the ecosystem. In the process of photosynthesis, they release molecular oxygen (O_2) into the water. Between 50 percent to 80 percent of the oxygen in the water and atmosphere is produced by phytoplankton with the rest being produced by plants. As a result, phytoplankton have been responsible for keeping the CO_2/O_2 balance since the Precambrian time (Prothero and Dott 2004).

The main innovation in the Silurian period was in the vertebrates. By this time, jawless fish were around for about 100 million years.

During the Silurian, they developed head shields and body armor, a slit-like mouth and no jaws (figure 23.8). This was the time when fish with jaws came to existence. The first fish with jaws are from the Silurian time and are called *acanthodians*. They had strong bites and fins along their bodies (figure 23.10; Prothero and Dott 2004).

The diversity of jawed fish increased around Devonian time (416–359 MYA). The first sharks with skeletons made from cartilage, fins, and flat skulls appeared during this time. There are two types of fish developed in the Devonian time that are of particular importance. First is the *ray-finned fish*, with their fins supported by bony spines. These are the ancestors of 99 percent of the bony living fish today. Second was the lobe-finned fish (figure 23.11). Their fins supported by bones that would give them the ability to eventually walk on the land. Lobe-finned fish diversified and developed lungs allowing them to breath outside water and eventually led to *amphibians* and other land vertebrates. At this point, about 359 MYA, the marine life was ready to test the land as its new habitat.

MIGRATION FROM WATER TO LAND

Figure 23.10. The fossil of an acanthodian fish from 400 MYA (during the Devonian time). These were the first fish with jaws, and it is believed their jaws evolved from their jaw-less ancestors. They became extinct 250 MYA.

Figure 23.11. A lobe-finned fish, probably the earliest known bony fish, lived 410 MYA (during the Silurian time). It has a combination of both ray-finned and lobe-finned features.

Around 500 million years ago, toward the end of the Cambrian period, life started to move from oceans to the land. The process was gradual with plants and animals having different histories. It is easy to imagine a scenario under which simple single cell organisms moved from the sea to the land. They could prosper where they have access to liquid water and could live underground to shield against the UV radiation from the sun. For the multicellular organisms, migration from sea to land was more challenging, as they needed the means to obtain water and mineral nutrients from their surrounding without absorbing them as they did in water. The formation of the ozone layer to protect them against the ultraviolet radiation from the sun and buildup of oxygen in the atmosphere were among the essential factors in the timing of the migration of multicellular organisms from sea to land.

The available fossil evidence suggests that plants were the first organisms to develop the ability to live on the land around 475 million years ago. The first plants moving onto the land were of the kind that did not have water-conducting tissues (called *nonvascular*) and algae. As a result, their height was limited to only a few centimeters as this was about the size water could reach to all the parts of the plant. Among the first plants settled on damp areas on land were algae. To be able to live on the land and face dry conditions, they developed thick cell walls. This happened during the Ordovician period (about 443 MYA). The earliest evidence for plants containing water-conducting tissues (called *vascular*) goes back to the Silurian period (about 400 MYA). They grew to large

Figure 23.12. Shows examples of extinct and modern arthropods.

plants during the warm climate of the Carboniferous period that followed. These plants had part of them searching for sunlight to produce energy and a part to acquire water from the soil, and hence the nutrition. Around 430 MYA a seedless vascular plant called *cooksonia* covered the land. The diversity of these plants increased during the Carboniferous period with the plants reaching large sizes. The earliest fossil evidence for seed plants goes back to 385 MYA, in the Devonian period when they diversified during the Carboniferous period. Formation of supercontinents during the Permian period (299–251 MYA) led to dryer climate and extinction of large (tree-size) seedless vascular plants. The dry conditions at that time favored drought-tolerant plants, the ancestors of today's pines and spruces (Prothero and Dott 2004).

The first animals started to move to land about 75 million years after the plants did. Land plants created a habitat for the first land animals, the *arthropods* (from Greek: *arthro* meaning "joint" and *podos* meaning "feet") that moved to the land during the warm Carboniferous period (figure 23.12). These are invertebrates (animals without the vertebrate column) with protective exoskeleton (external skeleton) that include insects, spiders, millipedes, and crustaceans. The earliest fossil evidence for arthropods is from 541 MYA (early Cambrian period). Fossils found from 419 MYA and terrestrial tracks from 450 MYA confirm that the arthropods were the first animals conquering the land. On the land, their joint exoskeleton protected them against extreme dryness and supported them against gravity, providing means for movement without relying on buoyancy of water (figure 23.12). It is not known as what was the last common ancestor of arthropods. However, it is likely that the earliest known arthropods had a number of segments in their body with unspecialized appendages that functioned as legs and extracted nutrition from mud (this definition resembles worms). The oldest insect fossils date as far back as 407 to 395 MYA and these were the insects with wings. The ability to fly, allowed this species escape enemies and to move to other regions in search of food, resulting this to become one of the most abundant and diverse creatures.

The ancestors of vertebrate land animals are lobe-finned fish (figure 23.11), first attempted to experience life on the land around 419 MYA. These had paddle-like fins with their lobe fins made of flesh, making it adaptable to change to limbs under nonaquatic conditions (figure 23.13). The lobe-finned fish were therefore the ancestors of the first *amphibians* (*tetrapods*—*tetra* meaning "four" and *pods* meaning "feet" in Greek) on dry land (figure 23.14). However, they were not well adapted to life on land and had to move to sea to reproduce. The transition of amphibians to life on land was a combination of both skeletal and biological changes. Their skeletons had to change to support their weight in the absence of water. The vertebrae evolved and became stronger and able to distribute the weight. Skulls disconnected from the body and necks evolved to allow

Eusthenopteron

Tiktaalik

Acanthostega

Figure 23.13. Evolution of the lobe-finned fish to land tetrapods over 20 million years. This took place around 450 MYA.

Figure 23.14. Examples of the Amphibians that were the first creatures to move from sea to land, around 450 MYA.

better mobility of the head. Bones were shifted to align with the limbs. Joints were rotated to allow four-legged crawling (figure 23.13). This way, the first tetrapods started to walk on land.

How did the transition from a fully aquatic life form of a lobe-limbed vertebrate (like *Eusthenopteron*) to four-legged tetrapods (like *Acanthostega*) that walked on dry land take place (Figure 23.13)? In 2006, scientists presented evidence of a Devonian fossil lobe-limbed vertebrate, called *Tiktaalik*, which had the intermediate characteristics of both the fins of a fish and the limbs of a terrestrial tetrapod (Figure 23.13). It appeared that the limbs needed to move a large fish forward and enable a land vertebrate to walk were developed in water. These fins helped the fish to hold the animal in shallow waters, allowing them to move their heads above water's surface. The limbs vertebrates used for movement on land were developed from the muscular fins of their aquatic ancestors. The tetrapods were among the first vertebrates that migrated from sea to land. This evolutionary process took place over a time of 15 million years, from 380 MYA to 365 MYA.

The tetrapods are divided into two distinct groups of land vertebrates: *amphibians* that remained in moist environments, and *amniotes* that adapted a drier habitat. The amphibians require moist environments because they rapidly lose water through their skin. The amnions exploited a range of environments. They developed thick skins with hair or with feathers, which reduced the loss of water from their skin. The amniotes were split into two major groups during the Carboniferous time: the *reptiles*, and the lineage that led to the *mammals*.

A major change during this transition was the adaptation to the air and the way they provide oxygen to their body. Initially they could breathe through their thin skins. Later, division of the heart to three chambers allowed blood flow into the body and the lung that played an increasingly important role. Some researchers proposed that through this process, the gills were converted to lungs. However, there are independent studies suggesting that the lung was formed from the fish's digestive system. In this scenario, the first tetrapods (figure 23.13), when leaving water, breathed by swallowing air and extracting oxygen in their gut. This evolved to special packet allowing more efficient way to absorb the air. Alternatively, it is likely that the swim bladder in fish, responsible for buoyancy in water, were converted to lungs. Other internal changes needed to take place for adaptation for life on dry land. For example, changes to the inner ear improved detection of airborne sounds with the eyes becoming protected against dryness by the eyelids. The change from a body living and navigating in the sea to a body being able to move and breathe on the land is one of the amazing evolutionary changes observed (figure 23.13). These evolutionary changes took 30 million years to adapt marine animals to those living on land. Around 360 MYA tetrapods could finally roam freely on the land and breathe the air as their descendants do today. The amphibians prospered during the warm climate of Carboniferous period and came in different shapes and sizes. At the same time, life in the sea continued and evolved for the plants and animals adapted to marine conditions.

EVOLUTION OF LIFE ON LAND AND CONTINUED EVOLUTION IN THE SEA

The migration from the sea to land was a gradual process taking place between 500–355 MYA (starting around late Cambrian and continuing to late Devonian periods). It is not known exactly why the animals moved from sea to the land. One possibility is that the seas became dry and they had to "walk" across the land to seek new habitats. The other possibility is that they escaped predators that were in the sea in abundance while the land was safer. During this 150 million years, significant evolution in both the structure and the biology of plants and animals took place to adapt them for life in dry land. Over the 100 million years (251–355 MYA) covering Carboniferous and Permian times, significant changes took place. While previous to this the world was dominated by oceans, tectonic changes raised much of the land above the sea level, making more land. Therefore, a transition took place from marine to land conditions. At the end of the Paleozoic era (around 251 MYA) the world was dominated by wide-ranging plants (forests and swamps) and animals (reptiles and giant insects). The land life was helped by the formation of the supercontinent *Pangea*. The connected continent provided a land large enough to accommodate a diversity of nonmarine life.

By late Carboniferous, insects dominated the world. The primitive insect groups had wings that could not fold along their back, turned to insects with folded wings by early Permian time (about 300 MYA). Among this family, *cockroaches, beetles, and grasshoppers* were particularly common. The largest of the animals during the Carboniferous were amphibians, developed into land predators with their flattened heads, long snouts and eyes on the top of their head (figure 23.14).

An important event during this period (300 MYA) was the emergence of the animals that laid their eggs on dry land, the *amnions* (from the Greek word *amnion* meaning "membrane surrounding the fetus"). This was the beginning of the reproduction on the land. Amnions are tetrapod vertebrates comprising reptiles, birds, and mammals. They are divided into *sauropsids* (reptiles and birds) and *synapsids* (mammals) and their ancestors. The first amnions looked like small lizards and evolved from the amphibians around 312 MYA (in the Carboniferous time). From this point onward the sauropsids and synapsids spread in the world and dominated the land, eventually becoming our ancestors (Prothero and Dott 2004) (Chapter 24).

A major mass extinction took place at the end of the Permian period (299–251 MYA), leading to the extinction of 90 percent to 95 percent of marine and over 75 percent of terrestrial species (Prothero and Dott 2004). When life started again in Triassic period millions of years later, there was a completely different landscape. The cause of the Permian extinction is likely to be cooling of the atmosphere. There were also evidence for both significant cooling (at high latitude) and drying (at the equator) on the land. This significant temperature gradient caused instability in the climate during late Permian period around 251 MYA. Another factor affecting

Earth's climate during this time was joining of the continents creating a single land—Pangea. This cut water circulation in the oceans, affecting the climate again.

After this major mass extinction extinguishing the land and marine life, reptiles dominated the sea and the land. On the sea floors abundant animals from the Permian time—crinoids, bryozoans, and brachiopods—disappeared, giving their place to *molluscs* that evolved rapidly to fill the seas. The molluscs consist of *gastropods* (snails and slugs), *bivalves* (clams), and *cephalopods* (squids), which all lived in oceans. Cephalopods such as squid, cuttlefish, and octopus are among the most neurologically advanced invertebrates (Prothero and Dott 2004). The clams and gastropods developed features that could escape shell-crushing predators (sea reptiles or mollusc eating fish) in the sea and hence survived during the Triassic, Jurassic, and Cretaceous (251–145 MYA). They developed the ability to burrow or swim fast and hence hide or escape from predators. Many developed spines or thickening in the shell to avoid predators to crush them or peeling the shells. However, during the Jurassic and Cretaceous time several groups of marine animals with shell-crushing teeth (fish and sharks) and claws (crabs and lobsters) appeared. Similarly, ammonites developed jaws able to crushing the prey. The abundance of predators at this time forced development of characteristics and skills for the prey to hide or escape including greater mobility, spiny and armored shells, and the ability to burrow. Sea life during the Mesozoic time (around 251–145 MYA) went through significant diversification from burrowing molluscs to shell crushing animals, fish, and marine reptiles.

Plant life also diversified at sea floors. This includes planktonic organisms converting nutrients and sunlight to living tissues to feed more complex creatures. Among the sea plants, there were microscopic diatoms with silicate rich shells that appeared during the Jurassic and Cretaceous periods. There were a huge radiation of life during the Cretaceous time, allowing diversification of nutrients for molluscs and fish to feed from and therefore, food for higher-level predators.

The evolution in the sea continued and around 251 MYA (early in the Mesozoic) there was a huge radiation of bony fish related to today's sturgeon (the fish producing caviar). By late Jurassic (145 MYA), the ancestor of all the

Figure 23.15. Huge xiphactinus fossil also showing another fish inside it, evidence that this was an active predator.

living bony fish, modern *teleost fish*, appeared. These had mobile jaws, giving them the flexibility to feed. Among the teleost fish family was the now extinct gigantic *xiphactinus*, a 4-meter-long predator living in the Cretaceous time (figure 23.15).

A large diversity of marine reptiles also radiated at the Triassic. These had short legs and short necks feeding from fish. An example is *placodonts*, which used their strong teeth to feed on molluscs (figure 23.16). By the Jurassic time, the placodonts had evolved to *plesiosaurs* with very long necks to enable them to catch fish (figure 23.17). The plesiosaurs became extinct by the end of the Mesozoic. Finally, among the sea reptiles with currently living relatives are huge marine turtles and the *mosasaurs* (related to lizards). The marine turtles were 4 meters in length, while the mosasaurs were land lizards adapted for life in sea and possessed flattened tails to allow underwater swimming and feet modified into flippers (Prothero and Dott 2004). The marine animals appeared during the Triassic were the ancestors of the animals dominating our seas today.

Figure 23.16. Placondonts with short legs and short neck. They used their legs as flippers to swim in the sea. The features (development of the neck and legs) show the transition from sea to land animal.

Figure 23.17. Plesiosaurs with long necks, allowing them to catch fish.

EVOLUTION OF PLANTS ON LAND

The seed plants grew during the Permian when also *gymnosperms* (vascular plants with seeds that are not enclosed in an ovary or fruit) in the form of seed ferns were prominent (Box 23.2). During the Triassic and Jurassic periods, two groups were outstanding. The first group is the *cycads* that look similar to the palm trees (except that they are gymnosperms), and they can be found today in tropical places. The cycads are either male or female with the pollen being released by male and carried by the wind to the female. The second group is the *ginkgo*. The oldest fossils for ginkgo trees were found from 270 MYA. During early Mesozoic (251 MYA) large trees covered the land as well as cycads and ferns (Prothero and Dott 2004). However, the vegetation at this time were slow growing, had mostly spiky leaves and were somewhat toxic. As a result, they were not suitable for feeding large dinosaurs that started to dominate the world in that time. In the absence of grasses that did not grow till mid-Cenozoic, only ferns could have provided rapid growth to feed the dinosaurs.Without a major evolution of plants on the land, able to provide fast growing and nutritious plants to feed large creatures, continuation of life on the land would be impossible (Box 23.2). We will return to this after we discuss emergence of dinosaurs in the next section.

BOX 23.2: HISTORY OF PLANTS ON EARTH

Plant life began in aquatic environments. The first known plants were multicellular photosynthetic eukaryates, known as *algae*. The first land plants were from 475 MYA. They had no roots, no leaves, and no flowers. They provided food for early land animals.

Plants were the first multicellular organisms to live on land. The earliest plants were nonvascular, meaning that they had no vessels to transport water and nutrients to their components. The vascular plants appeared afterwards when they developed vessels to conduct water from the soil. Since early land plants did not have the tissues to conduct water and nutrition to all of their parts, they only grew a few centimeters. The evolution of vascular tissue allowed land plants to transport water up their stem. As a result, they could access nutrients in the soil more efficiently, enabling them to grow taller than nonvascular plants. Ferns are the most important of vascular plants and were major players during the Carboniferous period (360–300 MYA).

The next step of plant evolution after the vascular plants was the appearance of the seeds. This was an amazing innovation made by nature—an embryonic plant with its own supply of water and nutrients. Seeds provided nutrients for the next generation of plants. There are two groups of seed-producing plants: *gymnosperms* (including pines, firs, and redwoods) and *angiosperms* (all the other flowering plants). Gymnosperm dominated the forests during the Mesozoic era about 160 MYA. They were the earliest plants to produce seeds. This property of gymnosperms made them the dominant plants of the Mesozoic era. After plants with seeds, flowering plants came on the scene. Flowers were used by insects and birds to spread the plant seeds during the Cretaceous period around 100MYA. The emergence of flowering plants roughly 35 million years after the seed plants led to a rapid increase in the number of plants since a flower has the plant's reproduction system.

THE EMERGENCE OF DINOSAURS

In the mid-Triassic time, two distinct groups of reptiles appeared. The first group included all the lizards, snakes, and plesiosaurs. The second group was *archosaurs* that dominated the world during 251–65 MYA (the Mesozoic time). The archosaurs consisted of crocodiles, dinosaurs, and flying reptiles. Anatomical study of the fossils of the earliest known dinosaur and those of the other animals in their close family has revealed that the common origin of dinosaurs goes back to a bipedal named *euparkeria*. This was a predator with forelimbs much smaller than hind limbs and a long tail to balance the body on two legs (figure 23.18). They lived 245 MYA and were among the first in the dinosaur family, sharing their ancestry with dinosaurs, crocodiles, and birds. All dinosaurs have

Figure 23.18. An artist's painting of euparkeria. These are believed to be the first to live in the dinosaur family, around 245 MYA.

defining characteristics in their skeletal structure, making them distinguishable from their other close relatives. These include the position their upper legs are formed and connected to their body (for dinosaurs the legs are right underneath their body providing support for the body and allowing easy and fast movement; Prothero and Dott 2004).

The first dinosaurs emerged around 200 MYA (late Triassic) and dominated the world for 150 million years after the last of synapsids and amphibians were extinct. Their fossils are found in all corners of the world indicating they were not concentrated in a single habitat. The largest dinosaurs, *prosauropods*, reached 9 meters in length. By the Jurassic time, the sauropods reached 23 meters in length and weighted 27,000 kilograms (Figure 23.19).

How could dinosaurs rule Earth for so long? This was due to a number of different facts. Breaking up of the large land, the Pangea, allowed more habitats with different conditions for them to live in. Also, changes in climate and adaptability to new conditions and evolution of their physical characteristics helped them to survive for such a long time. This allowed other species of these animals to develop and new descendants to grow, including modern birds.

Study of fossil records and DNA evidence shows that birds are likely to be descendants of dinosaurs, coming from the *theropod* group of dinosaurs. They first appeared during the Cretaceous time, about 100 MYA. A large number of feathered dinosaurs have been found, supporting the fact that birds are descendants of feathered dinosaurs. The feathers were mainly used to insulate their body during early Cretaceous. Also, in many features and structures of their body, theropods are indistinguishable from birds. The earliest of the birds found was the *archaeopteryx* (figure 23.20). The remaining fossils of them show that they still had teeth and a skeleton similar to theropods (Prothero and Dott

Figure 23.19. A reconstructed sauropod, the largest dinosaur group known.

Figure 23.20. Specimen of an archaeopteryx, the ancestor of today's birds. The wings and feathers are clear.

2004). The birds were diversified around the time of Cretaceous extinction that killed the dinosaurs. The birds survived in South America and then migrated through land to other parts of the globe, diversifying throughout their journey.

At late Cretaceous time, the air filled by insects, birds, and flying reptiles. The flying reptiles, *pterosaurs*, show similar skeletal features as dinosaurs. They reached their peak size around 65 MYA with one of the largest, *pteranodon*, having a wingspan of 7.5 meters. The largest of the flying reptiles or in fact the largest flying animal ever appeared on the planet is *quetzalcoatlus* with a wingspan of about 11 meters, the size of a small airplane (Prothero and Dott 2004).

Dinosaurs were completely disappeared around 65 MYA. The reason for their demise range from climate change, change in vegetation and their food supply, mammals eating their eggs and hence stopping reproduction and allergy to the pollen floating in the air. It is not possible to scientifically test these possibilities. The leading scenario for extinction of dinosaurs, for which there is strong scientific evidence, is collision of an extraterrestrial object with Earth around 65 MYA. This was discussed in chapter 17.

CHANGES IN WORLD VEGETATION AND FOOD SUPPLY FOR DINOSAURS

To feed all these huge dinosaurs, a significant supply of vegetation was required. Grasses were not around until the Cretaceous time, too late for dinosaurs to feed from. What sustained their growth and life was the appearance of a new kind of plants, the *flowering plants or angiosperms*. Compared to *gymnosperms* that rely on wind to carry the pollen to seeds, angiosperms developed flowers to attract insects and birds. The flowering plants were able to double fertilize by one pollen grain fertilizing the ovary (to produce more grains) and the other instigating the growth of nutrition, including fruits and nuts. This produces nutrient plants much more efficiently than the gymnosperms that were in action previously. The first angiosperms appeared around 100 MYA (mid-Cretaceous time). They had the ability to grow fast with a life-cycle of eighteen months. Their efficient production allowed them to diversify into different species over time that includes sycamore, magnolia, palm, oak, and walnut.

BOX 23.3: A WAY FOR MEASURING PAST TEMPERATURE

There is a relation between the size of the leaves on plants and the climate. Plants with large, thick, smooth edge leaves are found in tropical and warm climates. Plants in cool climates are thinner and smaller with rough edges that are shed every winter. In general, plants in climates with seasonal weather often do not have enough time to grow large and thick leaves. Therefore, by monitoring the shape and characteristics of fossils from plants, we could draw the climate map during the time of those plants.

Fertilization, evolution, and diversification of the flowering plants depended on the evolution of the insects. Different specialized angiosperms (flowers and fruits) attract different insects and required them to take their pollens to the same species. This led to rapid mutation and more efficient spread through a restricted gene pool. The insects that were most responsible for angiosperm were moths and bees that have their evolutionary history going back to late Cretaceous times.

CLIMATE AND THE EVOLUTION OF LIFE

The event that led to the demise of dinosaurs affected Earth's climate and therefore, life on land and the sea. During that time the change in the shape and configuration of the continents also affected the ocean circulation and the climate. The increase in greenhouse gas (CO_2) during the Paleocene (65–55 MYA) increased the temperature of the atmosphere. It became warmer during the Eocene (55–33 MYA), where tropical plants (palms and cycads) and animals (alligators and tortoises) were found in the Arctic zone.

Toward the end of the Eocene (about 35 MYA), the climate took a turn toward lower temperature, with the mean global temperature dipping by over 10 degrees Celsius. This produced large temperature gradients across the planet, with polar glaciers forming. The glacier build up at the poles continued to the Oligocene time (34–23 MYA), seriously affecting the land and sea life. The reason for this cooling is attributed to the change in the oceanic currents. For example, separation of Australia from the Antarctic caused the cold water to circulate around the South Pole, separating the South Pole waters from the warmer water at the equator. Apart from the oceanic currents, a possible decrease in the CO_2 level could have caused the reduction in the greenhouse effect and cooling of the atmosphere. The cooling of the climate continued to the Miocene (23–5 MYA) leading to the buildup of the Antarctic icebergs and as a result, lowering of the sea levels. Many parts of today's land came out of water at that time. The first evidence for Arctic ice cap appeared around 3 MYA during the Cenozoic time, when glaciers started to move to the northern land.

The change in climate has had direct effect on the life and its evolution. Changes in sea levels appeared during the Pleistocene (2.5 MYA to 11,000 years ago) as a result of glaciers forming and disappearing. This affected the coral reef in the sea and likely the aquatic food chain. On land, remains of plants and trees as far back as 30,000 years ago are found in areas that are dry and inhospitable today. Each change in climate led to extinction of groups of plants and animals and migration of animals to the climate they could best adapt to.

A BRIEF HISTORY OF LIFE ON EARTH

Life has had a complicated history since it began on our planet. It started in the sea and then moved to land. It was in the form of plants and then animals, who lived and evolved at the same time, both on land and in the sea. They affected each other's lives and evolutions as well as the environment and habitats. In the following, I summarize the main steps that led to the evolution of life on Earth.

Single-Celled Life: Early protocells emerged around 3.8–4.00 GYA. These prokaryotic cells are believed to be the first living things on the earth. The examples for these cells are photosynthetic cyanobacteria.

Photosynthesis: This is the process of converting light energy to chemical energy and to fuel an organism's activity. The chemical energy is stored in carbohydrate molecules synthesized from carbon dioxide and water. The ability to use water as the source for electrons in photosynthesis started with cyanobacteria. This took place about 2.4 GYA. Study of the archean sedimentary rocks indicate that life existed as far back as 3.5 GYA. It is not clear, however, at what point oxygenic photosynthesis evolved. Cyanobacteria were the first organisms to produce their required energy from photosynthesis. They dominated the landscape through to the Proterozoic Eon (2500–543 MYA). They were responsible for raising the oxygen level in the atmosphere and for nitrogen fixation and were major contributors to the chemical compounds needed for the cells.

Eukaryotes: These are the cells that have specialized organelles that contain the genetic material for the cell. Eukaryotic plant cells have chloroplasts. They reproduce both asexually and sexually (with the sex cells, i.e., gametes).

Multicellularity: These organisms contain more than one cell. They include all land plants and animals. They were formed at about the same time that the oxygen content in the atmosphere rose.

Precambrian Time: This covered 80% of the age of the Earth (3,500 to 550 MYA). During this time, cyanobacteria built up the oxygen level in the atmosphere and fixed its nitrogen content, resulting in the formation of multicellular organisms.

Cambrian Explosion: This occurred around 545 MYA and lasted for 45 million years. During this time, a burst of life took place, and the ancestors of modern plants and animals appeared. The known species of animals changed from a Precambrian estimate of 150 to nearly 400 at the time of the Cambrian explosion.

Marine Plants and Animals: Invertebrate marine animals with shells (branchiopods), coral reefs, planktons, nautiloids, and mollusks were among the first animals living in the Ordovician time around 450 MYA. The first vertebrate animals were jawless fish, living 443 MYA. Algae were the first marine plants.

Migration to the Land: Around 475 MYA plants were the first organisms that moved to land. Nonvascular plants were the first to move, followed by algae, and then vascular plants. The earliest seedless vascular land plants were Cooksonia (~430 MYA) appearing near rivers and streams. The diversity of this plant increased during Carboniferous time. The earliest fossil evidence of seeded plants goes back to 385 MYA.

The first animals started to move to land 75 million years after the plants. The arthropods were the first invertebrate animals that moved to a dry-land habitat (around 541 MYA), with most fossils found around 420–450 MYA. The ancestors of vertebrate land animals are lobe-finned fish. Between 380 MYA and 365 MYA their fins evolved into limbs; they developed thicker skins and a respiratory system that could function on dry land. They formed the amphibian family.

Evolution on the Land: Plants and animals adapted to dry conditions on land during the Permian period (250–300 MYA) with gymnosperms dominating the landscape. During the Triassic and Jurassic periods (250–150 MYA), cycad and ginkgo were found in abundance. Flowering plants also played a significant role, reproducing and increasing the diversity of plants.

Tetrapods evolved and were separated into two groups: amphibians and amnions. The amnions adapted and continued to live on land, eventually divided itself into two groups: reptiles and the branch that ended up with mammals.

Dinosaurs: They ruled the earth between 251 MYA and 65 MYA, forming a diverse population of some of the largest land animals. They are classified as reptiles and are ancestors to modern birds. They were extinct ~65 MYA by unknown events; it is very likely that the collision of an external object with Earth caused their extinction.

SUMMARY AND OUTSTANDING QUESTIONS

Study of the evolution of life on Earth is a complicated, multidimensional and nonlinear problem. One could divide different forms of life into two broad categories: the plant and animal life, in the sea or on land, each consisting of many widely different forms, all inter-related. For example, a source of nutrition is required for both plants and animals and this is different whether they reside in sea or on the land. To provide this, a constant food supply is needed. This food supply is directly affected by the sea level at any given time, the atmosphere, the climate and temperature, as well as catastrophic events that have many times wiped out the life. In the study of the evolution of life, all these parameters must be taken into account.

There is ample evidence that life first started from the sea. This was mostly because of the lack of an ozone layer in the atmosphere to protect Earth from the harmful ultraviolet radiation by the sun. In the sea, the protective layer was supplied by water. Around 2 GYA the oxygen accumulated in the atmosphere due to cyanobacteria that absorbed CO_2 gas and released oxygen to the atmosphere. This led to the appearance of stromatolites (made from cyanobacteria) in shallow waters. The evidence for the first aerobic organisms in the form of eukaryotic cells is from 2.8 GYA.

Combinations of an atmosphere rich in oxygen and formation of eukaryotic cells was the first step toward appearance of multicellular organisms around 1.4 GYA. A major development at this time was the emergence of specialized components to perform different tasks, leading to the birth and evolution of invertebrate animals in the sea around 630 MYA and a significant radiation of life—called the Cambrian explosion, which lasted for 40 million years. During this time some of the aquatic plants grew, for the first time, to reach outside water.

Around 480 MYA conditions became suitable for the development of vertebrate animals as a result of increase in the oxygen level needed to develop calcite in the vertebrate skeletons. The first ancestors of the vertebrate animals were jawless fish that lived around 480 MYA. The first fish with jaws are from 416 MYA and were divided into two groups—one being the ancestor of all the bony living fish today (ray-finned fish) and the other supported by bones that eventually gave them the ability to walk on land (lobe-finned fish) as amphibians.

The gradual migration of life from sea to the land started 500 MYA. Plants were the first to develop the ability to live on land. The first land-based plants were short to allow water to reach all their parts as they lacked water-conducting tissues. They developed the characteristics to be able to live in dry environments. About 400 MYA plants acquired the ability to conduct water and hence reached large sizes.

The first invertebrate animals moving onto the land were arthropods, migrating around 75 million years after the plants did, as confirmed by fossil evidence found from 419 MYA. They faced serious challenges in living on the land, including a dry environment, movement, and holding their weight against gravity, as well as developing the respiratory system needed outside water. The first vertebrates living on the land were the amphibians, likely descendants of lobe-finned fish. It took them until 360 MYA before they could freely venture the dry land. The rapid explosion of life and its diversity also led to different types of insects living on the land around 300 MYA. Almost at the same time, the first animals laying their eggs on the dry land lived, starting the new chapter of reproduction on the land.

While the evolution at the sea continued, the plants on the land evolved and by the appearance of vascular plants with seeds able to freely disperse, the rate of reproduction of plants increased. This provided food for some of the largest creatures ever known to live on the planet, the dinosaurs that ruled Earth for over 150 million years (between 251 and 65 MYA). The origin of dinosaurs goes back to bipedal creatures 251 MYA, with the first dinosaurs emerging around 200 MYA, growing in size to 23 meters and a weight of 27,000 kilograms. Study of the DNAs and skeletal features of the fossils shows similarities between the dinosaurs and birds. While dinosaurs were likely the ancestors of today's birds, they themselves were entirely wiped out 65 million years ago, due to collision of an exernal object with Earth.

There are a number of unanswered questions regarding the study of the evolution of plant and animal life on Earth. The time intervals between different fossils found (temporal resolution) is not enough to allow a detailed study of the evolutionary chain of events to find where the common origin of different species came from. This is often interpolated and could result in errors in identifying the evolutionary sequence. Through the prehistoric life and evolution, many different parameters affected the outcome. These included natural factors (temperature variations, change in proportions of gas in the atmosphere, formation of the ozone layer), the food supply (plants were needed to feed animals or nutrition required for aquatic creatures), mass extinctions (caused by temperature variations and cooling of the climate or collision with extraterrestrial objects), predators (some animals hunted others for food), and the general environmental conditions that allowed different forms of life to prosper. These affected sea and land as well as plant or animal life. The effect is therefore nonlinear. When studying the evolution of life on Earth, it is essential to consider the relation between different factors that affect life. This is a fascinating area for study.

REVIEW QUESTIONS

1. Why do scientists believe that the first life on Earth started in the sea?
2. What are stromatolites, and how were they formed?
3. When is the earliest evidence found for multicellular activity?
4. Explain in detail the characteristics of *acritarchs* and *trilobites*. When did they live?
5. Explain the reasons that led to the *Cambrian explosion*. How long did it last?
6. What are the first known ancestors of vertebrates? When did they live?
7. What creatures are responsible for keeping the CO_2/O_2 balance? And how?
8. Explain the characteristics of two types of fish developed during the Devonian time.
9. What changes needed to be made to the body of sea animals for adaptation to a life on land?

10. When did animals start to lay eggs on land? What were the first animals to reproduce on land?

11. What characteristics or skills did sea animals have to acquire in order to escape predators?

12. What were the main characteristics of sea animals during the Mesozoic time (about 251 MYA)?

13. What are *gymnosperms*?

14. Explain the two groups of plants that were prominent during the Triassic and Jurassic times.

15. What was the common origin of dinosaurs? What skeletal characteristics did this have?

16. The dinosaurs lived on Earth for 150 million years. What conditions were present to make it possible for them to live this long?

17. It is believed that dinosaurs are ancestors of today's birds. What is the evidence for that?

18. What is the difference between angiosperm and gymnosperm plants?

19. How could scientists estimate past temperatures by monitoring tree leaves?

20. Explain different steps of development of life from the first cells to Dinosaurs.

CHAPTER 23 REFERENCES

Hillis, D.M., D. Sadava, H.C. Heller, and M.V. Price. 2014. *Principles of Life*. New York: Freeman.

Prothero, D.R., and R.H. Dott. 2004. *Evolution of the Earth*. 8th ed. New York: McGraw-Hill.

FIGURE CREDITS

- Fig. 23.1: Copyright © Paul Harrison (CC BY-SA 3.0) at https://en.wikipedia.org/wiki/File:Stromatolites_in_Sharkbay.jpg.

- Fig. 23.2: Adapted from Copyright © Bdcarl (CC BY-SA 3.0) at https://en.wikipedia.org/wiki/File:Anabaena_circinalis.jpg.

- Fig. 23.3: Copyright © Vassil (CC BY-SA 3.0) at https://en.wikipedia.org/wiki/File:Trilobite_Ordovicien_8127.jpg.

- Fig. 23.4a: Copyright © Didier Descouens (CC BY-SA 4.0) at https://en.wikipedia.org/wiki/File:Liospiriferina_rostrata_Noir.jpg.

- Fig. 23.4b: Source: https://en.wikipedia.org/wiki/File:Haeckel_Bryozoa.jpg.

- Fig. 23.5: Copyright © Toby Hudson (CC BY-SA 3.0) at https://en.wikipedia.org/wiki/File:Coral_Outcrop_Flynn_Reef.jpg.

- Fig. 23.6: Copyright © Dlloyd (CC BY-SA 3.0) at https://en.wikipedia.org/wiki/File:Nautiloid_trilacinoceras.jpg.

- Fig. 23.7: Source: https://en.wikipedia.org/wiki/File:Cypraea_chinensis_with_partially_extended_mantle.jpg.

- Fig. 23.8: Copyright © Zsoldos Márton (CC BY-SA 3.0) at https://en.wikipedia.org/wiki/File:Eudontomyzon_mariae_Dunai_ingola.jpg.

- Fig. 23.9: Copyright © Uwe Kils (CC BY-SA 3.0) at https://en.wikipedia.org/wiki/File:Tomopteriskils.jpg.

- Fig. 23.10: Copyright © FunkMonk (CC BY-SA 3.0) at https://en.wikipedia.org/wiki/File:Diplacanthus.jpg.

- Fig. 23.11: Copyright © ArthurWeasley (CC BY-SA 3.0) at https://en.wikipedia.org/wiki/File:Guiyu_BW.jpg.

- Fig. 23.12: Copyright © Pter Halsz/Nobu Tamura/Guy Haimovitch/Wpopp/Marshal Hedin/John Kratz (CC BY-SA 3.0) at https://en.wikipedia.org/wiki/File:Arthropoda.jpg.

- Fig. 23.13a: Copyright © Nobu Tamura (CC BY-SA 3.0) at https://en.wikipedia.org/wiki/File:Eusthenopteron_BW.jpg.

- Fig. 23.13b: Copyright © Nobu Tamura (CC BY-SA 3.0) at https://commons.wikimedia.org/wiki/File:Tiktaalik_BW.jpg.

- Fig. 23.13c: Copyright © Nobu Tamura (CC BY-SA 3.0) at https://en.wikipedia.org/wiki/File:Acanthostega_BW.jpg.

- Fig. 23.14: Copyright © Froggydarb (CC BY-SA 3.0) at https://en.wikipedia.org/wiki/File:Amphibians.png.

- Fig. 23.14a: Copyright © Froggydarb (CC BY-SA 3.0) at https://commons.wikimedia.org/wiki/File:Litoria_phyllochroa.JPG.

- Fig. 23.14b: Copyright © Ryan Somma (CC BY-SA 3.0) at https://commons.wikimedia.org/wiki/File:Seymouria1.jpg.

- Fig. 23.14c: Copyright © Patrick Coin (CC BY-SA 2.5) at https://commons.wikimedia.org/wiki/File:Notophthalmus_viridescensPCCA20040816-3983A.jpg.

- Fig. 23.14d: Copyright © Franco Andreone (CC BY-SA 2.5) at https://commons.wikimedia.org/wiki/File:Dermophis_mexicanus.jpg.

- Fig. 23.15: Copyright © Spacini (CC BY-SA 3.0) at https://en.wikipedia.org/wiki/File:Xiphactinus_audax_Sternberg_Museum.jpg.

- Fig. 23.16: Copyright © Ghedoghedo (CC BY-SA 3.0) at https://en.wikipedia.org/wiki/File:Placodus_gigas_2.JPG.

- Fig. 23.17: Copyright © Kumiko (CC BY-SA 2.0) at https://en.wikipedia.org/wiki/File:Plesiosaurus_in_Japan.jpg.

- Fig. 23.18: Copyright © Nobu Tamura (CC by 2.5) at https://commons.wikimedia.org/wiki/File:Euparkeria_BW.jpg.

- Fig. 23.19: Copyright © Tadek Kurpaski (CC by 2.0) at https://en.wikipedia.org/wiki/File:Louisae.jpg.

- Fig. 23.20: Copyright © H. Raab (CC BY-SA 3.0) at https://en.wikipedia.org/wiki/File:Archaeopteryx_lithographica_(Berlin_specimen).jpg.

THE ORIGIN OF MAMMALS AND PRIMATES

CHAPTER LEARNING OBJECTIVES

This chapter will cover:

* The origin and evolution of primates
* Origin of bipedalism
* The *Homo* genus
* Evolution of Homo brain
* The origin of Homo Sapiens and modern humans
* Migration out of Africa
* Origin of modern human
* Origin of speciation and diversity
* Origin of consciousness

"We can allow satellites, planets, suns, universe, nay whole systems of universes, to be governed by laws, but the smallest insect, we wish to be created at once by special act"

- CHARLES DARWIN

"Everything that is made beautiful and fair and lovely is made for the eye of one who sees"

- RUMI

In the aftermath of the event that led to the extinction of dinosaurs 65 MYA, a new landscape opened. The mammals (from the Latin word *mamma*, meaning "breast"), defined as animals with body hair or fur who nurse their baby, give birth to alive young, have more developed brains (neocortex) and three middle ears, soon became prominent and dominated the world. This is a category of living things that ends up with humankind on the top and for this reason, study of their evolution directly relates to where we are today. The mammals were hiding under the ground for most of the time the dinosaurs ruled the Earth (and hence were not preyed by them) before going through an explosive radiation of life. Adapting to new conditions of life, they diversified, the effect of which we see today.

In studying the history of life on Earth, as we get closer to the present time, we find more detailed fossil evidence as how life was evolved. The Cenozoic time (65 MYA to present) is therefore a time with abundant information about the origin and evolution of mammals. The radiation of life that took place during the Paleocene (65–55 MYA) and Eocene (55–34 MYA) times took around 15 million years and was one of the most dramatic times for diversification. The process of the evolution of mammals started before that and by the end of the Eocene time, the form, shape, and characteristics of the living things significantly changed. The evolutionary process has been gradual and affected by many factors including the climate, movement of the continents, presence of predators, migration, and the availability of nutrition in nature to name a few. Mammals were the first creatures that started to involve in more complex tasks and learned to live together in communities.

The most important subtype of mammals, the primates (from the Latin word *primus*, meaning "the first or best of the group"), include the human genus. Within the primates group, the process of evolution and natural selection has led to significant changes in their bodies and the size of their brain, allowing them to make decisions and change their living conditions. This was an important factor in their survival over the last millions of years, in finding new ways to feed themselves, in forming communities and cultures, in protecting themselves and their communities and eventually exploiting whatever the planet could offer to them to improve their living conditions. Mammals were our distant ancestors. This is an important page in the history of life.

This chapter presents a study of the origin of mammals and primates and their evolution to the present time. It then discusses the places and conditions under which they were developed, diversified, and adapted to the existing living conditions. The chapter explores the transition from mammals to primates, the evolution in their brain size and origin of consciousness. Finally, the migration of our ancestors across the world will be studied.

THE EMERGENCE OF MAMMALS

In the previous chapter I discussed the division of land animals into two distinct groups: amphibians and amniotes. The amniotes themselves are divided into two broad groups: *synapsida* and *reptilia*. The reptilia are the lineage that eventually led to turtles, lizards, dinosaurs, and birds. The synapsida are a pre-mammalian vertebrate group that, after 300 million years of evolution led to the mammalian lineage. They dominated the world during the Permian and Triassic times and were extinguished by the Permian-Triassic mass extinction around 252 MYA, with only the mammalian line surviving to the present.

The early mammals were mouse sized, descended from the last *synapsids* (a group of animals that contains mammals or are closely related to mammals than other creatures). The fossil evidence suggests that first mammals appeared about 200 MYA as small *nocturnal* (active at night) animals, evolved from a group of synapsids during the Triassic time. For about 150 million years, during which dinosaurs ruled the Earth, they lived under the ground and hid in bushes. When dinosaurs disappeared around 65 MYA, they came out and replaced them, diversified, and soon dominated the habitat. They fed by capturing insects, as is evident from the shape of their teeth, revealed from the remaining fossils. Relative to the size of their body, they had larger brains and used it partly for more complex tasks. They evolved to give birth to young live mammals (not laying eggs).

Mammals were initially *amniotes* (animals that lay their eggs on land or retain the fertilized egg within the mother) differing from birds and reptiles by their developed brain (a neocortex part of the brain exclusively possessed by mammals). They possess hair that keeps their body temperature constant and were highly active (relative to other members of the animalia kingdom). The activity is possible because of their constant body temperature, with an efficient respiratory system separating the nasal and mouth cavities allowing them to breath while eating. They also developed four-chambered hearts, separating the oxygenated and deoxygenated blood. Female mammals have mammary glands that provide milk to feed newly born babies. The skeletons of mammals show that their limbs are located beneath their body rather than in the sides, allowing flexible movement and maneuvering of the body. Mammal teeth are also different from others in the animalia kingdom. The teeth are specialized to carry certain tasks allowing them to eat a variety of foods.

By mid-Cretaceous (about 100 MYA), mammals evolved into two (still surviving) groups: *marsupial* and *placental* mammals. The marsupials are mammals whose off-springs are born as premature embryos. They are positioned in the mother's pouch to complete their development. This group includes opossum, kangaroo, wallaby,

BOX 24.1: MAMMALS VS. PRIMATES

The mammals were evolved to primates while they diversified within their own population. The main characteristics of primates setting them apart from mammals are as follows:

- Opposable thumbs, making it easy to grasp tree limbs
- Primitive five-digit feet to allow arboreal life (life on trees)
- Stereoscopic vision that is needed for judging distances, particularly important for arboreal animals

bandicoot, and koala. The placental mammals carry the embryo in mother's womb until they are well developed and ready for birth. During the Paleocene time the placental mammals went through a huge diversification, splitting into different types that included the ancestors of almost all the land animals seen today. The mammals in these two categories are likely to have had different origins and evolutionary histories.

The majority of mammals are placental. Recent genomic studies have revealed that there was an early split in the placental mammals that coincided with the breaking of the continents during the Mesozoic time (250–65 MYA). When the continents settled at the end of Mesozoic, the mammals radiated independently in different regions and habitats. For example, when South America and North America connected around 3 MYA, the groups that had evolved independently in two separate lands, started to move between these continents. The main reason for rapid increase in the number and size of the mammals was the extinction of dinosaurs. Large groups of grazing herbivores (mammals that eat plants as food) fed from grasslands whereas browsers fed on shrubs and trees.

Some of the placental mammals finally moved to their original habitat—the aquatic environments. The completely aquatic mammals like whales and dolphins evolved from even-toed-hoofed (animals whose weight is equally divided between their third and fourth toes) ancestors. The sea lions and seals also moved to the sea with their limbs converting to flippers. The most important subgroup of the placental mammals is *primates*. This is the group that includes humans. On the land, the movement of the continents and their connection to one another affected where the mammals eventually ended up. For example, by late Cretaceous, marsupial and placental mammals started to evolve separately while both being in Americas. However, for some reason there are more marsupials in Australia than placental. As a result the marsupials evolved independently in Australia resulting in unique animals not existing in the Americas where their ancestors lived side-by-side millions of years earlier. Although they look similar to their counterparts in the northern hemisphere, there are distinct differences between the two. Fossil records show that they likely originated in the Americas and migrated to Australia via the Antarctic during the Cretaceous time when the two continents were still connected. Similarly, lack of marsupials in New Zealand indicates that the island had already separated from Australia before the marsupial's arrival.

THE ORIGIN OF PRIMATES

One lineage of the placental mammals went through extensive evolution becoming primates (Box 24.1). The fossil evidence indicates an origin for the primates as early as 55 MYA in North America and Europe when these continents were still connected. The early primates were mammal-like and *quadrupedal,* with a size smaller than a cat and legs adapted for climbing on trees (figure 24.1). The fossil remains of the oldest primates indicate that around the time they lived, the climate was warmer and more humid, resulting in an adaptive radiation of first primates. The fossils found from these early primates indicate stereoscopic vision with reduced snout and larger brains. Anthropologists believe that early primates divided into two lines: one the ancestor of today's *lemurs* and *lorises* and the other the lineage leading to *anthropoids* that include apes and monkeys (Larson 2011). I discuss this in the next section.

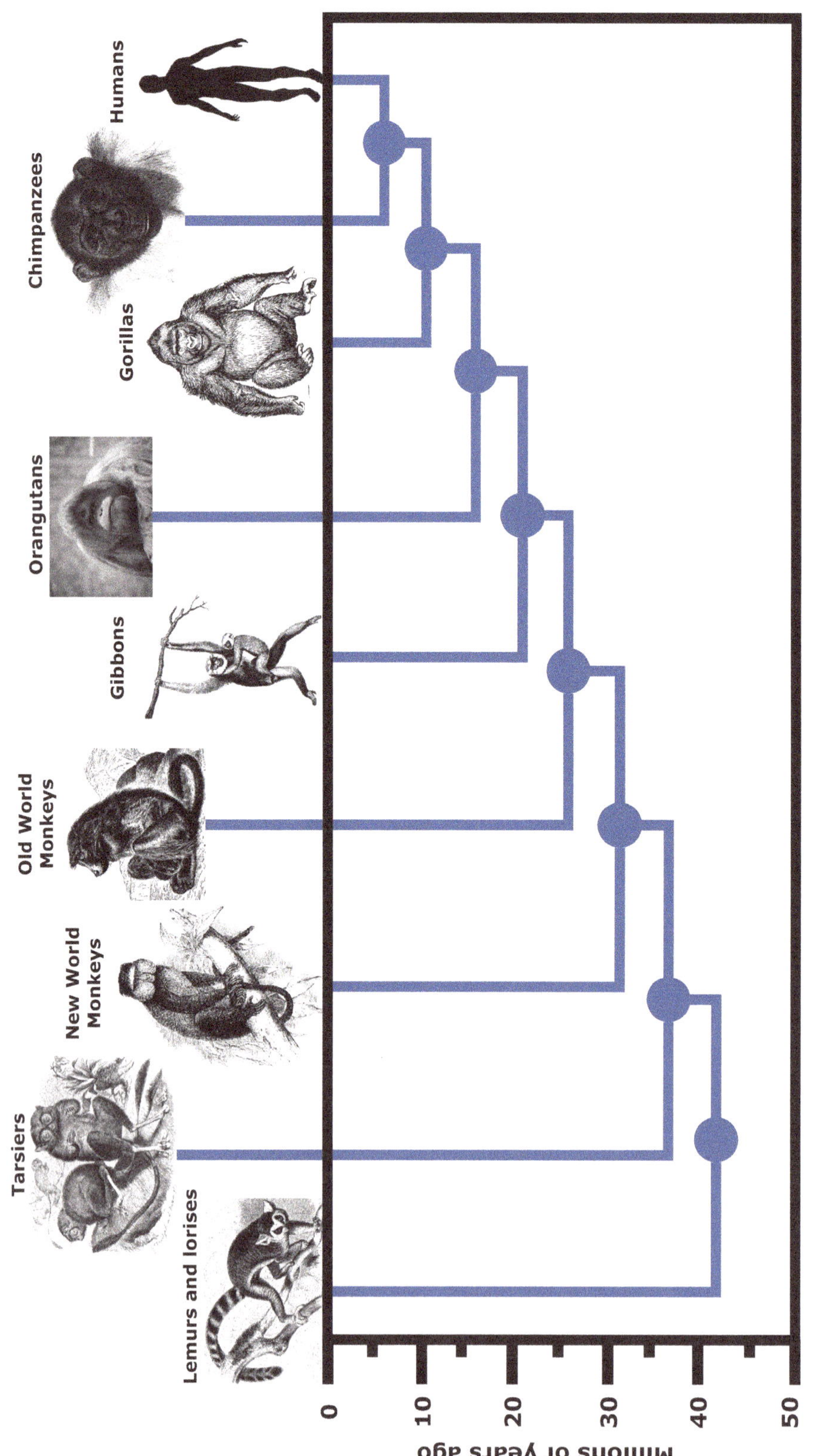

Figure 24.1. The evolutionary tree for primates, dating back to 55 MYA. The diagram shows division of primates to prosimians (lemurs and lorises) and anthropods (monkeys, apes and humans).

EVOLUTION OF PRIMATES

As I mentioned above, the earliest primates looked like shrews and squirrels with their fossils dating back to 55 MYA (Paleocene time; figure 24.1). They had grasping feet and opposable toes with a nail rather than a claw, the signatures that separate primates from mammals (Box 24.1). This feature allowed them to grasp objects more easily. During the Eocene time (55 to 33 MYA) primates were evolved into two distict groups: *prosimians* and *anthropods* (Haviland et al 2014).

Prosimians include lemurs and lorises that today live in Africa and tropical Asia. Anthropods include tarsiers, new world monkeys, old world monkeys, and apes that began to diversify in Africa after the mass extinction

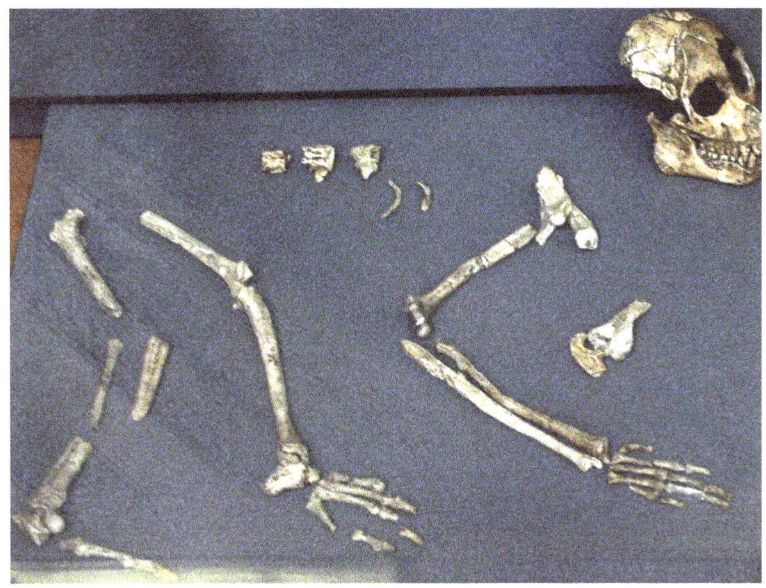

Figure 24.2. Shows an example of a Proconsul fossil dating to 14 million years ago. This indicates a transition state between monkeys and apes (from the National Museum of Natural History, Paris).

at the end of Cretaceous (figure 24.1). New old monkeys were developed from old world monkeys and apes and moved from Africa to South America when these continents were still connected, explaining the origin of the new world monkeys found today in South America. The earliest fossil evidence for the apes is found in Egypt and is from the Oligocene time (34–23 MYA). Sometime around 35 MYA the lineage that contained modern apes was separated from the old world monkeys (Figure 24.1), becoming a distinct population that eventually ended up in widely scattered areas such as Europe, Asia, and Africa between 22 and 5.5 MYA. The remaining fossils from these apes show features indicative that they were related to *hominids*. The gibbons and orangutans are descendants of these apes, separating from them around 12 to 18 MYA (for gibbons) and 12 MYA (for orangutans; Haviland et al. 2014).

One genus of particular significance is the *proconsul* that lived in Africa between 21 and 14 MYA. The proconsul fossils show a mixture of monkey and apelike features (figure 24.2). Like modern apes and humans they did not have a tail but show limb proportions similar to monkeys. Their arms and hands are monkey-like, while their shoulders and elbows are like apes. The proconsuls are transition between monkeys and apes (figure 24.2).

The primates appeared at the time when a mild climate allowed the spread of tropical forests over much of Earth. Primates evolved from arboreal (tree-living) mammals and could adapt to life in forests. The features found in primate fossils support the fact that they were indeed arboreal. The ability to live on trees opened the door to a significant source of food supply like leaves, fruits, flowers, and bird eggs as well as allowing them to escape predators. Natural selection allowed the survival of those who could, for example, judge depth accurately (when

BOX 24.2: EVOLUTION OF BRAIN SIZE

Fossil records show that the brain size has increased from the smallest in the oldest hominid (350 cm³) for chimpanzee to the largest in the Homo sapiens (1,450 cm³) for humans.

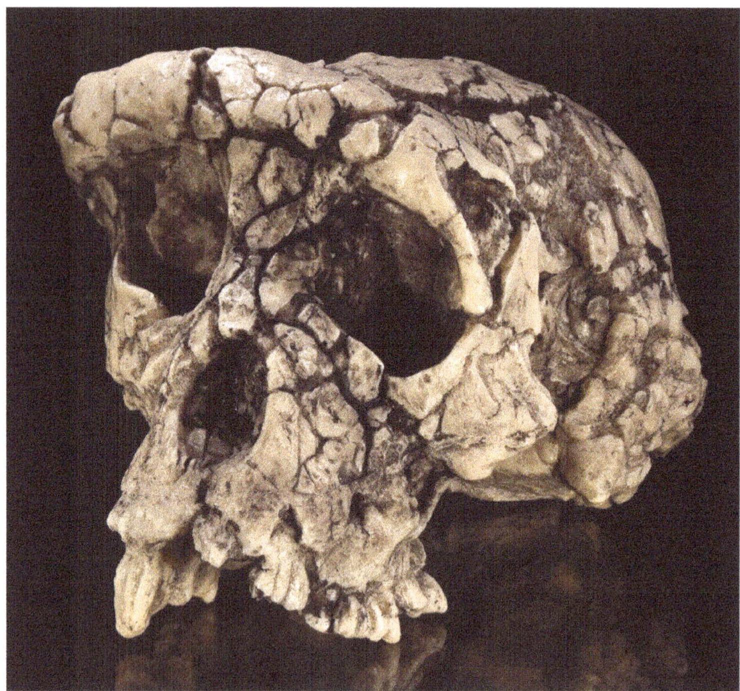

Figure 24.3. The cranium fossil of a *Sahelanthropus tchadensis*. This dates back 7 MYA and is close to the time of human-chimpanzee lines diverged. The fossil was found in western Africa.

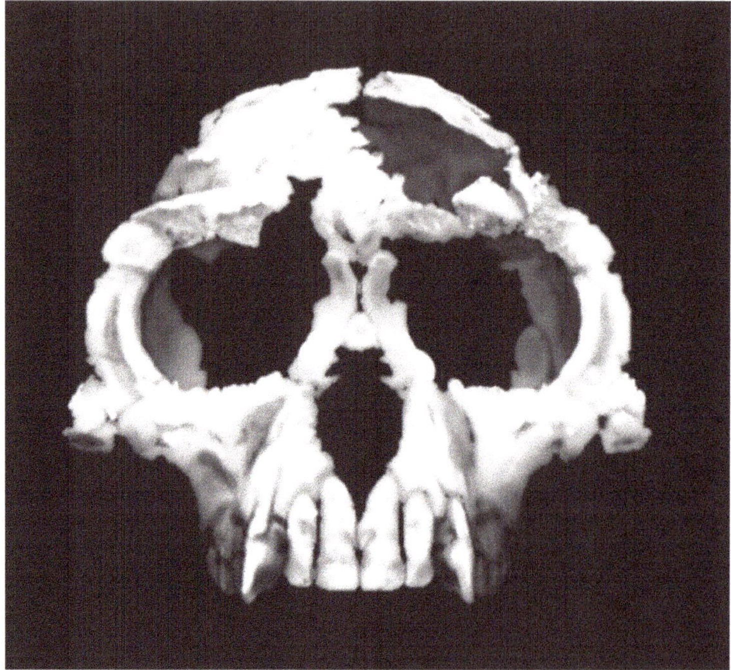

Figure 24.4. The fossil of an *Ardipithecus ramidus* who lived 4.4 MYA. This is considered to be one of the earliest fossils showing chimpanzee-like characteristics with bipedal mobility.

jumping between tree branches), and those who survived passed their traits to their off-springs. At some point in their life, they transformed to *diurnal* (active in the day) animals (Jurmain et al. 2013).

There is fossil evidence that the apes and human lineages separated around 7 MYA. These studies reveal the first recognizable ancestor of the lineage leading to humans (figure 24.1). All the fossils are found in central and eastern Africa implying that the first hominids started and evolved there. Evidence from cranial capacity (a measure of the brain size) is used to study the closeness of the hominine to the human (Box 24.2).

Molecular evidence confirms that around 6 MYA first the gorillas and then the chimpanzee split off from the line that eventually led to humans (figure 24.1). The closeness of human and chimpanzee families is supported by the study of their DNA, showing 98 percent similarity. The earliest fossils belonging to human lineage was found in Chad, western Africa, called *Sahelanthropus tchadensis* (named after "Sahel" the place it was found in the southern Sahara desert). It is 7 million years old and has chimpanzee-like features such as large brow ridges and small brain as well as human-like features like flattened face and enlarged cheeks (figure 24.3). An important feature of *S. tchadensis* was that the place where the spinal cord exits the braincase is directly below the skull, showing that they had upright figures. This provided the earliest evidence and the time when the ape-human split took place and the first hominid fossils (Box 24.3). In the same area, a number of other hominid fossils were found (Park 2013).

The next youngest fossil after *S. tchadensis* was discovered in the Tugen Hills of Kenya, dated between 5.72 and 5.88 million years ago. It is called *Orrorin tugenensis*. They have limb anatomy indicating bipedal posture but large and pointed teeth with the arms and finger bones indicating a tree-living animal, likely ancestral to the chimpanzee today (Larson 2011).

BOX 24.3: PRIMATE CLASSIFICATION

Study of the past history of organisms and their relation to one another is known as the **evolutionary tree**. The lines of descent from one common ancestor is called **lineage**. New genetic and biological analysis has been used to classify the primates as prosimians (including lemurs, tarsiers, and lorises) and anthropoids (monkeys, apes, and humans). They are also sub classification of these groups. The **hominid** includes apes (gorillas and orangutans), chimpanzees, and humans. The **hominine** only includes gorillas, chimpanzees, and humans. The **hominin** includes all members of the genus *Homo* and their close relatives.

The two fossils discovered in Ethiopia—named *Ardipithecus ramidus* and *Ardipithecus ramidus kaddaba* (in the local Afar language, *ardi* means "ground" and *ramid* means "root")—are from 4.4 MYA and 5.8 MYA respectively, with both apelike and human-like features (figure 24.4). The ardipithecus's big toe is opposable, more like an ape and unlike human. However, their foot has a rigid structure and does not show the flexibility to grasp the tree limbs or moving on the trees but rather appropriate for moving bipedal. These observations confirm that the ardipithecus were adapted to life both on trees and on the ground. Their bipedal ability was finally transferred to the hominids. These confirm that ancestors to modern humans were bipedal and "upright" walkers as early as 5 million years ago (Larson 2011; Park 2013).

The fossils from the first member of our ancestral genus, *Australopithecine*, were originally found in Africa and were 3.9 to 4.2 million years old. This was the start of the hominin line (Box 24.3). Some australopithecines were small and slender, called *gracile*, and some were robust and powerful, called *robustus*. The youngest fossil of this genus was found near lake Turkana in Kenya and is called *Australopithecus anamensis* (*anam* meaning "lake" in the Turkana language). The most complete fossil skeleton was discovered in Ethiopia, of a young female from 3.5 MYA and was of gracile type. This is assigned to the species *Australopithecus afarensis* and is known as "Lucy" (after the song "Lucy in the Sky with Diamonds"; figure 24.5). The *A. afarensis* had brain sizes 500 to 600 cm³ with their hands and feet showing clear bipedal features (Haviland et al. 2014; Larson 2011).

The two distinct groups of hominids are found in a few locations in south and east Africa and provide the link to the *Homo* genus. *Australopithecus gracile* had small cheek teeth and centrally located spinal opening (and hence

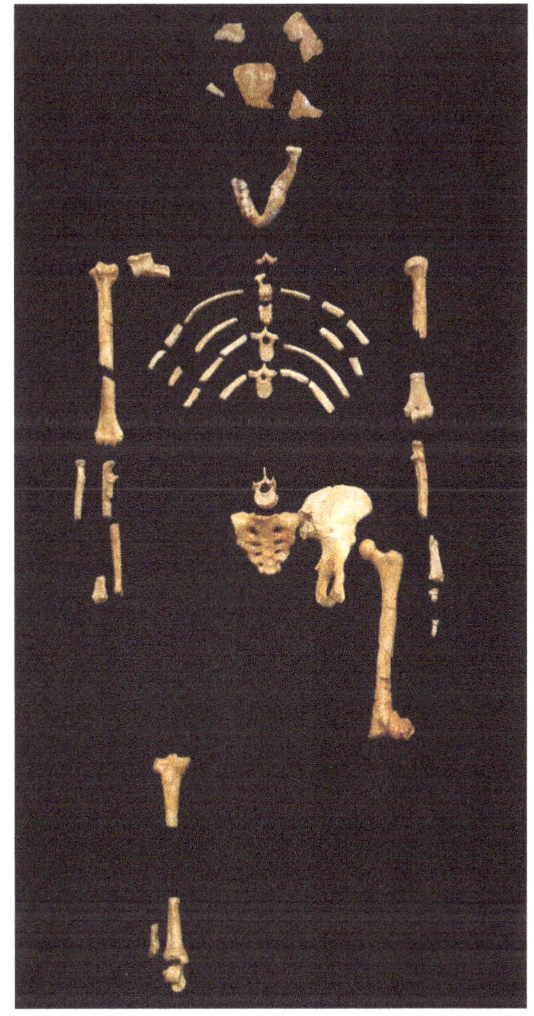

Figure 24.5. A nearly complete skeleton of *Australopithecus afarensis*, also known as Lucy. The fossil is of a female from 3.5 MYA and was discovered in Ethiopia.

being walkers) with a brain size of 450 cm³ dated 3.0 to 1.5 MYA. *Australopithecus robustus* had large teeth and massive jaws and a skull size of 530 cm³, dated 1.9 MYA. One of the most robust of its kind was *A. boisei* with an age of 1.2 to 2.2 million years. The early hominids of different groups lived side-by-side about 2 MYA and evolved to the *Homo* genus, of which we are a part.

THE ORIGIN OF BIPEDALISM

One of the earliest traits first hominins acquired was the ability (possibly because of the necessity) to walk with two legs. Fossil evidence of the hominins indicates that bipedalism preceded the growth in brain size or changes in their face or teeth. Bipedalism dates back to about 4.2 MYA compared to the expansion of the brain that started around 2.5 to 2 MYA.

To find about the origin of bipedalism, we need to first explore how our ancestors moved. There is convincing evidence that our common ancestors were knuckle-walkers like the apes. They then evolved in two different lineages—one line continued to knuckle-walk and this is the lineage that led to the apes and the other line evolved to move on two legs and that led to the hominins. The environmental factors were also important when studying the origin of bipedalism. The forests started to disappear because of the change in climate, resulting in more grasslands and savannas. When coming down the trees and living on flat lands, early hominins had to look out for predators and to hunt because an up-right hominin could see further, allowing an increase in the depth of the vision. By providing these characteristics, bipedalism allowed early hominins to survive. Furthermore, by becoming bipedal, the early hominins freed their hands to carry tools or infants, increasing the survival rate and population growth.

The origin of bipedalism is also related to more efficient movement and long-distance travels as well as an energy efficient way to carry out tasks. As the forests shrank and grasslands grew, the early hominins had to travel longer distances in search of food while looking out for predators. Bipedalism therefore evolved because of natural selection. It seems that there was no single reason why hominins became bipedal but a combination of different needs.

THE RISE OF THE *HOMO* GENUS

Apart from the australopithecines, there is another largely contemporaneous hominin that is even more closely related to us. Their fossils are found in east Africa and are assigned to the genus *Homo* (Box 24.4). The earliest appearance of genus *Homo*, found in 1960s, dates before 2 MYA and found in the Hadar area in Ethiopia. These had significantly larger brains than the australopithecines and were given a new classification, *Homo habilis*, meaning "handy man." The name comes from the fact that their fossils were found along with tools used for hunting. These comprise the early *Homo* fossils found in Olduvai and near lake Turkana in Tanzania. There is good evidence that the oldest *Homo* found in East Turkana, dated back to about 1.44 MYA, coexisted with other *Homo* species for several hundred years in exactly the same area (Larson 2011; Haviland et al. 2014).

The *Homo* genus has a number of characteristics separating them from the australopiths. These include: (1) a larger cranial size, with the early *Homo* (such as *H. habilis*) having 45 percent larger skulls (775 cm³) compared to australopiths (442 cm³); (2) the early *Homo* genus had smaller, more human-like, cheek teeth; and (3) the shape and anatomy of their skulls is different from other hominin.

An entirely different kind of fossil was discovered in East Africa, classified as *Homo erectus*, with an age dating back to 1.7 MYA (figure 24.6). These show robust body structure and an increase in body size compared to previous species, with a weight of 100 pounds and height of over 5 feet. They have a cranial capacity (directly proportional to the brain size) of 700 to 1250 cm^3 (compared to 500 cm^3 for early *Homo* species) (Box 24.4). Due to the increased brain and body size, they show a distinct cranial shape and large brow ridges above the eyes. These characteristics indicate that the *H. erectus* are closer to modern humans than their ancestors and are the result of a major adaptive shift in hominin evolution toward a more human direction (figure 24.6). It is becoming clear that some type of early *Homo* was evolved to *H. erectus* in East Africa around 2.0 to 1.8 MYA. They are the first fossils that were also found out of Africa and as far away as eastern Asia and China. There were larger differences in the fossil characteristics among the *H. erectus* in Africa and in Europe and Asia. The population in Europe appeared to be more similar to the original population in Africa than the one in Asia. As a result researchers classified the population in Asia as *H. erectus* and the one in Africa as *H. ergaster*. The *H. erectus* evolved nearly 2 MYA in Africa, from an earlier species, *H. ergaster*. By 1.8 MYA it had expanded as far away as to Indonesia and by 500,000 years ago to northern China and Europe. They lived, as a known species in China and Africa until 250,000 years ago. They were the first species to build tools and showed interest for learning and culture.

The shift from *H. erectus* to the first premodern humans took place over 780,000 years ago. To study the origin of modern human, scientists look for fossil remnants that show similar features as modern humans. The earliest premodern humans identified, share some basic characteristic with *H. erectus*, including large face, low forehead and projected brows. They had larger brains, more rounded braincases and more vertical nose. The premodern humans emerged around 850,000 years ago and extended to about 200,000 years ago and are categorized in the *H. heidelbergensis* group (named after a fossil found in Germany in 1907). This is a transitional species between *H. erectus* and the *H. sapiens* that contains the human species. In Africa, the *H. heidelbergensis* were evolved to modern *H. sapiens* while in Europe, *H. heidelbergensis* evolved to *Neanderthals* (named after the Neanderthal Valley in Germany).

Neanderthals lived in Europe and west Asia for about 100,000 years. They were heavily built with brain sizes (1,520 cm^3)

Figure 24.6. The skull fossil of a *H. erectus*, likely to be ancestors of the premodern human from 780,000 MYA.

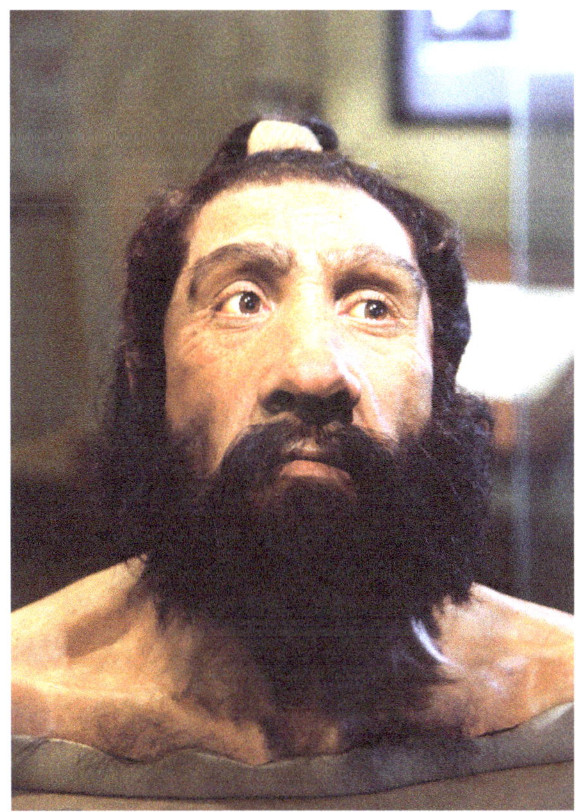

Figure 24.7. The reconstructed image of a Neanderthal man.

even bigger than the modern humans—*H. sapiens* (1,400 cm³; figure 24.7). They developed technical tools for hunting, language, and culture. DNA sequencing of three female *Neanderthals* from 38,000 to 44,000 years ago when compared with modern human has revealed that they share 99.84 percent of modern human's genome. Study of the genetic differences has also shown that *Neanderthals* and modern humans diverged from a common ancestor between 270,000 to 440,000 years ago. These dates combined with the geographic distribution points to the fact that *H. heidelbergensis* were the common ancestor. There was also some interbreeding between *Neanderthals* and modern humans, with 1 percent to 4 percent of the ancestry of modern humans out of Africa coming from *Neanderthals*.

EVOLUTION OF THE HOMO BRAIN

It is clear from the fossils and the size of the skulls found by the archaeologists that the *Homo* genus brain size has increased by a factor of three between 2.5 MYA (when the first evidence for *Homo* genus was discovered) to 200,000 years ago when it reached the size of modern humans (figure 24.8). Starting from the plant-eating Australopithecus with a cranial capacity of 310 to 530 cubic cm, the brain size increased to the range 580 to 752 cubic cm for the *Homo habilis,* the earliest known meat eaters in East Africa, to a cranial capacity of 775 to 1,225 cubic cm for *Homo erectus*, who acquired their food by hunting (figure 24.8).

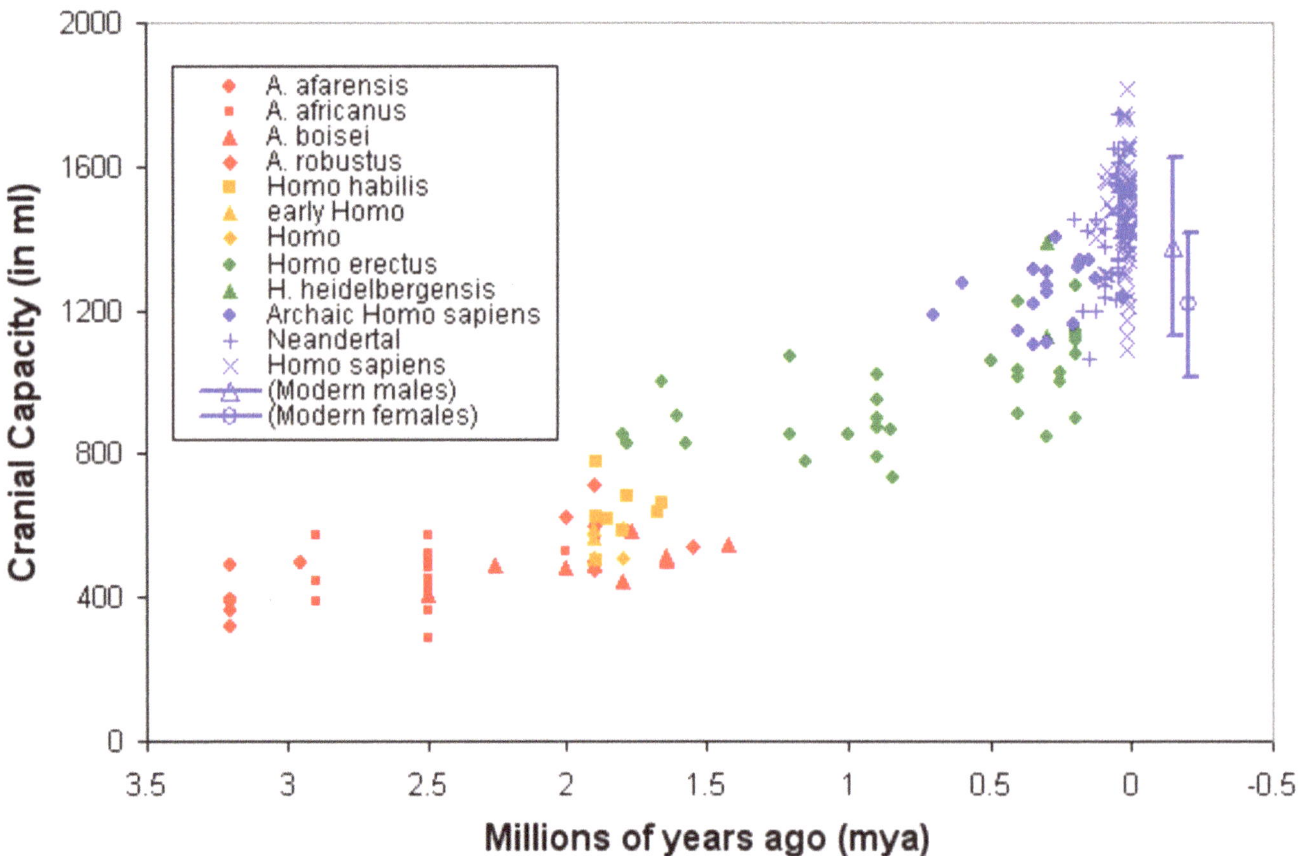

Figure 24.8. All measurements of the hominids cranial capacity or skulls. Only adult specimens are used (de Miguel and Henneberg 2001).

The nerve tissues from which brains are made require more energy to function compared to tissues in other parts of the body. Larger the size of the brain, higher is the energy requirement. For example, for a modern human up to 25 percent of the energy of the body is consumed by the brain. This demands special diet to produce the energy of the brain. Plants could provide some of that energy but it could most efficiently be provided by meat. This is the reason the increase in the brain size of the *Homo* genus went hand-in-hand with their ability to consume meat as nutrition. Furthermore, archaeological evidence as far back as 2.5 MYA indicates a direct relationship between cultural ability (like toolmaking) and increase in brain size.

ORIGIN OF CONSCIOUSNESS

Consciousness is the degree with which our neurons respond to input from the senses. It emerges from neural circuits with greater degrees of consciousness associated with increasingly complex circuitry. Unlike feeling and emotion that can be associated with specific regions of the brain, consciousness depends on many areas working together. This is the reason study of circuitry connections are needed to explain consciousness. This is a combination of sensory perception, memory, learning and language. A lot of work is still needed to understand the phenomena of consciousness and its origin however, it is clear that consciousness is a process and not a single thing. Also, it is not a single process but a combination of several processes such as those associated with seeing, touching, thinking, emotion and language.

What give the nervous system its learning and memory ability are the synapses. These are the junctions where nerves exchange electrical signals through biochemical switches (neurotransmitters) (figure 22.9). A small number of proteins in single celled organisms form the ancient synapses that are responsible for simple behavior. By the evolution of single-celled organisms to invertebrate and vertebrate animals the number of these proteins increased, leading to more complex behavior for animals at higher levels of consciousness (like birds and mammals). The number and complexity of proteins in the synapse started to increase around three billion years ago at the time when multicellular organisms emerged. The next rapid increase took place when vertebrates appeared around 500 million years ago.

During this evolutionary process of protein generation, a number of proteins appeared that made different parts of the brain specialized to specific tasks such as the cortex, cerebellum and spinal cord. The presence of big synapses and their subsequent evolution and growth of complexity may have been responsible for the emergence of large brains found in primates, mammals and vertebrates.

THE ORIGIN OF MODERN HUMAN

The origin of modern human can be investigated within the context of the following two questions:

1. Where and when did the first modern human appear?
2. How did the transition to modern human take place?

Before addressing the above questions one needs to clarify what "modern human" means in this context. Anthropologists define *modern* based on a number of distinctive anatomical features different from the earlier hominids. Modern humans are defined to have a high, vertical forehead, a round skull, small teeth and small brow ridges. Fossils with these characteristics are considered to be analogous to modern humans—*Homo sapiens*. There are distinct differences between the fossil remnants for archaic (early) *H. sapiens* and modern *H. sapiens* from which one could follow their evolution. The earliest *H. sapiens* appeared around 350,000 years ago in Africa, Asia, and Europe.

There is evidence from the fossil remains that archaic *H. sapiens* and *H. erectus* had a number of common features including massive brow ridges, small teeth and brain size as well as cultural complexity. This confirms that early *H. sapiens* evolved from the *H. erectus*. The archaic *H. sapiens* lived in Africa until 200,000 years ago and in Asia and Europe until 130,000 years ago. The hominids found around this time showed a continuous increase in brain size, reduced tooth size and less robust skeletal features. The shape and morphology of the *H. sapiens* found in western Asia (Middle East) and Europe show distinct differences from those in Africa, indicating both regional variations and adaptation to a colder climate. These new features define the characteristics of *Neanderthals* that include projecting face, long and low skull, large front teeth, wide body, and short limbs. The earliest evidence for *Neanderthals* were found in Amud, Kabara, and Tabun in Israel and date back to 55,000–40,000, and 60,000 years ago for those found in Amud and Kabara respectively. The skulls found in Tabun are recently estimated to date back as far as 170,000 years ago. The *Neanderthals* found in Europe lived between 130,000 and 32,000 years ago (Larsen 2011; Park 2013).

How did the early *Neanderthals* evolve to become modern human? The earliest *H. sapiens* skulls that resemble modern human were found in Ethiopia's Middle Awash Valley in Eastern Africa. These dated back to 160,000 to 154,000 years ago and show large cranium, a vertical forehead and small brow ridges. These findings suggest that modern human first emerged in Africa before moving to Europe and Asia. The earliest modern *H. sapiens* in Europe are found in Pestera Cu Oase in Romania and date back to 35,000 years ago, with similar remains resembling modern human found in Czech Republic dating to 35,000 to 26,000 years ago. Fossils are found in other parts of Europe also confirm these dates. There is evidence that the first cultural activities for modern *H. sapiens* were in Africa.

What happened to Neanderthals? Were they eliminated by the first Homo Sapiens that arrived in Europe? This is a population known as *Cro-Magnon*, named after the site in France where their first specimens were discovered. Did *Neanderthals* interbreed with the Cro-Magnons and if so, are we modern descendants of the Neanderthals? Reliable clues concerning the fate of Neanderthals and the origin of modern humans can be acquired from the study of mitochondrial DNA (mtDNA; Box 24.5). Comparison of mtDNA between remnant fossils from Neanderthals with early and modern humankind show no similarities indicating that no gene flows appeared between them. In this case, *Neanderthals* disappeared with no evidence of interbreeding between them and early modern humans. More recently, however, doing detailed DNA analysis geneticist constructed detailed genome of *Neanderthals* and discovered that between 1 percent and 4 percent of the nuclear DNA are in common between Eurasians and *Neanderthals* while there is no common DNA between *Neanderthals* and Africans. This strongly suggests that some genetic flow between these two populations took place when *H. sapiens* moved out of Africa (Larson 2011), with little interbreeding between Neanderthals and Eurasians.

IN SEARCH OF OUR GRANDMOTHERS

Using molecular biology techniques, biologists studied changes in mitochondria DNAs (mtDNA) among different populations in multiple locations worldwide (Box 24.5). mtDNA is only inherited by maternal lineage and therefore, is not affected by recombination (mixing of parent's DNAs) and can only be changed by mutation. The mtDNA passed from mother to daughter and down the line is affected by mutation. The populations that have been around the longest have had their mtDNAs undergo more changes (mutation) and are therefore more diverse. By looking at changes in mtDNAs among different populations, it has been found that the variation in mtDNA among the modern Africans is the most. This means that they have been around the longest and, in other words, we all originated in Africa.

By studying the rate at which mutation occurred in mtDNA, biologists were able to estimate the elapsed time since the common ancestor of modern humans lived. Applying this technique to data from Africa, anthropologists found that the common origin of all humankind goes back to 200,000 years ago in Africa. Given that this is based on

It is possible to use techniques in molecular biology to determine the time of the split between humans and apes as well as the evolution of the human genus. Inside the nucleus of every cell, there are DNA sequences that encode all the proteins in our body. There are two copies of nuclear DNA and they are packaged as chromosomes (with 23 chromosomes existing in each cell nucleus). There are also DNAs in other parts of a cell, like mitochondria. Each cell has many mitochondria with each containing many DNAs (mtDNA); the mitochondria have their own genome because they are evolutionary remnants of the first complex cells, i.e., a bacterium that a single-cell organism swallowed. Therefore, mtDNAs are much more abundant than nuclear DNAs and can more easily be extracted and examined. The most important property of mtDNA however, is that it is only produced by our maternal side through the egg she produces (sperms do not carry mtDNA). The implication of this is that mtDNAs are not affected by genetic recombination (mixing of the father and mother DNAs resulting in new DNAs). Therefore, unlike nuclear DNAs that are affected by both recombination and mutation, the mtDNAs are *only* affected by mutation. This is because mtDNA only exists in one copy, which means that they could not recombine. This property of mtDNA is used to study the timeline of evolution. Therefore, by studying differences in the mtDNA sequences between people from different parts of the world, we can investigate if they shared a common ancestor.

Populations that show the largest difference among their mtDNAs have been around for the longest. This is because they have coexisted for a long enough time to allow more mutations and therefore, more changes in the mtDNAs from mothers to daughters and so on. This has led to a larger variation of mtDNAs in their modern descendants.

Performing the mtDNA analysis in various people around the world, it was found that the most divergent mtDNAs were among the Africans. This means that Africans are the oldest group in the world, and by extension, our species originated from Africa.

mtDNA that is carried by our mothers, this must be the great, great ... grandmother of all human beings alive today. Of course, this is not one individual person. The transfer of mtDNAs only takes place through our maternal lineage, restricting its distribution down the line (e.g., the transfer stops at a family if they only have male members). Also, the transfer from ancestors to descendants stops if people do not survive long enough to transfer the gene.

EARLY MIGRATION FROM AFRICA

All the early *Homo* fossils prior and including the *H. erectus* were found in Africa. The fossils dated back to 1.7 MYA and classified as *H. erectus* were found in East Turkana, Kenya, and other sites in East Africa. The earliest of the *H. erectus* fossils come from the same area where the *australopith* and earlier *Homo* fossils were found. It is now confirmed that a similar population of *H. erectus* species were living around 1.8 MYA in southeast Europe and around 1.6 MYA in Indonesia. Based on these findings, it is logical to develop a hypothesis in that the *H. erectus* first originated in East Africa and then rapidly migrated to other continents (figure 24.9). Indeed all the sites where the *H. erectus* were found out of Africa show dates later than the ones inside Africa. All these travelers who left Africa have characteristics similar to the *Homo* genus. Also, since they were living in East Africa, they were close to the Middle East and likely they went through that route to Europe and East Asia. The time took the *H. erectus* to travel from East Africa to Southeast Asia is around 200,000 years as suggested by dating of the fossils and comparing their characteristics at different geographical locations (figure 24.9).

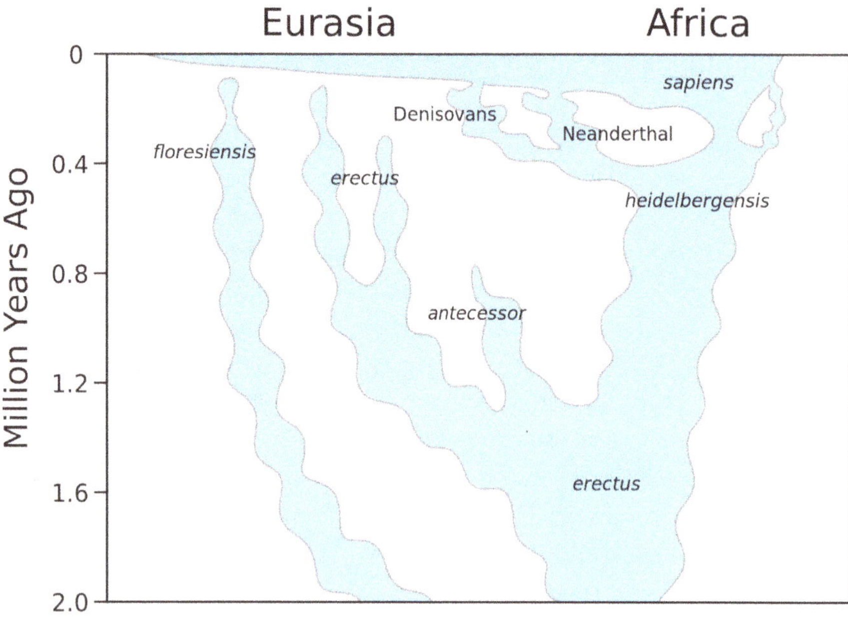

Figure 24.9. Human tree evolution showing the place and time each species dominated.

The species *Homo neanderthalensis* was widespread in Europe and Asia between 500,000 to 28,000 years ago. *Neanderthals* were short and strongly built with a large brain size. Early modern humans (*H. sapiens*) expanded out of Africa around 60,000 to 70,000 years ago and around 35,000 years ago, *H. sapiens* co-lived with the *Neanderthals* in Europe and western Asia (figure 24.9). *Neanderthals* disappeared around 28,000 years ago, likely by the early humans. As also mentioned in the last section, study of the genetic similarities between the *Neanderthals* and us indicates that there was some degrees of interbreeding between the two. In particular in modern humans with European ancestry, about 1 to 4 percent of the genes in their genomes are likely derived from *Neanderthal* ancestors (Larsen 2011; Jurmain et al. 2013).

For the modern humans, there are a few models explaining their movement out of Africa. The *African replacement model* states that *H. sapiens* started in Africa as a new species, splitting from *H. heidelbergensis* about 200,000 years ago. Some of the members of this population then started to leave Africa around 100,000 years ago and spread throughout the Old World (figure 24.9). In this scenario, any human outside Africa became extinct and therefore, are not part of the ancestry of modern humans. A competing scenario is the *assimilation model* implying that the initial modern human developed in Africa but these then spread to other populations outside Africa through gene flow. The genes of non-African population were then assimilated into the genes of the modern human population rather than being replaced. In the African replacement model the ancestors of modern humans lived in Africa about 200,000 years ago while in the assimilation model not all the population lived in Africa (Relethford 2013; Jurmain et al. 2013).

SPECIATION AND DIVERSITY

A group of organisms that can mate with one another and reproduce offspring is called *species*. The divergence of biological lineages and the emergence of reproductive isolation between these lineages are called *speciation*. This is at the heart of the diversity observed in nature. A lineage is an ancestor-descendant series of populations followed over time. Each species starts with a speciation in which one lineage is split into two and ends either with extinction or with another speciation event, at which time the species produces two descendant species (Relethford 2013).

The speciation could happen as a result of geographical separation when a population is divided by physical barrier. This is called *allopatric speciation* (*allos* means "other" and *patria* means "homeland"). The physical barrier separating the species could be water or mountain range or dry land for aquatic organisms. The initial populations divided by these barriers are large and once divided, sometime (and not always) evolve differently through mutation, genetic drift, and adaptation resulting different sister species that breed among themselves, producing a different population from where it started. Allopatric speciation also happens when members of one population cross an existing barrier to establish a new isolated population.

Speciation without physical barrier is called *sympatric speciation* (*sym* meaning "together" in Greek). This occurs when some form of disruptive selection takes place when individuals with different characteristics have a preference for a certain habitat where the mating takes place. For example, a form of sympatric speciation results from duplication of a set of chromosomes within individuals—either from chromosome duplication within an individual or the combining of chromosomes of two different species.

The diversity also happens through *mutation*, that is a change in DNA (Chapter 23). There are many types of mutation but the most widely used is point mutation that is the substitution of one DNA base for another. To have an evolutionary consequence, point mutation must occur in sex cells since mutation must be passed from one generation to the next. For any given trait, mutation is low and it cannot be seen in a small population. Mutation has significant effects when combined with natural selection. The only way to produce a new gene is through the mutation (Relethford 2013).

It is known that people who live and interbreed in a single place for a long time will have the highest genetic diversity. This is because they have had sufficient time for many mutations to produce the genetic diversity observed today. Such genetic diversity has been observed in Africa more than anywhere else. This, combined with DNA sequencing confirm that the origin of the modern human, *H. sapiens*, going back to two hundred thousand years ago, was in Africa, very likely from *H. heidelbergensis*.

SUMMARY AND OUTSTANDING QUESTIONS

The mammals lived under the ground for 150 million years to protect themselves from dinosaurs. They went through a major radiation of life and rapidly diversified soon after dinosaurs disappeared. The evolution of mammals was directly affected by the geographical separation and climate. When continents were connected, the mammals freely roamed between the lands and when climate became colder (in North America and Europe), they moved to warmer regions (Africa). This explains the distribution and the origin of mammals in different parts of the world.

One lineage of mammals led to primates around 55 MYA. Early primates had mammalian features as well as some primate characteristics (opposable toes and grasping feet). Primates were divided into two groups: prosimians (lemurs and lorises) and anthropods (tarsiers, old world monkeys, and new world monkeys). The line eventually leading to apes separated from anthropods about 23 MYA. The transition from monkeys to apes took place between 21 and 14 MYA and this is evident in the fossils of proconsuls showing combined monkey and apelike features.

Our own ancestors branched out from the apes around 6 MYA. The oldest species of primitive hominins were classified in the genus Australopithecus. These were bipedal with ape-size brains and stereoscopic vision. One species of Australopithecus evolved into the *Homo* genus sometime between 2 and 2.5 MYA. The earliest species that closely resembled modern anatomy was *Homo erectus* that first appeared in Africa around 2 MYA. They were bipedal with a larger brain size than previous species. *H. erectus* were the first hominin that moved out of Africa and into Europe and Asia. They discovered fire, hunted, and built tools.

The transitional species between the *H. erectus* and *H. sapiens* is the *H. heidelbergensis* that appeared around 800,000 years ago, occupying parts of Africa, Asia, and Europe. They had large brains (almost the same size as modern humans), made stone tools and hunted. In Africa they evolved to *H. sapiens* (around 200,000 years ago) while in Europe they evolved to *Neanderthals* (around 150,000 years ago).

Homo neanderthals lived in Europe and South East Asia from 225,000 to 28,000 years ago. This group had features between *H. heidelbergensis* and *H. sapiens*. There is evidence that *Neanderthals* formed communities and developed culture. The *H. sapien* fossils, first found in Africa, lived there around 200,000 years ago before spreading to other parts of the world over the next 100,000 years. The modern human dispersed, geographically reaching Australia around 60,000 years ago and the New World around 20,000 to 15,000 years ago. There is evidence that the *H. sapiens* interbred with *Neanderthals* as indicated from the common genes they have.

With the invention of agriculture and new hunting methods, modern human population rapidly grew and evolved to the present time. The increase in the brain size led to invention of new hunting methods by premodern and modern human. This allowed more meat consumption, providing the energy needed for the brain to function.

It has been hypothesized that *H. Heidelbergensis* are the common origin of the *Neanderthals* and *H. sapiens*. The size of hominids brain increased by a factor of four over the last 3 million years. However, not much is known about the *H. Heidelbergensis* themselves. The fossil evidence about their structural characteristics is sparse. More fossil samples and research is needed to confirm the role of *H. heidelbergensis* in the evolution of *Homo Genus*. It is now well accepted that the first humans started life in Africa. From there our distant ancestors migrated and conquered the world. It is not clear however, why all this started in Africa. Did climate play a role, with early premodern humans escaping the cold climate in search of warmer climate in Africa? The are two competing scenarios regarding the replacement of early Homo Genus. The African replacement model where the ancestors of modern humans lived in Africa around 200,000 years ago, and the assimilation model proposing that not all the population lived in Africa. Another outstanding question is, how has the variation in *H. sapien* fossils been interpreted? And what are the key evolutionary trends in the early *Homo Genus*? With new techniques based on mtDNAs and data analysis, one could now study the variation of mtDNAs in different geographical locations. This could also be done in temporal coordinates to locate the places where the maximum diversity exists as well as the time when mtDNAs diverged.

REVIEW QUESTIONS

1. When did the first mammals appear, and with what group they were most closely associated?
2. Describe the two categories of mammals.
3. What are the main reasons for the wide distribution of mammals in the world?
4. Explain the differences between mammals and primates.
5. What were the first primates like? And what were their main characteristics?
6. Explain classification of primates and the timeline for the appearance of different classes of primates.
7. What are the proconsuls, and what is their significance?
8. How did the ability to live on trees save the first primates?
9. At what time did the ape and human lineages diverge? What were the animals that first separated from the line that eventually led to humans?
10. What was the earliest fossil belonging to human lineage? How far back does it extend? What feature in those fossils indicated the human had an upright figure?
11. Explain the origin of bipedalism. When is the earliest evidence for bipedal movement of *Homo genus*?
12. Where was the first fossil from the genus Homo found? What is its name, and how long ago did it live?
13. What were the first fossils found "out of Africa"?
14. Who were the Neanderthals? How are they related to modern humans?
15. The size of the human brain has increased over the past 2.5 million years, where the first evidence for the genus Homo was discovered. Could brain size increase indefinitely into the future? Explain the reasons.

16. What is the definition of speciation?

17. Explain the evidence that the genus Homo first lived in Africa.

18. Discuss and compare the two prevailing models describing migration of the early humans out of Africa.

19. How mtDNA helps to study the variation in genomes?

20. Explain different ways diversity takes place in nature.

CHAPTER 24 REFERENCES

de Miguel, C., and M. Henneberg. 2001. "Variations in Hominids Brain Size: How Much Is Due to Method?" *Homo* 52 (1): 3–58.

Haviland, W.A., D. Walrath, H.E.L. Prins, and B. McBride. 2014. *Evolution and Prehistory: The Human Challenge*. 10th ed. Belmont, CA: Wadsworth.

Jurmain, R., L. Kilgore, W. Trevathan, and R.L. Ciochon. 2013. *Inroduction to Physical Anthropology*. Boston: Cengage Learning.

Larsen, C.L. 2011. *Our Origins: Discovering Physical Anthropology*. 2nd ed. New York: Norton.

Park, M.A. 2013. *Biological Anthropology*. 7th ed. New York: McGraw-Hill.

Relethford, J.H. 2013. *The Human Species: An Introduction to Biological Anthropology*. 9th ed. New York: McGraw-Hill.

FIGURE CREDITS

- Fig. 24.1a: Source: https://commons.wikimedia.org/wiki/File:Lemur_catta_-_Brehms.png.

- Fig. 24.1b: Source: https://commons.wikimedia.org/wiki/File:Koboldmaki-drawing.jpg.

- Fig. 24.1c: Source: https://commons.wikimedia.org/wiki/File:An_introduction_to_the_study_of_mammals_living_and_extinct_(1891)_(20544138118).jpg.

- Fig. 24.1d: Source: https://www.flickr.com/photos/internetarchivebookimages/20585753528.

- Fig. 24.1e: Copyright © Matt Biddulph (CC BY-SA 2.0) at https://www.flickr.com/photos/mbiddulph/4353478083.

- Fig. 24.1f: Source: https://pixabay.com/en/ape-wild-sitting-mammal-hairy-47790/.

- Fig. 24.1g: Source: https://commons.wikimedia.org/wiki/File:Schimpanse-drawing.jpg.

- Fig. 24.1h: Source: https://pixabay.com/en/boy-human-male-man-people-person-2025115/.

- Fig. 24.2: Copyright © FunkMonk (CC BY-SA 3.0) at https://en.wikipedia.org/wiki/File:Proconsul_nyanzae_skeleton.jpg.

- Fig. 24.3: Copyright © Didier Descouens (CC BY-SA 4.0) at https://en.wikipedia.org/wiki/File:Sahelanthropus_tchadensis_-_TM_266-01-060-1.jpg.

- Fig. 24.4: Copyright © T. Michael Keesey (CC by 2.0) at https://en.wikipedia.org/wiki/File:Ardi.jpg.

- Fig. 24.5: Copyright © 120 (CC BY-SA 3.0) at https://en.wikipedia.org/wiki/File:Lucy_blackbg.jpg.

- Fig. 24.6: Source: https://en.wikipedia.org/wiki/File:Homo_Georgicus_IMG_2921.JPG.

- Fig. 24.7: Copyright © Tim Evanson (CC BY-SA 2.0) at https://commons.wikimedia.org/wiki/File:Homo_neanderthalensis_adult_male_-_head_model_-_Smithsonian_Museum_of_Natural_History_-_2012-05-17.jpg.

- Fig. 24.8: C. de Miguel and Macej Henneberg, from "Variation in hominids brain size: How much is due to method?" Homo, vol. 52, no. 1 pp 3-58. Elsevier B.V., 2001.

- Fig. 24.9: Copyright © Chris Stringer (CC BY-SA 3.0) at https://commons.wikimedia.org/wiki/File:Homo-Stammbaum,_Version_Stringer-en.svg.

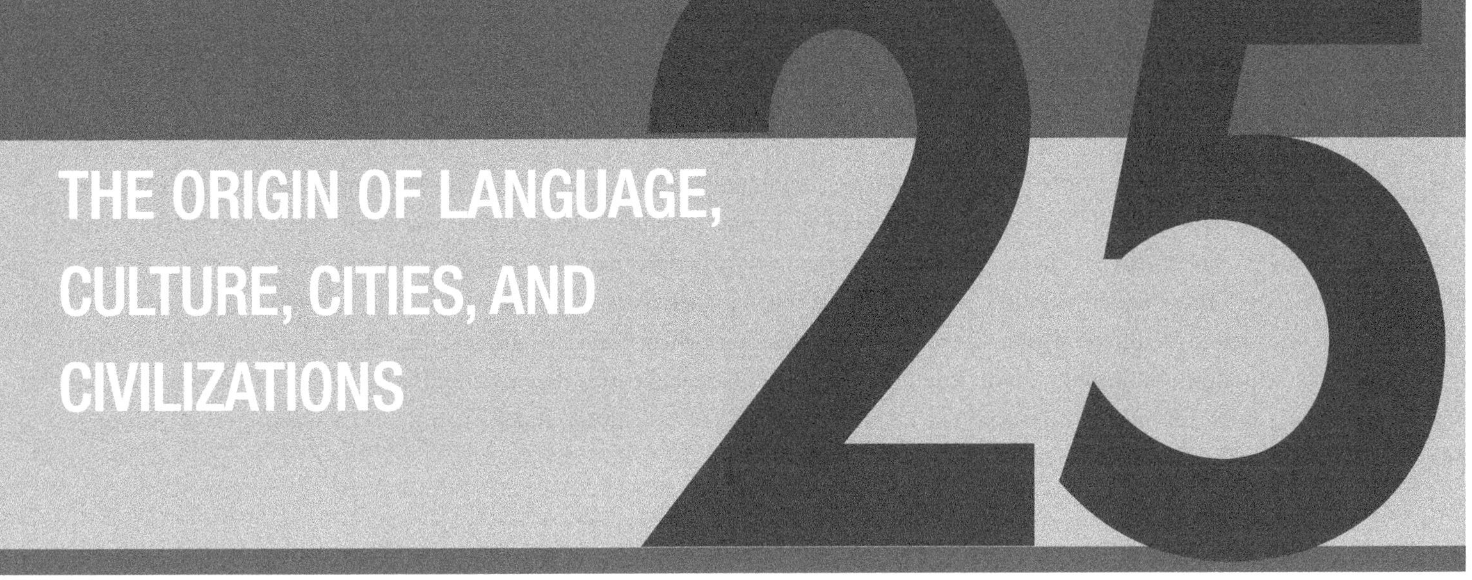

THE ORIGIN OF LANGUAGE, CULTURE, CITIES, AND CIVILIZATIONS

CHAPTER LEARNING OBJECTIVES

This chapter covers:

- The first tools used by humans
- Discovery of fire
- The origin of language
- The origin of agriculture
- The first cities
- The origin of civilization

In the last chapter I reviewed the evidence regarding the origin and evolution of primates and mammals, leading to modern humans. Bringing the findings together from different aspects of evolution, I argued that when their brain size grew, as it did for evolving primates, their ability to carry out more complex tasks increased. This resulted in a better understanding of their environment, inventing new tools defending themselves, looking for means of communication and forming cities and societies. Eventually, this led to the creation of civilization and culture. These all happen as a result of biological evolution as well as natural selection and the influence by the environment and later, by social interactions.

When accommodating the entire history of the universe in one year (figure 1.3), the emergence of societies, civilization, and culture occupied the last few seconds of that year, showing how recent this has been compared to the age of the universe (this corresponds to less than one-millionth of the age of the universe). Nevertheless, this is when the world's communities, cities, diverse cultures and civilizations formed. As the size of the communities and their population increased, so did the need for the means to feed them and to provide the necessary nutrition. This required new methods for hunting and development of agriculture. The invention of agriculture led to the formation of new societies and cities. As more efficient ways for food production developed, leading to food surplus, the early human found more free time to spend on other activities. At this point new specializations appeared (like making tools, pottery, and craftsmanship), leading to first civilization and cultural

"Civilization began the first time an angry person cast a word instead of a rock"

- SIGMUND FREUD

"Try to learn something about everything and everything about something"

- THOMAS HUXLEY

centers. When early humankind shaped communities, they needed means of communication. This led to the development of language, which resulted to enhancement of brain activities. The result of all these was the formation of modern societies. The earliest evidence for this dates back about eight thousand years ago.

This chapter presents a short review of the main activities of *H. sapiens* after they migrated out of Africa and settled in different parts of the world. At this point their brain had grown to its current size, enabling them to perform multiple and more complex tasks. The chapter explores the origin of the first tools our distant ancestors made for hunting. The origin of language, invention of agriculture, and formation of the first cities and civilization will then be discussed.

THE FIRST STONE TOOLS AND THE USE OF FIRE

The increased brain size enabled early *Homo* to perform more complex tasks while bipedalism freed their hands to carry or build new things. The archaeological evidence shows that the first tools were made about the same time the growth in the brain size started and about the time the *Homo* became bipedal. First stone tools were found in association with *H. habilis* around 2 MYA (the name *H. habilis* means "handy man"). These were simple chopping tools with rough cutting edge. The tools were made from materials such as flint or quartz. They were used for cutting meat and bones or scraping wood. These tools were first found at Olduvai Gorge in Tanzania, east Africa, and are called Oldowan tool tradition. They signify the beginning of the *Lower Paleolithic* time that covers the Old Stone Age spanning from 2.6 MYA to 200,000 years ago. More recently, tools of similar age (and probably older) were found in Gona, Ethiopia, dating to about 2.6 MYA (Haviland et al. 2014).

Manufacturing of these tools needed highly skilled workforce to produce sharp-edged tools with the limited facilities they had. They also had to develop the idea of what the end product would look like and plan different steps, from going out and searching for the raw materials, carrying them to a place to build the tools, and building them. This required a combination of a well-developed brain and bipedalism to be able to carry materials, design and manufacture tools. The fossil evidence confirms that the first tools were made in Africa. It was also in Africa that the *Homo* genus originated and later became bipedal. About 1.8 MYA the genus *Homo erectus* started to move out of Africa and spread to other places in the world, as far away as in China, India, western Europe, and Russia. They started to make more complex tools. They invented the first hand-axe (now found in Africa) around 1.6 MYA while those found in Europe are 500,000 years old. They then used tools that resembled cleavers (hand-axe with sharp and long edges) and knives for hunting purposes.

After the *Homo* genus moved out of Africa and to China, Europe, and Asia around 780,000 years ago, they were exposed to colder environments. To survive, they needed to warm themselves. There is evidence that the *H. erectus* used controlled fire to warm themselves against the colder climate or to protect themselves against predators. The discovery of fire-cracked basalts and bones in a 700,000-year-old rock shelter in Kao Poh Nam, Thailand, confirms the use of fire by the *H. erectus*. There is evidence for earlier use of fire in South Africa going back 1 to 1.3 MYA where fossil remnants and bones show exposure to high temperatures. Discovery of the use of fire for cooking allowed production of better nutrients needed for growth, and led to the evolution of heavy jaws and sharp teeth as well as reduction in the size of digestive tracks, all characteristics of modern humans.

The discovery of fire led to other activities. For example, it was found that clay hardens when exposed to high temperatures produced by fire. This allowed our ancestors to make tools with clay that included pottery containers around 35,000 years ago. Evidence found in Yuchanyan Cave in China shows some of the earliest pottery dated back fifteen to eighteen thousand years ago.

THE ORIGIN OF LANGUAGE

Evidence for the linguistic ability of the *Homo* genus comes from the unlikely source of the first tools made by the humankind. There is evidence that most of the stone tools recovered today, were manufactured by right-handed individuals. This implies development of the specialization process in the brain of *H. habilis* and *H. sapiens*, indicating that the evolutionary specialization needed for language was well underway at that time. The *hypoglossal canal*—the canal through the skull that hosts the hypoglossal nerve in control of the movement of the tongue, essential for speaking, acquired its contemporary size around 500,000 years ago (Haviland et al. 2014). The reduction in the size of the jaws and teeth, as well as the need for spoken as compared to gestural language led to the development of the ability to "talk."

For the early *Homo* genus (2.5 to 0.8 MYA), it is likely that the change in the shape and size of the skull at the bipedalism stage (developed around 3.5 MYA in Australopithecines) affected their vocal track. The shape of the track and a low-lying larynx (compared to the apes) are essential for generating many of the sounds human make. Furthermore, the structure of the vocal cord acts as a "bandpass filter" to modify the sound, allowing only selected frequencies to pass. This is indicated by the length and shape of the vocal cords and modified by the tongue, lips and the palate. The first in the *Homo* genus that managed to make controlled vocalization were *H. heidelbergensis*, developing early forms of vocal language. However, study of the stone tools made by the early *Homo* species found that for 2 million years since *H. habilis* emerged and to the Neanderthal time, there was little change in the stone tools they made. This indicates that perhaps the functional activity of the *Homo* brains, including the *Broca region*, which is responsible for languages (after Paul Broca, who in 1861 identified the region of the brain that can produce speech; figure 25.1), were not well developed. Therefore, while the size of the brain played an important role in human ability to speak, neurological developments inside the brain were as important (Jurmain et al. 2013).

The center of language faculty in most people is the left hemisphere of the brain. In particular, two regions play the main role: the Broca area in the left frontal lobe and *Wernicke area* in left temporal lobe (figure 25.1). The Broca area is located in the motor cortex close to the region that controls muscle movement in face, lips, and larynx. The task of Broca region is to receive information and to organize it for communication. Once done, it sends instructions to adjacent areas to activate muscles responsible for speech. Wernicke area is close to the part of the brain that is responsible for sound reception and interpretation. Its task is to process the words that we hear when spoken or analyze language input. The auditory information related to language is transmitted from Wernicke's region to the Broca area by a bundle of nerves connecting the two regions. The cooperation and coevolution of these two areas of the brain have provided us with the ability to interpret and understand language and to respond to what we receive. In other mammals, these regions are less well developed and as such, they are unable to develop language and speech abilities (Jurmain et al. 2013).

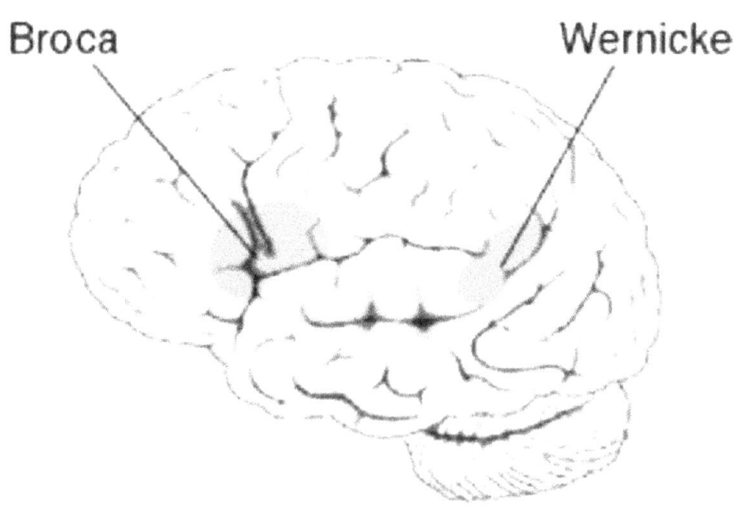

Figure 25.1. The relative locations of Broca and Wernicke areas of the brain. These two areas are connected by a bundle of nerves, allowing understanding and interpretation of language.

The ability to communicate using language requires both biological development as discussed above, and development in structural language itself. The exact origin of language is not known. Most likely, physiological evolution of the *Homo* genus combined with social and cultural needs were responsible for the development of language. Furthermore, by being able to communicate, our ancestors could inform one another about the dangers in their surrounding and proximity of predators. Therefore, natural selection must have played a significant role in the development of language.

Fossil remnant of a Neanderthal found in the Kebara Cave in Israel shows evidence that the *hyoid bone* (the C-shaped bones associated with speech muscles in larynx) have a similar shape to that of modern human, indicating that they had the necessary anatomy to make controlled vocal sounds. The Neanderthals are also found to have undergone the necessary neurological development in their brain to process spoken language ability. Also, their expanded upper body consisting of thoracic vertebrate canal could control breath needed for structured speech. It is therefore likely that Neanderthals who lived around 60,000 years ago were able to speak and use language as means of communication.

Was there a single language with all the rest of the languages originating from that? If so, how did the *H. sapiens* come up with so many different languages? This is a difficult study as most words evolve too rapidly and do not preserve their ancestry. By studying the frequency with which words are used in everyday speech (the number of times a word is repeated) among seven Eurasia languages, linguists identified a set of "conserved" words that evolved from a common ancestor around 15,000 years ago (Pagel et al. 2013). They find that some widely used words have retained their forms since last ice age (Box 25.1).

Tracing back the history of the most frequently used words between seven widely different languages and similarities between the sounding of the words, linguists concluded that languages evolved gradually over time as human population diverged. They found that the change in some words happened so slowly that we are still using the same words our ancestors were using ten thousand years ago (estimating an average half-life of two thousand to four thousand years for any random word and ten thousand to twenty thousand years for most frequently used words) (Haviland et al. 2014). This points toward the existence of a single ancient language, a linguistic super-family tree that unites all current languages. With this, linguists are now able to trace back to the "mother" of all languages that used by our early ancestors (Pagel et al. 2013) (Box 25.1).

In summary, the origin and development of language depend on many different factors. They are in parallel with the development of the brain, the physiology of the mouth, the tongue and the position of the larynx and the coordination of all these organs, as well as the social need to be able to communicate and to invent the words to do that. It is likely that *H. heidelbergensis* used language around 600,000 years ago. It first started with hand gestures and sounds that resembled natural events, gradually evolving into a protolanguage.

Some linguists believe that language (used by modern humans) has a recent origin. By developing a list of 100 basic lexical concepts found in all languages (words like "I", "two", "sun" etc), they estimated the rate at which these words would have changed as new dialects of language were formed. They found a rate of change of 14% in every 1000 years. When they compared the list of words spoken in different parts of the world today, they estimated that between 10,000 and 100,000 years ago all the inhabitants of the earth spoke the same language (Kolb & Whishaw 2009). The diversification then appeared as soon as everyone started to speak the same language. The question with this scenario is how early humankind, spread around different parts of Africa with no means of communication, could speak the same language. Also, given that hominids lived around 4 MYA, why they did not start to speak before the time they did (i.e. 100,000 years ago)?

THE ORIGIN OF AGRICULTURE

Until about twelve thousand years ago, humankind acquired most of their required food through hunting and fishing and by collecting large variety of plants. This changed when the cold and dry Pleistocene epoch came to an end, being replaced by warmer climate of the Holocene epoch. A significant consequence of this climate change was for the humans to take control of the plant and animal growth cycles, a process called *domestication.* They replaced their diet from wild to domesticated plants and animals, seriously changing their life style and habit. This period is called *Neolithic.*

The Neolithic revolution, as it is called, happened gradually throughout thousands of years. Through domestication, early humans could impose selective breeding to eliminate for example, plants with thorns, bad tastes, or toxins. Archaeologists can easily distinguish between domesticated and wild plants and animals. Domesticated animals often have a different size (in terms of the skeletal structure and the size of their horn) than wild animals. They can also identify domesticated and wild seeds. Therefore, by studying the age and sex distribution of animals in archaeological sites and the variation of these between different sites they can show if domestication was practiced (Box 25.2).

The Neolithic revolution coincided with people forming communities and small villages. As the size of the population increased, so did the need for developing new ways to feed them. Hunting and collecting wild plants were no longer efficient ways to produce the food needed. On the contrary, domestication could provide the largest amount of food per unit area of land than could other means of producing the food. Also, with the invention of agriculture, humankind took control of the type and amount of the food with the option to store the food.

Agriculture started sometime around 11,500 years ago when people started to make simple observations that some seeds falling on the ground result in new plants. They observed the conditions needed for the plants to grow, like the need for water or protection from animals eating the plants. They observed if seeds are planted in the ground, they result in new plants that could be harvested and used as food. This is common sense today but a great discovery about 12,000 years ago.

The earliest form of plant domestication took place in the area from the Upper Nile (currently Sudan) to the Lower Tigris (currently Iraq). This area is called *Fertile Crescent* (figure 25.2). There is evidence that humankind first domesticated rye around 13,000 years ago around Aleppo in Syria. At almost the same time, plant domestication happened in other parts of the world. For example, the earliest domestic rice is from China around 8,000 years ago. Archaeologists have found other sites of early plant domestication of corn in Mexico (9,000 years ago), banana in New Guinea (7,000 years ago), squash and sunflower in North America (6,000 years ago), potatoes and sweet potatoes in South America (5,250 years ago), and sorghum and yams in Africa and the Saharan Desert (4,500 years ago). Similarly, domestication of wheat and barley spread from southwestern Asia to Greece around 8,000 years ago. Starting from these centers, the domestication idea spread to other places on the globe (Larsen 2011).

BOX 25.2: BEGINNING OF DOMESTICATION

The change from forging (search for wild food resources) to farming happened around ten thousand years ago, and this is one of the most significant events during the life of the *Homo* genus. For over 7 million years of human evolution, they ate all kinds of plants but had never grown any. The start of domestication affected biological evolution, brain growth, genetic changes, and social aspects of their lives.

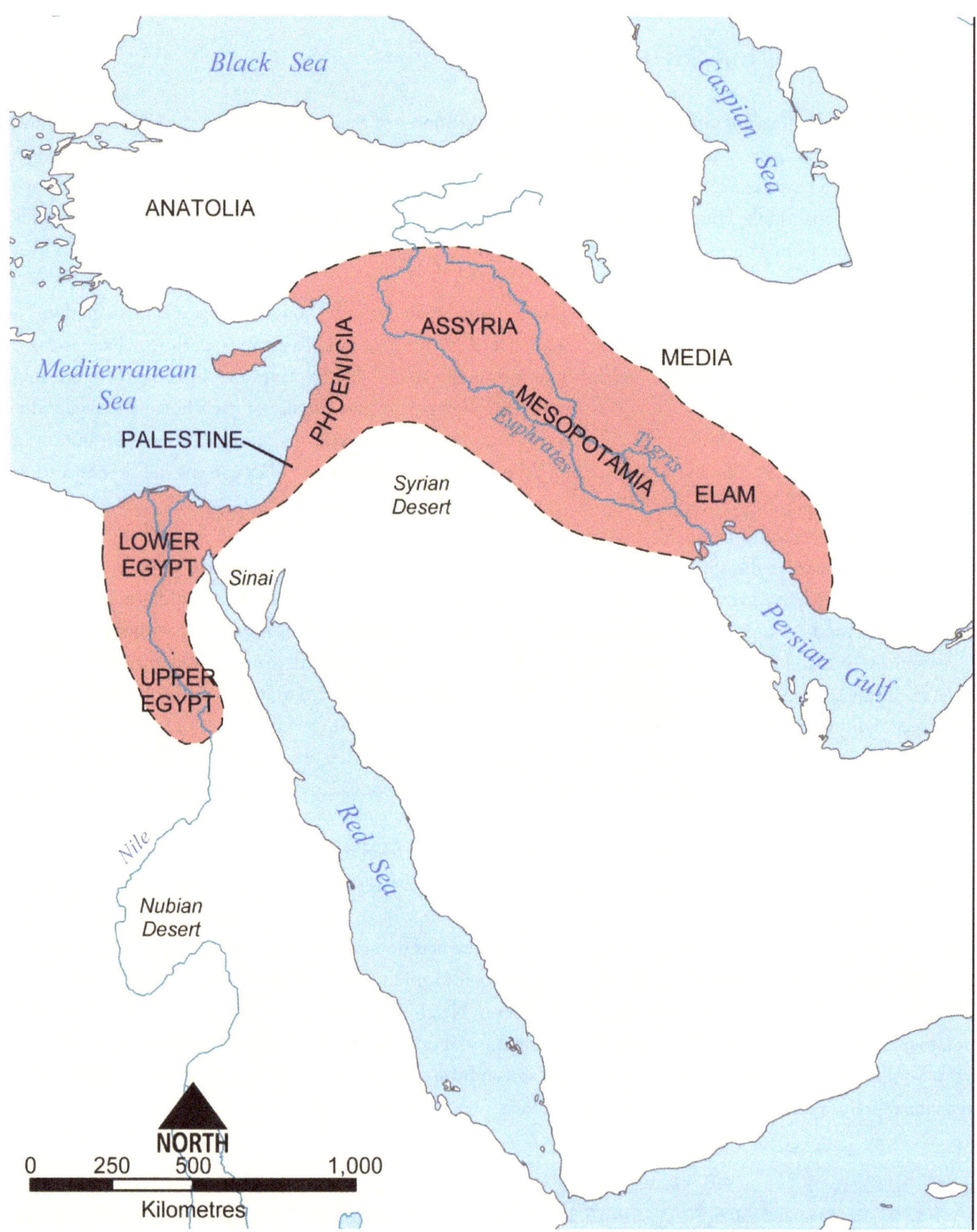

Figure 25.2. The geography of the Fertile Crescent (the brown region), one of the first places agriculture started.

The first animals to be domesticated were dogs around fifteen thousand years ago. This followed by goats, sheep, and cattle around seven thousand to eight thousand years ago. Archaeological evidence points to southern

Turkey, northern Iraq, and Zagros mountains in Iran as the main sites of animal domestication. The main reason was the environmental diversity of these regions. The herds of sheep and goat needed to graze in the fields and because of the variety of ecological conditions different kinds of plants were available in different regions throughout the year. The animals fed from plants and were hunted by humans for food. Archaeological evidence shows that at first animals of all sexes and ages were hunted but later, there was a decrease in female animals that were killed for food as they were kept for breeding.

The discovery of agriculture was the very first step towards establishment of civilized societies. "Agriculture is not crop production as popular belief holds—it's the production of food and fiber from the world's land and waters. Without agriculture it is not possible to have a city, stock market, banks, university, church or army. Agriculture is the foundation of civilization and any stable economy"— Allan Savory.

Figure 25.3. Remains of Jericho (meaning the "city of palms") located in what is now Israel. This is regarded as the oldest city in the world, dating back over 7,000-9,000 years.

THE FIRST CITIES

The invention of agriculture led to human population growth and formation of Neolithic villages (Stone Aged communities built around 10,000 years ago) that later developed to cities and complex societies. Early cities were developed in a number of different regions. The oldest known city in the world is *Jericho*, located in the West Bank region of Middle East (currently Israel). The history of the city goes back to fourteen thousand years ago when it was constructed, with a population reaching three thousand people around nine thousand years ago, living in an area of about 10 acres. Jericho was the first known example of a large-scale construction project surrounded by large walls made from stones (figure 25.3). Excavation of burial sites in Jericho shows signs of social structure and order and cultural activities that define cities.

Other first cities were founded in *Mesopotamia* (in old Greek meaning "land between rivers"), located in present-day Iraq (figure 25.4). The city was established between sixty-five hundred and eight thousand years ago by movement of population from foothills of the Zagros Mountains (present-day Iran) and

Figure 25.4. Mesopotamia (meaning in old Greek "the land between rivers") in Ur. The history of the city goes back about eight thousand years.

after the Neolithic revolution. Around 6,300 years ago, with the development of civilization in Mesopotamia, some of the first cities were developed. These include the first true cities of *Ubaid, Uruk, Ur,* and *Eridu.* There is evidence of temple structures in Ubaid, probably serving as administrative centers indicating that these cities were being developed into urban centers. Around fifty-eight hundred years ago, one of these cities, Uruk, grew to a size and density that could be considered as the first genuine city. The structures in Uruk were more centralized with more standard structures implying the presence of a central authority to set the rules. This area is considered as the *cradle of civilization* where many of the branches of science first developed. It is believed that wheels were first invented here, as well as mathematics and astronomy.

Early cities were also developed in what is now Egypt. These cities appeared alongside the Nile river—a narrow river through the otherwise dry and lifeless desert. Central communities were formed around fifty-five hundred years ago. These cities often competed for resources and for the control of more land and people. Their locations were strategic to provide routes for trade and access to mineral resources found in desert. There is evidence for social structures in these cities found from the burial sites.

Two of the most ancient cities in the world were developed in the Indian subcontinent in the *Indus Valley* around forty-five hundred years ago traced to a region in western Pakistan called Baluchistan mountains. *Mohenjoo-daro* was one of those cities with forty thousand inhabitants (figure 25.5). Archaeological sites have unearthed the presence of a citadel with a temple, a granary, and a bathhouse. The other city was *Harappa* with a similar planning. The city had streets and a brick-covered sewer trench with connection to different residential areas, an indication of the urban planning forty-five hundred years ago. Hundreds of farming villages were surrounding these cities.

What was the reason for development of cities? Why were cities formed in certain places in the world and not others? The main reason behind the development of cities was agriculture as the history of formation of first cities coincides with the Neolithic revolution and invention of agriculture (figure 25.6). Agriculture brought hunters to settle and produced food that made larger communities possible. Efficient farming led to food surplus, helping trade and the economy of cities. Once the food supply was sufficient, people had more time to spend on other things instead of spending all their time in search of food. This led to development of civilization and culture. Stable climate conditions and warmer weather also helped formation of cities. It is clear that the first cities were located in Fertile Crescent, a suitable land for farming (figure 25.6). Existence of water that would support growing plants and animals was another factor for formation of the first cities.

Some of these first cities still exist. The geographical advantages, right climate and right conditions for agriculture were the main reasons for their longevity and sustained occupation by the inhabitants (figure 25.6). The archaeological excavations have revealed rich and diverse cultures in these cities from which the first seeds of civilization were planted.

Figure 25.5. Remains of Indus Valley in Mohenjoo-daro, located in present-day Pakistan and northwestern India. A civilization formed forty-five hundred years ago on the banks of the Indus river.

Figure 25.6. The oldest cities in the world (green band), located in Mesopotamia, compared to the location of new cities.

THE ORIGIN OF CIVILIZATION

What is the definition of *civilization*? There is no simple answer to this question. This is further complicated by the fact that this definition could change with time. A general definition for civilization should embody a society with food and labor surplus, social stratification, formal governing authorities, densely populated settlements, and specialization of labor (Haviland et al. 2014). In search for the first civilizations, it makes sense to start with the first cities as discussed in the last section. After all, civil societies could only emerge after cities are established. Here the focus of our discussion is not the description or names of early civilizations but how and under what conditions first civilizations developed (Box 25.3).

BOX 25.3: THE CAUSES OF CIVILIZATION

The main cause of acceleration in the development of civilization was innovation in agriculture. Once early humans invented efficient methods for producing food, less people were needed to work in the farmlands and therefore, they had more time to spend on other things like cultural activities, art, and trade.

Figure 25.7. Sumerian cuneiform written on a clay tablet. This is from Shurppak in Iraq and dates back to 2500 BCE.

The *Sumerians* who lived in the city of Sumer in southern Mesopotamia (modern-day Iraq) -(figure 25.6) around 5500 to 4000 BCE (Before Common Era) were the first civilization in the world. They invented writing with the earliest texts produced in the city of Uruk, dating back to 3300 BCE (figure 25.7). They developed trade and industries including masonry, pottery, and metalwork. Sumerians invented irrigation and canals to control water, allowing more efficient farming and food production methods. This created a higher population density in the region allowing other activities but agricultural. Many of the disciplines that form the basis of modern science originated from this part of the world.

Around five thousand years ago in the Uruk city in Mesopotamia, a new style of writing was invented. The first writing tools used in Mesopotamia was wedged-shaped reed stylus on tablet of damp clay. Each marking stood for a word (figure 25.7). Recent archaeological findings point to the Henan province of China as the first place where writing originated around eighty-six hundred years ago.

Because of their size and increasing complexity, large cities needed a central authority to govern. The evidence for central authority is often found in the form of monuments, temples, and palaces, as well as large sculptures. Examples of such monuments are the Great Pyramid for the tomb of Egyptian pharaoh, Khufu. This contains 2.3 million stone blocks with an average weight of 2.5 tons. Building of such huge structures needs central authority to administer, coordination to apply engineering skills and the labor force to build.

The first legal system was developed during the reign of the Babylonian king Hammurabi who lived in Mesopotamia around 3,700 to 3,950 years ago. It contained law applying to perjury, false accusation, loans, and debts. It indicated the rate to be charged in various trades and was designed to protect the rights of individuals.

Figure 25.8. Shows the Step-Pyramid of Djoser. This was the first pyramid made by Egyptians, around forty-seven hundred years ago.

Civilizations were also developed in other parts of the world based on the same concepts as for the Mesopotamia. In Egypt, cities were formed alongside the Nile River around fifty-five hundred years ago. Excavations discovered the city of Hierakonpolis showing series of houses made of mud-bricks with as many as ten thousand people living there. There is evidence that the power in these cities were in the hands of the elites. Dryer climate led to the manufacturing of irrigation systems (first developed by Sumerians), controlling the water flow and the food production. With the growth of villages and cities, an elite class emerged, food surplus was produced and as a result, new specialties were developed. Different cities alongside the Nile were finally united under King Narmer, the ruler of Hierakonpolis, around fifty-one hundred years ago. The successor to King Narmar was King Djoser, the first pharaoh of Egypt. With all the power in his hand and all the villages and cities under his authority, he started the construction of the first Egyptian Pyramid around forty-seven hundred years ago (figure 25.8).

On the banks of the Huan River in the northern Hunan province of China is the ancient city of Yin, dating back to thirty-seven hundred years ago. A number of small settlements were located close to this, providing one of the earliest cultural centers in East Asia, called the *Shang culture*. Remaining structures of this city show palaces and evidence for pottery, stone tools, and art work. The Shang culture developed from the Neolithic culture in the region. They grew domesticated rice and wheat and raised animals including sheep, pigs, and chicken. Large buildings and monuments do not appear in abundance around the Shang sites but it still fits the definition of civilization. Different burial sites and house sizes reveal social stratification. There is evidence for different specializations such as bronze metallurgy and stone sculpture as well as evidence for early writing that included five thousand characters. In many respects this civilization, the first in South East Asia, had many things in common with civilizations independently developed in other parts of the world.

Meanwhile another civilization was developing far away from Mesopotamia, Egypt, and China, where is now South America. The *Olmec civilization* was the oldest in South America, located in south-central Mexico. The Olmec civilization emerged around 4,500 years ago as the first of the Mesoamerican civilization and was the foundation of many other civilizations that followed. Their roots come from the farming cultures of Tabasco that started around 7,100 to 6,600 years ago. The Olmecs are known for their artwork showing "colossal heads" (figure 25.9).

The next major civilization in the Mesoamerica after the Olmec was the *Maya civilization*. This was developed in an area covering southern Mexico, the entire Guatemala and Belize and western Honduras and El Salvador. The earliest villages were formed around four thousand years ago when agriculture was developed producing maize, squash, and chili pepper. The first cities were developed around twenty-seven hundred years ago and this extended to a series of cities connected to one another through a trade network with large monuments built in them. There was clear structure in the architecture of the city, remnants of which still exist. The Olmec and Maya civilizations were the first and the oldest in the western hemisphere.

Figure 25.9. Artwork from the Olmec civilization showing "colossal heads."

SUMMARY AND OUTSTANDING QUESTIONS

The increase in world population since agriculture revolution has been enormous. Around 10,000 years ago when agriculture started, the population of Earth was 2 million to 3 million. This increased to 250 million to 300 million around 2,000 years ago and 1 billion in the eighteenth century. Today there are over 7 billion people in the world. The increase in the world population soon after the agricultural revolution (10,000 years ago) was 0.01 percent per year. This increased to 0.3 percent and 0.6 percent per year during the eighteenth and nineteenth centuries respectively and currently is at 2.0 percent per year (Jurmain et al. 2013).

As brain size increased and the *Homo* genus became bipedal, the ability of our ancestors to carry out more complex tasks increased as well as their ability to use their hands in making crafts. The first stone tools were made by the *H. habilis* around 2 MYA. These needed both a thinking process and free hands to make them. They were used for cutting meat and woods. With the discovery of fire in Africa around 1 MYA and the understanding that it could be used for cooking, a more nutrient diet was discovered. This sped up the evolution of the digestive system.

There is evidence that around the time the human made the first tools, the specialized parts of the brain started to develop. With the growth of language centers in the brain to control muscle movements on the face and mouth, changes in the anatomy of the larynx and development of the center in the brain to interpret the sound, early human started to communicate with producing sounds. It is not clear exactly when and how the communication using language started but it is believed that Neanderthals living 60,000 years ago were able to communicate by language. An important question is whether all languages started from a common super-language and then diversified or languages were developed locally? There are different hypothesis concerning this.

The first civilization started by the Sumerians around 5500 to 4000 BCE in southern Mesopotamia (modern-day Iraq). They developed writing, trade, industry, and invented irrigation systems. It was around the region rich in agriculture. Civilization then appeared in other parts of the world, including China (3,700 years ago), Egypt (5,500 years ago), and South America (4,500 years ago). The main reason for civilization appearing where they did was the development of agriculture. Once they appeared, the culture, art, science, and trade followed.

Civilizations form, evolve, and sometimes collapse and disappear, leaving behind their remnants in the form of monuments, burial sites, and crafts in the form of art or writings. From these archaeologists decipher details about each civilization. We discussed the main facts behind development of civilizations, but why did civilizations collapse? And why is nothing left today from many of the richest and most prominent civilizations?

In his book *Collapse: How Societies Choose to Fail or Succeed*, Jared Diamond (2005) attributes the fall of civilizations to five factors:

1. Environmental degradation caused by practices in the society
2. Climate changes that affect food production
3. War between different states
4. Failure to acquire resources beyond the immediate geographical borders
5. Failure of the society to respond to the above items.

Amazingly, many of these points apply to today's civilizations. As civilizations penetrate into the lives of individual citizens, we should hope that we could handle these items better.

1. When and at what part of the world were the first tools discovered?
2. When was fire discovered, and what is the evidence for the first fire? What was fire used for?
3. What are hypoglossal canals, and when did they acquire their modern size?
4. Which of the *Homo* genus first developed the ability for vocal communication?
5. What observation did archaeologists use to conclude specialization of different parts of the brain?
6. Explain the tasks of the Broca and Wernicke regions of the brain.
7. Explain the Neolithic revolution.
8. When and how did agriculture start?
9. Where is the Fertile Crescent, and what is its significance?
10. Where is the cradle of civilization considered to be?
11. What are the first true cities that satisfy the modern definition of a city?
12. Why were cities developed in some regions and not others?
13. Sumerians were the first known civilization. What is their contribution to the world?
14. What was the main reason for building large constructions like the Pyramids?
15. What were the oldest civilizations in the Western Hemisphere?

CHAPTER 25 REFERENCES

Diamond, J. 2005. *Collapse: How Societies Choose to Fail or Succeed*. New York: Penguin.

Haviland, W.A., D. Walrath, H.E.L. Prins, and B. McBride. 2010. *Evolution and Prehistory: The Human Challenge*. 9th ed. Belmont, CA: Wadsworth.

Jurmain, R., L. Kilgore, W. Trevethan, and R. Ciochon. 2013. *Introduction to Physical Anthropology*. 4th ed. Boston: Wadsworth/Cengage Learning.

Larson, C.S. 2010. *Our Origins: Discovering Physical Anthropology*. 2nd ed. New York: Norton.

Pagel, M., Q. Atkinson, A.S. Calude, and A. Meades. 2013. "Ultraconserved Words Point to Deep Language Ancestry across Eurasia." *Proceedings of National Academy of Sciences* 110 (21): 8471–76.

FIGURE CREDITS

- Fig. 25.1: Source: https://commons.wikimedia.org/wiki/File:BrocasAreaSmall.png.
- Fig. 25.2: Copyright © 92bari (CC BY-SA 3.0) at https://commons.wikimedia.org/wiki/File:Fertile_Crescent_map_it.PNG.
- Fig. 25.3: Source: https://commons.wikimedia.org/wiki/File:Jerycho8.jpg.
- Fig. 25.4: Copyright © Danyelflorea (CC BY-SA 3.0) at https://commons.wikimedia.org/wiki/File:Zigurat_Ur.JPG.
- Fig. 25.5: Copyright © Comrogues (CC by 2.0) at https://commons.wikimedia.org/wiki/File:Mohenjo-daro-2010.jpg.
- Fig. 25.6: Copyright © Goran tek-en (CC BY-SA 3.0) at https://commons.wikimedia.org/wiki/File:N-Mesopotamia_and_Syria_english.svg.
- Fig. 25.7: Source: https://commons.wikimedia.org/wiki/File:Sumerian_account_of_silver_for_the_govenor_(background_removed).png.
- Fig. 25.8: Copyright © David Broad (CC by 3.0) at https://commons.wikimedia.org/wiki/File:Step_Pyramid_of_Djoser_at_Saqqara_-_panoramio_(1).jpg.

- Fig. 25.9: Copyright © ·Maunus· (CC BY-SA 3.0) at https://commons.wikimedia.org/wiki/File:OlmecheadMNAH.jpg.

- Fig. 25.8: Copyright © David Broad (CC by 3.0) at https://commons.wikimedia.org/wiki/File:Step_Pyramid_of_Djoser_at_Saqqara_-_panoramio_(1).jpg.

- Fig. 25.9: Copyright © ·Maunus·l· (CC BY-SA 3.0) at https://commons.wikimedia.org/wiki/File:OlmecheadMNAH.jpg.

CONCLUDING REMARKS

This book is about the origin of almost everything in the physical world and how they came to be the way they are today. By the very meaning of the word *origin*, we make the implicit assumption that everything we observe has had a beginning. In other words, nothing has been here forever. If this is the case, then, how did they come to existence? And what was their state before that? This book attempted to address these questions using scientific arguments. However, when exploring frontiers of knowledge or unveiling the unknowns in the physical world in search of truth, we are constantly reminded of a philosophical question: Is there an absolute truth waiting to be discovered, and what are the ways to uncover it? The *principle of sufficient reason* developed by the seventeenth-century philosopher Leibniz states that for anything that exists, there must be a reason for its existence, and for any truth there must be a reason why it is so and not otherwise. This is the challenge—to find the reasons for the existence of things and for why they exist the way they do. The question then is: would mankind ever be able to know everything about the physical world? In *Feynman Lectures on Physics* (volume 1), Richard Feynman refined this in his elegant statement: *"Each piece, or part, of the whole nature is always merely an approximation to the complete truth, or the complete truth so far as we know it. In fact, everything that we know is only some kind of approximation, because we know that we do not know all the laws as yet. Therefore things must be learned only to be unlearned again or, more likely, to be corrected"* (Feynman et al. 2014). The conclusion therefore is that there is truth to be discovered but once discovered, more refined form of truth would still remain to be uncovered and the search goes on.

The two most fascinating and at the same time intricate intellectual challenges are the questions of the origin of the universe and the origin of life. Could they both have started spontaneously from absolute nothingness? A few chapters in this book were dedicated to study these questions. We are better placed to explore the origin of life, as we can perform controlled tests and experiments to decipher it. Study of the origin of the universe is more speculative and much harder to verify through experiments. We know the life started from life and did not come about spontaneously from nothing. However, we do not yet know details of how

> "My only wish would be to have 10 more lives to live on this planet. If that were possible, I'd spend one lifetime each in embryology, genetics, physics, astronomy and geology. The other lifetimes would be as a pianist, backwoodsman, tennis player, or writer for the 'National Geographic' "
>
> - JOSEPH MURRAY

> "When it is obvious that the goals cannot be reached, don't adjust the goals, adjust the action steps"
>
> - CONFUCIUS

life really started. For the universe, the situation is more challenging. We do not have the necessary tools and techniques to study the conditions of the very early universe. We may come up with some models for the universe at that time but to confront theoretical concepts with observational facts is a serious problem as it is hard to come by experiments to test those theories. Nevertheless, as shown in this book, we have made significant progress in understanding the physical universe and how it came to be. Questions that, until a few years ago, were in the domain of philosophy could now be addressed by science and be verified experimentally. This is what I tried to show in this book—to present answers to the most fundamental questions in the boundary of humankind's knowledge without speculating. Lets pose one of those most fundamental questions: when did space and time come to existence? To take this a step further: did the universe start from absolute nothingness? Or, was it something before it all started? In the following, I look at these questions through some philosophical arguments.

In his book *Why Does the Universe Exist?* Jim Holt (2012) contemplates the question of *why there is something rather than nothing*. Let's explore what is meant by "nothing." The world of absolute nothingness (if such world ever exists) still requires an observer to think about it. Otherwise, how do we know about the world and that it has "nothing" in it? This makes such a discussion self-contradictory, as at least one observer must exist to observe that there is nothing in that world. By implication, every world must at least have one conscientious observer (Holt 2012). The next question is: does a universe without conscious observer physically exist? Imagine the physical constants were slightly different from what they have been in our universe. In this case, the evolution of life would likely not have happened with the universe being filled with an unknown form of matter (depending on the physical constants in that universe). By the above logic of the observer argument, such a universe would be impossible since there would be no observer to observe it. Therefore, the reality of the world may be synonymous to our own consciousness.

How could the concept of space and time fit into this picture? Were they the first things that came to exist? If so, how did they appear? Indeed, one of the most fundamental challenges in both philosophy and science is to explain the nature of space and time—whether there were *something* or *nothing* in the early universe. If you empty everything from the universe—all matter, planets, stars, galaxies, and everything else—you are still left with the space (or the space-time) that accommodated them. The point then is, if the presence of space is the prime necessity of any possible reality (Holt 2012).

There are two views about the nature of space-time. One goes back to Newton, supposing that space is "real" with its own properties and geometry. In this scenario, the space would continue to exist even if all its contents are removed. The second scenario is due to Leibnitz that implies the space is just the result of the relation between the stuff in it. In this scenario, space would not exist independent of its content (if the contents of the universe are removed, the space would vanish, leaving nothingness). The first view (Newton's hypothesis) does not accommodate "nothingness" but the second (Leibnitz's hypothesis) does. Now, consider Newton's view that the space exists on its own. If this is the case, its characteristics and geometry ought to exist alongside the space itself. Therefore, a finite and unbounded space (like the surface of a balloon) would shrink once its radius is reduced, eventually vanishing into "nothingness" once its radius is zero (Holt 2012). The conclusion is that whatever is the nature of the space and time, they can be reconciled with "nothingness" at some point in the distant past. It is therefore possible that the "something" we observe in the universe today, all started from "nothing." In the language of modern physics, "nothingness" is when all the energy from a volume of space is removed, reducing it to the state of its lowest energy—"vacuum." Once this energy depletion continues, it gets to a stage that the actual energy in the space becomes "negative" or there is less energy than "nothing." At this point a bunch of virtual particles appear and disappear and appear again. This occupies the absolute empty space, making "something" out of "nothing." This is indeed how scientists believe it all started.

Conditions responsible for the early evolution of the universe some 13.8 billion years ago and for its subsequent evolution, left imprints that can be observed and studied today. By confronting these observations with theories, scientists have produced charts of the history and evolution of the early universe, including the timeline of the events. The present expansion of the universe, production and abundance of light chemical elements, formation of structures, and the presence and spectrum of the microwave background radiation all provide the observational evidence confirming a widely accepted framework for the evolution of the universe during its first 300,000 years. There are scenarios that are more theoretical and waiting experimental verification like the universe before and during the Planck time. To test these needs progress in theories to predict the phenomena that can subsequently be observed. There are also things that exist in the universe but we have no explanation for them or no information about their nature—like dark matter and dark energy. These will be understood by acquiring more extensive data and with future advancement in technology and by developing new theories to explain present observations. There are also natural limits to what we can do or accomplish. For example, no material body can move with a speed faster than the speed of light, one cannot break the conservation laws in nature, reverse the arrow of time or simultaneously measure the position and speed of a particle with infinite accuracy. No matter how much we advance in technology, we cannot break the above barriers. Without these laws, we would not have been here to begin with.

Would we ever be able to know the answer to all the questions? In his book *The Island of Knowledge: The Limits of Science and the Search for Meaning*, Marcelo Gleiser discusses that *"the unknowns are not a reflection of our ignorance or limited tools or exploration. They express nature's very essence"* (Gleiser 2014). There are some scientific predictions that submit themselves to tests by experiments and observations. As a result, we may find out the facts about them although we may never know why. For every question we answer, many unknowns appear. This has been the trend and what is responsible for the progress in science.

In 428 BCE Plato proposed his model that Earth was at the center of the universe. For almost seventeen hundred years, people believed that, until Copernicus removed Earth and put the sun at the center of the universe (Copernican revolution). Although still wrong, this was a big leap forward in attempting to understand the world. Lets put this in the context. On December 17, 1903, the Wright brothers made history when they managed to fly the first controlled heavier than air aircraft. Just over 100 years from that historic event, today we have conquered space and sent missions to solar system planets and beyond. It took seventeen hundred years before Copernicus rejected Plato's view of the universe by replacing Earth by the Sun at the center of the universe, and only one hundred years for the humankind to go from the first plane to exploring the space. This example shows the amazing speed with which science and technology progress. Today we have generated human's genome, have started to understand working of the human brain and the interconnections between the neurons, have developed theoretical and experimental tools to explain the internal structure of atoms and have observed and studied the first generation of stars and galaxies in the universe. All these were done over the last half a century. Compare this to the seventeen hundred years it took for the mankind to replace earth by the sun at the center of the universe. The scientific progress has become nonlinear and is not clear where it would all lead to in the next decades.

Today we apply our knowledge of the natural laws to the very beginning of the universe to explain its turbulent birth and early history. Many of the things we experience today, including our own existence, are the result of the conditions at that time. The fact that *things are like this and not like that* are all the result of the events very early on. The four forces of nature gained their distinct identity that they preserve until today, the coming together of quarks to form protons and neutrons and formation of the light chemical elements, the interaction between matter and radiation and the decoupling of the two 300,000 years after the beginning, are all explained and confirmed by our observations. After the universe became matter dominated, the force of gravity governed its subsequent evolution, forming the structures from super clusters and clusters of galaxies to individual galaxies, stars, and planets. This continued until nine billion years after the beginning of the universe (just over 4 billion years ago) when dark energy took over, accelerating the rate of expansion of the universe to an infinite future.

It is not clear if the formation of our own planet, the way it did, was a deterministic or chaotic process. What is clear, however, is that many things came together in the right place and at the right time to form the Earth. Once Earth was formed in the habitable zone of a main sequence star (our sun), a sequence of events led to the development of conditions to make it the planet it is today. A unique aspect of our home planet, not seen anywhere else in the universe, is its habitability. It has formed and for billions of years sustained life. How the life was originated on Earth is the most fundamental question. We are now at the position able to scientifically address this.

A number of different (and independent) events had to happen in the right order for life to originate and thrive on Earth. Failure of any of these events in the sequence would have seriously affected the life, as we know it today. Chemical elements essential for building amino acids must have been produced and cooked in stars and distributed in the interstellar medium. Water needed to exist to protect primitive life deep in the oceans from the UV radiation from the sun. The ozone layer must form and that needed oxygen, which was produced by cyanobacteria. All these, and a lot more, needed to be in place before life could start and prosper. The transition from chemistry to biology is most fundamental in exploring the origin of life. Once life started, it again went through a series of sequences before it could form complex and multicellular organisms. This again depends on the sustained conditions through the history of the Earth. For example, how could the oxygen level in the Earth's atmosphere be so constant over the last 500,000 years? How many parameters that make the Earth the planet it is today, have been fixed within the narrow range allowed to sustain conditions for life on the Earth? How has life survived and adapted again after mass extinctions? The answers to these questions are not clear but are fundamental to our very presence here.

There is evidence that some form of life existed over 3 billion years ago. The zircon crystals (the oldest minerals found on Earth) were found to contain water molecules, implying the existence of water very early in the history of Earth, about the time the crust of Earth was solidified. The life at that time was in the form of prokaryotic cells. The step from simple prokaryotic to more complex eukaryotic cells about a billion years ago was a huge evolutionary leap. At this point, with the development of cellular organelles and nucleus, the cells became larger and able to perform more complex functions. Each cell became specialized in performing specific tasks, allowing the development of multicellular organisms. The ability to grow hard shells around 570 million years ago led to the emergence of new species of animals that could protect themselves against the harsh environment and therefore, survive. All these took place in the bottom of the oceans where water protected the organisms against intense ultraviolet radiation from the sun. When Earth's atmosphere was formed through volcanic activities and the oxygen released by the cyanobacteria produced the ozone layer, the land was finally protected against intense UV radiation from the sun. This happened around 500 million years ago when plants and animals started to move to land around the time of Cambrian explosion. This led to development of species of land plants and animals. The vascular plants were developed and grew leaves that helped photosynthesis process and generation of oxygen. The increase in oxygen level in the atmosphere provided the means to support metabolic activities of larger multicellular animals by allowing more efficient energy production. This led to the emergence of different species of reptiles, birds, and mammals on Earth 200 –300 million years ago. The demise of dinosaurs 65 million years ago allowed the mammals (our very distant ancestors) to roam freely on the surface of Earth and to evolve if adapted to the existing conditions. At about the same time the land distribution on Earth took its present form with all the continents in place. Finally, the first evidence for the life that led to our species found in Africa around 4 million years ago. The evolution from the very primitive state to our own species, the Homo sapiens, has been the result of many different steps in a long chain of events—all needed to go the way they did. This has been a complicated process but is known relatively accurately given the number of fossils that are found.

At some point around one hundred thousand years ago, our ancestors started to move out of Africa and to other continents in the world. The development of agriculture, domestication of animals, communities, cultures, and cities then followed. These events are very recent compared to the history of the Earth.

An important factor in the development of human species is the growth in the size of their brain. This has increased by a factor of four (from 400 to 1600 *ml*) over 3 million years. Given that the brain consumes around 25 percent of the total energy of the body, this must be related to how efficient metabolic activities could produce energy. The fascinating question here is how and when the mammal's brain started to think and what the thought process was. As we studied in this book, we are made from the chemical material formed in stars and then scattered in the interstellar medium through supernova explosion when the stars die. Once cooled down, they formed complex molecules that eventually led to life. It is amazing that we use our brains, itself made up of the material cooked in stars, to unlock the puzzles of the origin of all these material and hence the very origin of the brain itself and how this powerful machine is able to think, discover, and invent. Max Planck, the German physicist, once quoted: *"Science cannot solve the ultimate mystery of nature. And that is because, in the last analysis, we ourselves are a part of the mystery that we are trying to solve"*

As I am writing these final sentences, I am at the Honolulu airport in Hawaii waiting for my flight back to California. I was here to make observations on the Keck telescopes on Mauna Kea mountain, currently the most powerful and technologically advanced ground-based telescopes in the world. My aim (with my students) was to search for and perhaps confirm the most distant galaxy candidates in the universe, over 13 billion light years away. This is to look for the first generation of galaxies in the universe and to push observations to the boundaries of the observable universe. I have been coming here to make observations for many years. Apart from the science about which I am so fascinated, I am excited about what I experience every time I visit Hawaii. I use the best technology could offer to see deep in space. These all are located at the top of the Mauna Kea volcano, the highest mountain in the world if measured from its base at the bottom of the Pacific Ocean. Itself, this mountain is a result of millions of years of geological evolution. These are all part of the story that provided the conditions for us to study how things all came together at a grand scale, as I describe in the following paragraph.

The Hawaiian Islands were formed as the crust of the Pacific Ocean floor moved above a heat source ("hot spot") beneath the crust. The magma released by this hot spot erupted through the crust and to the ocean floor. This built up the volcanoes that, with more eruptions, grew and moved above the sea level eventually forming the islands. The hot spot has been stationary over the last 45 million years. As the oceanic crust moved above the hot spot (to the northwest), new volcanoes and islands were created, resulting in a chain of islands away from the hot spot. For the last 1 million years, the hot spot has been under the Big Island (where our telescopes are located and we make our observations of the cosmos). If this hypothesis is true, then the age of the islands and volcanic material on them must increase as one moves away from the Big Island that is at the edge of the Hawaiian Island chain. This is what is indeed observed when analyzing and measuring the age of the rocks on those islands. The oldest of the Hawaiian Islands are 65 million years old while the youngest ones are only 1 million years old. As the islands are eroded by ocean waves, wind, and rain, the older islands are reduced in size and will eventually submerge in the ocean. Therefore, newer islands are expected to be bigger and have taller volcanoes. This is confirmed by the large size of the Big Island and the tall Mauna Kea, with an age of 0.375 million years and a height of 4,250 meters (14,000 feet), the youngest and tallest volcano in Hawaii. The height of this volcano makes it the best observing sites on the planet—with its thin and dry atmosphere—where we get the best data by looking through our telescopes, where I was last night and have been visiting for the last three decades. The above story indicates as how many different things have come together to allow us to address the most fundamental questions about the depth of space and time. As we saw in this book, all the changes on Earth, formation of continents, oceans, and mountain ranges took place in only a tiny fraction of the history of the universe. The humankind, whose presence on this planet only goes back to 4 million years, has evolved and became intelligent enough to study and discover the laws of nature. Using those laws, we decipher the reasons behind what we observe and experience as where all these came from. Exploiting these laws, we develop technology. With that technology, we build our powerful telescopes, make computers to process and analyze massive amounts of data and use them to explore the distant universe where it all started or look for planets that resemble our own home planet when it had just formed. And tonight, I submit myself to the trust I have on the laws of nature and the technology, to fly on a modern aircraft over half the Pacific Ocean to get back home. This is how things came together, we came to be here, and now, we have exploited what nature has had to offer to go back and discover all that came together to get us where we are today.

There are many concepts discussed in this book that are based on facts verified by experiments. There are also many points presented here that may turn out to be incomplete or even wrong. These will be replaced by new and possibly more complete concepts. This is the way science progresses and evolves. There is no preconception of what is the truth, but only the search for the truth. We are lucky enough to be living at a time when we can explore so much and have so many fascinating questions to find answers to. This is the purpose of humanity and the most valuable heritage we leave for next generations. This is done by asking fundamental questions and seeking the answers for them. The quest for discovering the origin is one of those attempts that brings together different disciplines and allows us to know ourselves better, as well as our position in the world. The search continues, as it indeed should.

CHAPTER 26 REFERENCES

Feynman, R.P., R.B. Leighton, and M. Sands. 2014. *The Feynman Lectures on Physics, Vol. I: The New Millennium Edition: Mainly Mechanics, Radiation, and Heat*. Vol. 1. New York: Basic Books.
Gleiser, M. 2014. *The Island of Knowledge: The Limits of Science and the Search for Meaning*. New York: Basic Books.
Holt, J. 2012. *Why Does the World Exist?* New York: Norton.

INDEX

Sahelanthropus tchadensis, 290
Saint Augustine of Hippo, 28
Satellites, 76, 160
   COBE, 76
   Kepler, 18, 97, 150, 152, 153
   Planck, 38, 57, 76, 108, 319, 321
   WMAP, 76, 108
Sauropsids, 273
Schleiden, Matthias, 22
Schrodinger, Erwin, 20
Schwann, Theodore, 22
Schwarzschild, Karl, 119
Seafloor spreading, 169, 170, 174
Shang culture, 313
Singularity, 56, 57, 62, 63
Socrates, 17
Space-Time, 28, 29, 30, 318
Speciation, 298, 299
   allopatric, 298, 299
   sympatric, 299
Stars, 86, 87, 88
   Giant stars, 140
   Main sequence, 120, 121, 122, 123, 124, 320
   Neutron star, 124, 132
   Population I and II, 114
   Population III, 79, 80, 87, 88, 112
   White dwarf, 130, 131, 132, 133
Stellar nucleosynthesis, 114, 135, 136, 140
Steno, Nicolas, 21
Stromatolites, 182, 183, 228, 246, 254, 263
Stromatoporoids, 266
Subduction zones, 174, 181
Supernovae, 106, 113, 132, 133, 140
Synapsids, 273, 277, 286

T

Thales, of Miletus, 16, 17
Tiktaalik, 184, 272
Trilobite, 184, 188, 265
Tycho Brahe, 18

U

Universe
   Age of, 21, 59, 74, 77, 83, 95
   Inflation, 58, 59, 60, 62
   The faith of, 23

V

Valance shell, 205, 206, 207, 208, 209
van Leeuwenhoek, Antonie, 21, 23
Vesalius, Andreas, 21
Volatile material, 156
von Humboldt, Alexander, 21
von Weizacker, Carl Friedrich, 120

W

Wallace, Alfred, 22, 263
Watson, James, 22
Wegener, Alfred, 165, 166
Wilson cycle, 173, 174, 175
Wilson, Robert, 20, 54, 76
Wilson, Tuzo, 173
WIMPS, 105
Wormhole, 33
W+ W, 43, 45, 48, 50

Z

Z0, 42, 45, 48, 50, 61
Zircon, 157, 158, 180, 320

CPSIA information can be obtained
at www.ICGtesting.com
Printed in the USA
LVHW06s1959021018
592169LV00004B/24/P